ENERGY MEDICINE

in Therapeutics and
Human Performance

ENERGY MEDICINE
in Therapeutics and Human Performance

James L. Oschman, PhD

Nature's Own Research Association
Dover, New Hampshire
www.energyresearch.us

Foreword by
Karl Maret, MD

Amsterdam   Boston   London   New York   Oxford   Paris
San Diego   San Francisco   Singapore   Sydney   Tokyo

An Imprint of Elsevier Science

The Curtis Center
Independence Square West
Philadelphia, Pennsylvania 19106

ENERGY MEDICINE IN THERAPEUTICS AND   ISBN 0-7506-5400-7
HUMAN PERFOMANCE

NOTICE

Complementary and alternative medicine is an ever-changing field. Standard safety precautions must
be followed, but as new research and clinical experience broaden our knowledge, changes in treatment
and drug therapy may become necessary or appropriate. Readers are advised to check the most current
product information provided by the manufacturer of each drug to be administered to verify the recommended dose, the method and duration of administration, and contraindications. It is the respon-
sibility of the licensed prescriber, relying on experience and knowledge of the patient, to determine
dosages and the best treatment for each individual patient. Neither the publisher nor the editors assume
any liability for any injury and/or damage to persons or property arising from this publication.

Library of Congress Cataloging-in-Publication Data
Oschman, James L.
   Energy medicine in therapeutics and human performance / James L. Oschman; foreword by Karl Maret.
      p.; cm.
   Includes bibliographical references and index.
   ISBN 0-7506-5400-7
      1. Energy—Therapeutic use. 2. Bioenergetics. I. Title.
   [DNLM: 1. Energy Transfer. 2. Complementary Therapies. 3. Energy Metabolism. QU
34 O81 2003]
RZ999.O836 2003
615.8'9—dc21   2002043687

Acquisitions Editor: Inta Ozols
Developmental Editor: Karen Morley
Project Manager: Peggy Fagen
Designer: Mark Bernard

Printed in the United States of America

Last digit is the print number: 9 8 7 6 5 4 3 2 1

Dedication

This book is dedicated to the living spirit of the great pioneers in science and medicine, especially John McDearmon Moore (1916-2000) and Albert Szent-Györgyi (1893-1986).

Our present and future science, medicine, and evolution as a species depend upon the pioneers who look beyond the way things seem to be, to the way things can become. The distinguished individuals to whom this book is dedicated are from opposite ends of the biomedical spectrum: an osteopathic clinician and a basic scientist. What they had in common, in addition to their dedication to life, was not what they knew but what they knew they did not know. John McDearmon (Mack) Moore, DO, Kirksville College of Osteopathic Medicine, became a country doctor, a loving witness and an aid to countless births, sufferings, and passings. Albert von Szent-Györgyi Nagyrapolt, MD (Budapest), PhD (Cambridge), Nobel Laureate, known to his colleagues as Prof or Albi, was one of the great scientific pioneers of the twentieth century. One of our continuing challenges is to connect advances in basic science with day-to-day clinical practice

The way of the pioneer is seldom easy, and we will always wonder what our world would be like if we could learn to embrace new ideas for the opportunities they offer.

John McDearmon Moore (Photo courtesy Agnes Moore)

Albert Szent-Gÿorgi (Photo courtesy Dr. Benjamin Kaminer)

Foreword

by Karl Maret, MD

Thomas Kuhn's 1962 book *The Structure of Scientific Revolutions* made us keenly aware of the principle of a "paradigm shift." Paradigm, literally "an example that serves as a pattern or model," is an existing belief system that anchors consensual reality. "Normal science" is predicated on the assumption that the scientific community knows what the world is like. Scientists take great pains to defend that assumption and to this end often suppress fundamental novelties because they may be seen as subversive to the existing set of beliefs. The scientific community cannot practice its trade without some set of received beliefs that forms the continuity of modern Western scientific thought. In turn, the scientific educational system is founded on a process of rigorous, sometimes rigid, preparation that ensures that the transmitted beliefs exert a "deep hold" on the students' minds.

One existing paradigm in contemporary medicine is that we are primarily biochemical beings. Consequently, the use of pharmaceutical agents, or alternatively herbs and vitamins, ought to be the most effective and widely used method to bring healing to an ailing organism. The paradigm shift that is beginning to occur, and to which this book so eloquently speaks, is that we are not only biochemical beings, but even more fundamentally, energetic and informational beings with sophisticated, high-speed communications systems mediating a complex information flow within our bodies.

No biochemist or physicist will deny that all chemical interactions fundamentally depend on energetic, informational, or structural conformational interactions in molecules. But just because molecules such as ATP are mediators of energy exchange, we should not assert that these mediators are the cause of these energetic interactions.

For almost a century, physics has lived with an emerging new vision of energetic interactions within matter by invoking such strange concepts as quantum mechanics, soliton waves, particle entanglements, non-locality, and action-at-a-distance, to name only a few. This new physics describes a world that is increasingly paradoxical and often counterintuitive and confusing to our fundamental sense perceptions. In that context, the modern biological sciences and medicine utilize concepts that are more analogous to the more traditional ideas of Newtonian physics.

Nearly 100 years ago, when this revolution in physics was occurring, American medical practice and education were also being fundamentally altered. The 1906 Pure Food and Drug Act declared that electrotherapy was scientifically unsupportable and should be legally excluded from clinical practice. This was followed in 1910 by the Flexner report sponsored by the Carnegie Foundation that reformed American medical education. The standardization of medical education resulted from Flexner's recommendation when he stated, "The curse of medical education is the excessive number of schools. The situation can improve only as weaker and superfluous schools are extinguished."

From these beginnings came a growing emphasis on a surgical and drug-oriented therapeutics, while effectively discouraging the use of the widespread electrical and magnetic treatment modalities that had found favor with many American consumers up to that time. Robert Becker and Andrew Marino estimated that by 1884, 10,000 physicians in the United States were using electricity every day for therapeutic purposes. Some of these ubiquitous "energy medicine" devices were sold by American retailers such as Sears through their catalog.

Although historians may one day determine whether the Carnegie Foundation–sponsored Flexner report had a self-seeking economic and political agenda to reform American medicine, one thing is for certain: the potential value of energy medicine treatments was effectively removed from American medical practitioners' consciousness. Over the ensuing years, research and education in electromagnetic medical approaches was actively discouraged if not outright shunned or ridiculed by the academic establishment.

Only in the waning years of the 20th century was there a renewed interest in this field, driven in part by consumer demand in complementary and alternative medicine that is facilitating a paradigm shift in health care. With U.S. levels of health care spending (or, truthfully, should we call it disease-care spending?) in 2000 reaching $1.3 trillion, and showing no signs of abating, we are now spending over $4500 annually for every American man, woman, and child to provide health/disease care. Only a minuscule percentage of this spending is for energy medicine diagnostic and treatment modalities, even though these have demonstrated many beneficial effects.

Diagnostic modalities such as the key EKG, EEG, and MRI, even though relegated to subspecialties of medicine such as cardiology, neurology and radiology, are essentially energy medicine modalities. Great promise also lies with newly emerging energy medicine therapeutic devices ranging from microcurrent stimulators, low level lasers, bone growth stimulators to broad-spectrum multiple frequency Tesla coil devices used for healing and performance enhancement. Although the genie is barely out of the bottle in the emerging field of energy medicine, yet already health-care benefits with lower delivery costs look very promising.

What Jim Oschman invites us to do is to journey together into the mysterious world of energy inside living systems. He pursues this path like Sherlock Holmes looking for clues, and we, the reader, collectively become the amazed Dr. Watson being led into the "heart of the matter." The reference to heart mentioned here is not without forethought.

One of the 11 definitions of heart in the *American Heritage Dictionary* is the "most important or essential part," as in "to get to the heart of the matter." To get to the heart of the current paradigm shift, we are invited to reexamine our current Newtonian medicine in the light of energy medicine involving the principles that belong to the quantum physics and information theory. A useful starting point might be to examine the actual heart in our bodies and discover what new revolutionary findings have emerged here over the last quarter-century.

The heart is not simply a muscular organ that pumps blood around the body and lungs of vertebrates. It has always been seen as a vital center and source of one's being, emotions and sensibilities. New research has shown how the heart is in intimate dialogue with our brain, body and the world at large. Dr. J. Andrew Armour, one of the pioneers of the newly emerging field of neurocardiology, introduced the concept of the functional "heart brain" in 1991. The heart's complex intrinsic nervous system contains tens of thousands of sensory neurites which detect circulating hormones and neurochemicals and sense heart rate and pressure information. Aggregates of the heart's neural structures have direct connections with the emotional-cognitive part of the brain, the limbic system, that can be called the old mammalian brain. An ongoing dialogue takes place between the heart and the brain through these ganglionic connections.

About 20 years ago it was also discovered that the walls of the atria of the heart had the capacity to secrete atrial natriuretic factor (ANF), a very potent blood pressure lowering agent which also affects sodium excretion by the kidneys and lowers blood volume. This hormone also affects many other organ systems, including the immune and reproductive systems, with hormone receptors found in many cells ranging from immune cells such as lymphocytes to male and female gonads. Thus we could call the heart an endocrine gland since it has the capacity to secrete this powerful hormonal peptide.

From an energy medicine perspective, however, the heart is the most powerful electromagnetic organ of the body having an important regulatory function. With a power output of around 2½ watts for each contraction by the coordinated cells of the heart, this organ sends a complex frequency pattern throughout the entire tissue matrix and creates a potent electromagnetic field that radiates out some 12 to 15 feet beyond the body itself. Any doctor can confirm the strength of this field since an electrocardiogram, with a voltage that is about 50 times greater than the EEG signal from the brain, can be picked up from anywhere on the body with ease.

All living processes in the body depend on the transfer of charges to conduct energy and support life. The entire watery matrix of our bodies is interconnected by complex charge-coupled fields that receive around sixty pulsations of electromagnetic energy from our beating heart each minute. From that perspective, our heart becomes a synchronizing electromagnetic master clock influencing the entire living matrix. Every cell in the body is in intimate electromagnetic contact with the toroidal-shaped magnetic field of the heart. Consequently, the heart constantly mediates between our thoughts, feelings, and actions. The body's heart rate variability is an instantaneous measure of our entire autonomic nervous system function, essentially the balance between our sympathetic nervous system stress response and the parasympathetic nervous system relaxation response.

One of the seminal contributions that this book makes to the establishment of the new paradigm of energy medicine is Dr. Oschman's attention to the living tissue matrix. It is a great tragedy that medical students and healers are not taught that the largest organ of the body is the connective tissue matrix. In the past this matrix was often simply seen as the stuff that "glues" the important organ systems of the body together. We now are realizing that it is here that mysteries of life can be most fruitfully explored. The Austrian professor of histology and embryology at the University of Vienna, Alfred Pischinger, MD, in his 1975 book entitled *Matrix and Matrix Regulation: Basis for a Holistic Theory in Medicine,* already presented the concept about the Ground Regulation System and the communications that exists within the connective tissue matrix. Unfortunately, little mention is made of this important work to today's medical students.

By his heartfelt listening to the stories of body workers, massage therapists, Rolfers, healers, clinicians, and research scientists, Dr. Oschman has done us all a great service to translate their insightful observations and life experiences into the articulate language of science. Throughout that process he maintained an attitude of open-minded inquiry and wonder that is characteristic of a master practicing with "beginners mind."

It must never be forgotten that all the scientific method can ever do is to disprove various hypotheses using the experimental method and then construct the most plausible theories from these observations. Unfortunately, modern science has all too often been held hostage to political, philosophical, or economic interests leading to what might be called scientism rather than true science. Many of the true breakthroughs in science have been made by questioning minds that thrive on anomalous observations and derive novel insights from them.

Inspired by seminal thinkers and teachers, such as Nobel laureate Albert Szent-Györgyi, who was a personal mentor to him, Dr. Oschman never forgot that the scientific process can be filled with fun and adventure, hard thinking and heart-felt listening. In an age of specialization, with more experts arising in an ever-increasing number of scientific disciplines, it is rare to find generalists who can bridge between various fields in biology, physics and medicine to inspire a new vision of our future.

An inspired vision of this emerging medicine emphasizes the importance of the living tissue matrix with its unique properties of semiconducting proteins, liquid crystalline holographic structure, vicinal water domains conducting Davydov solitons, and almost instantaneous, faster than nerve conduction, information transfer. It reads like science fiction while simultaneously having the feel of a good detective novel. Yet this is modern medical science at the new frontier of energy medicine and quantum healing.

Here cells are seen as fractals embedded in a holographic energetic matrix where everything is interconnected and capable of influencing any other part of the matrix. Information can be communicated through many, potentially redundant, modalities including photons of ultraviolet and visible light, phonons of sound, multiple resonant cellular vibrations, charge density waves and quantum potentials. The body becomes visible as a living, energetic whole even though composed of specialized organ systems and cellular aggregates. Ultimately, a picture of an electromagnetically unified matrix containing a self-organizing blueprint with innumerable feedback loops begins to emerge.

The ancient Yogic, Tantric, and Taoist inward-directed sciences of the East have studied this holistic blueprint for centuries. The Western esoteric traditions reexamined these more hidden doctrines and gradually awakened a renewed interest through fields such as Theosophy, Anthroposophy and the Arcane School.

More recently, modern genetic engineering research into cloning is also rediscovering that the energetic spark must be critically applied at the outset in order to create the fusion of genetic materials that result in the new clone. Before we get too excited about the wonders of modern genetic engineering and the emerging science of genomics, it is sobering to note that in Russia Dr. Chang Kanzhang was creating genetically modified organisms simply by manipulating the fields existing between living organisms and without resorting to the slicing and dicing of DNA fragments in test tubes and sequencers.

Energy medicine represents a vast promise to bring better health and wellness to humanity. Yet its greatest potential may be that it is capable of creating a bridge from the spiritual and social sciences to the modern life sciences and medicine. Imagine the impact of seeing our bodies as essentially whole, interconnected and with all parts actively communicating with each other to create a unified field of unique, individual self-expression. This model, if taught to future generations, might also inspire us to see our planet as an interconnected whole, with all of us as fractal, vibrant cells in the matrix of Gaia, our conscious Earth. Then biology and the life sciences might take the place of physics as the primary foundational discipline of our scientific consciousness. Perhaps then the grip of materialistic thinking might loosen enough to allow a shift to a more balanced energetic perspective that is characteristic of a deeper dialogue between the head and the heart.

It is my sincere hope that this book will be read widely and will find a broad audience throughout the world. May it be shared, discussed, and debated to ultimately inspire a new generation of scientists, clinicians, healers, athletes, and seekers of truth who will embrace paradox, mysteries of life, and the wonder of creation. May all who read these chapters enter into their body wisdom with a spirit of open-minded inquiry that is the hallmark of all true pioneers and explorers.

Michaelmas 2002
Santa Cruz, CA

Preface

It could be a plot for a detective novel, but it is all true. As in a Sherlock Holmes thriller, we gather a set of strange and disconnected clues that seem to make no sense by themselves. In fact, those who have discovered these clues are suspected of being way out on a limb or "over the edge."

As usual, Watson is bewildered. Our challenge is to fit these peculiar clues together to solve the mystery. When we are finished and all the pieces are in place, Sherlock wonders why it took us so long.

■ While on a camping trip in England, one of the world's foremost biochemists watches a startled kitten jump straight up in the air. Most of us have seen this, but this scientist thinks something new: the response is far too rapid to be explained by neuromuscular signals from the eyes to the brain and then to the muscles. He spends years in the laboratory researching an alternative explanation. The conclusion: living matter has a previously unsuspected high-speed electronic system for conducting energy and information. Water is an essential part of the circuitry.

The scientific community responds to these ideas by suggesting that, although this scientist did some good work in the past, he should now consider retirement.

■ An International Business Machines (IBM) researcher follows up on the water clue. He begins an experiment requiring the world's largest and fastest computers, running 24 hours a day, 365 days a year. Within the computer, he constructs a virtual model of one turn of the deoxyribonucleic acid (DNA) double helix. The molecule vibrates 3,000,000 times per second, as any molecule does at body temperature. He brings a water molecule into the virtual space within the computer. As the water molecule, with its electrical polarity, approaches the DNA, the structures of both the DNA and water interact via their energy fields and their internal structures change, requiring repeated recalculation of their structures. One at a time, he brings 447 water molecules up to the DNA helix, allowing each to settle into the region where the forces are about equalized or balanced.

When he is finished, he steps back to view the pattern his computers have been constructing and sees something incredible. Chains of water molecules extend along the DNA strands and form bridges or filaments connecting each helical turn with the next. Remarkably, it is water that holds the double helix together!

■ A British scientist studies an important protein in the membrane of red blood cells. The protein is responsible for a variety of blood group antigens and is the place where viruses attach. He uses a radioactive labeling procedure to determine whether the protein is on the inside or outside surface of the cell membrane. Interestingly, the protein shows up on both sides. Other investigators doubt this claim, but further work reveals that he was correct. This is a surprise, but it soon is discovered that most cells have comparable proteins extending from their interiors, passing through the membrane, and connecting to neighboring cells or the extracellular matrix. Further research reveals that these proteins are vital links in cellular communications.

■ One of the leading theoretical physicists studying crystals becomes fascinated with the enormous voltage (some millions of volts per meter) that can be measured across the membranes of living cells. He realizes that the huge electrical tension on the molecules in cell membranes should make them vibrate intensely at body temperature; therefore, they should emit coherent or laserlike light. Researchers look for, and find, these photonic signals, and they suspect the signals regulate living processes.

■ Bodyworkers interested in connective tissue dissect a cadaver and notice tiny threads running through the connective tissue sheets, called *fascia,* surrounding muscles. They suspect these fibers correspond to the acupuncture meridians of traditional Chinese medicine.

■ All of these pieces of the puzzle are presented at a seminar at an acupuncture school in New England. A classical scholar describes her translations of ancient texts revealing that the Chinese had observed that the fascial sheets are shiny, like metals, and that the fascia conduct energy, referred to in Chinese as *Ch'i.* Perhaps these shiny fibers are the substrate for the proposed high-speed electronic system that conducts energy and information within the body.

■ Scientists at a medical school in California insert needles into acupuncture points on the side of the foot. These points have been used for thousands of years to correct vision disorders. Remarkably, the needling quickly elicits nerve impulses in the visual cortex of the brain, as measured with a sophisticated tool called *functional magnetic resonance imaging (fMRI).* Suddenly there is scientific support for an ancient healing method that seemed to defy normal logic. After all, how could inserting a needle into the foot affect something taking place in the brain?

Further research reveals that the stimulus to the feet reaches the brain far too quickly to be explained by nerve conduction. There appears to be a nonneural communication system in the body capable of conducting signals extremely rapidly from the feet to the brain. Sherlock sees that the plot is getting thicker.

■ A team of athletes from East Germany wins an unexpected and unprecedented number of gold medals at the Olympic Games. One of their secrets: repeated mental rehearsals of their events. Soon researchers discover that mentally rehearsing any activity, without moving at all, vastly enhances performance. Neurophysiologists discover that thinking about a pattern of movements sets up a corresponding pattern of electrical activity in the nervous system, even though no actual movements are taking place. It is theorized that mental rehearsals and the electrical signals they create strengthen pathways for energy and information flow that can be called upon when it comes time for the real performance.

■ A dancer asks a University of California Los Angeles (UCLA) physiologist to measure the electrical activity of nerve-muscle connections during a "trance dance." Remarkably, the electrical signals cease, even though the dancer is still moving.

■ Astounded by these results, the physiologist observes another performer who has studied a "primitive" dance form in Haiti. This time an extraordinarily vigorous and acrobatic dance routine is correlated with a *decrease* in heart rate and blood pressure. The inescapable conclusion from studying these two dancers: there is a way of moving that is different from the textbook neuromuscular mechanisms.

■ The dancer mentioned above has an extraordinary and totally unprecedented idea. Perhaps the primal movements, sounds, and breathing patterns she learned in Haiti could help people with spinal cord injury. She meets Barbara, who has spent 11 years in a wheelchair after an automobile injury that rendered her paraplegic. They begin years of work together using breath and sound. Movement slowly begins to return below the site of the injury, gradually spreading to the pelvis, legs, and feet. Eventually, Barbara is able to crawl, an extraordinary accomplishment for someone with a severed spinal cord. Work with others with spinal cord injury confirms that it is possible to regain some degree of movement in individuals thought to be permanently paralyzed.

■ A sensory neurophysiologist and dedicated flute enthusiast publishes a theory about how the same cellular systems can be responsible for both movement and sensation. He and others conclude that *sensation is movement in reverse,* and *movement is sensation in reverse.* Bacteria and protozoa can sense their environment and move, using their cilia and flagella, even though they lack any form of nervous system. Cilia are part of virtually all sensory cells throughout the animal kingdom. Hence a comparable "primitive" sensory/movement system may exist in higher organisms and may explain the other pieces of the puzzle.

■ A scientist at Boston University School of Medicine studies a film of two people having a conversation. For a year and a half he carefully analyzes a 4.5-second piece of conversation. After watching 100,000 sequences, the film is worn out, so he makes another copy and wears it out as well. After analyzing 130 copies of the same film and making millions of observations, he reaches his conclusions. The speaker and the listener are performing a synchronous "dance" consisting of micromovements of different parts of their bodies. Each part of each word has a cluster of movements associated with it, and the listener moves in precise shared synchrony with the speaker's speech. Remarkably, there is no discernible lag between movements of speaker and listener.

■ Two Austrian PhD students, destined to become major figures in biology, get together from time to time in Viennese coffee shops. They "milk" each other's minds about systems concepts they are working on from entirely different perspectives. Their conversations lay the foundation for a "general systems theory" with implications for all of science and medicine. They both conclude, from their different perspectives, that the development of an organism, and other complex living processes, cannot be understood as the sum of the behaviors of isolated parts.

■ Leading biochemists gather at the New York Academy of Sciences to discuss a mystery related to how energy moves from place to place in the cell. Many creative ideas are presented, but the mystery is not solved.

■ A Ukrainian biophysicist attending the New York meeting responds by developing a new concept for energy transfer in living systems. Called the *soliton* or *solitary wave,* it is a coherent or laserlike wave resembling the oceanic tidal wave or tsunami. The soliton is ideally suited for energy and information transfer in living systems because, unlike other waves, it does not lose energy, dissipate, or disperse as it travels through a medium. A laboratory in Japan sends soliton data through a fiberoptic coil the equivalent of 4500 times around the Earth without any loss of information. Engineers incorporate soliton technology to create high-speed intercity fiberoptic connections for the Internet.

■ Physicians and therapists occasionally observe extraordinary "spontaneous" healings of very serious injuries or diseases. It is a rare occurrence but quite impressive when it happens. We all would like to know what triggers these remarkably rapid healings. Perhaps there is an aspect of living systems that is right in front of us but we cannot see. If we had the tools to see this "new stuff," many more pieces of the puzzle of life would fit into place.

■ Beyond the rarefied atmosphere of academic biomedicine, most people can recall special moments that tell them that there is more to life than usually meets the eye. During emergency or life-threatening situations, for example, "time seems to slow down" to allow quicker responses than usual. Athletes and other performers describe a similar kind of clarity and coordination during peak performances. The practice of certain martial arts techniques conditions the body to respond extremely rapidly without thought. For jazz musicians, highly synchronized yet unplanned improvisational music is commonplace. All performers agree, however, that if they start to think about what is happening, they lose their focus.

Let us document each of these fascinating clues. We will "connect the dots" by describing an energy and information system in the body that is the "missing link" for many phenomena that have seemed hopelessly inexplicable in the past. It is a system that is responsible for extraordinary feats of perception, movement, and healing. It is a *wet* system, consisting of all of the material parts the body is made of, without exception. Sometimes referred to as the *living matrix,* it forms the body's "operating system" that works quietly in the background, coordinating our every activity. It is a system that is more fundamental than any of the other systems in the body because it gives rise to and regulates all of them. Research from a variety of perspectives and a variety of disciplines describes its anatomical, biophysical, and behavioral properties.

James L. Oschman, PhD
Dover, New Hampshire
May 31, 2002

Acknowledgments

I am fortunate to have access to a source of immediate answers to complicated questions: my precious wife Nora. Although the usual method of research is to go to the library and dig through books, this often has proved unnecessary because the answers, Nora says, are to be found in nature. I am fortunate to live with a person who knows and lives the answers to what the rest of us consider to be the most perplexing problems of all kinds, scientific or not. I also am blessed that she is willing to live with someone who more often than not is lost deep within a cell, molecule, or biophysics manuscript. Thank you, Nora, for all of your years of dedicated support and nourishment.

Through the kind and generous efforts of Kevin and Deb Reichlin, we met Jeff Spencer, who is a key medical support person for Lance Armstrong, a repeat winner of the Tour de France. Jeff immediately incorporates information from energy medicine into his sophisticated medical practice and the "bag of tricks" he uses to keep his cyclists in the race or to keep the other famous performers he has worked with in top form. His inspiring work on human performance and rapid recovery from injuries helped me to appreciate how much energy medicine contributes to the training, performance, and injury recovery for athletes and other performers who are taking their bodies to and beyond the edge of what seems possible.

Emmett Hutchins asked a key question that required more than a year to answer. Major portions of the answer came from Emilie Conrad and Valerie Hunt.

Lorin Kiely, founder of *Earthfutures Communications,* introduced us to the profound concepts and teachings of Dr. Herbert V. Guenther, Professor Emeritus of Far Eastern Studies at the University of Saskatoon, Saskatoon, Canada. Dr. Guenther is the foremost scholar/translator of Buddhism of the twentieth century. In the 1950s, when the Chinese invaded Tibet, the Lineage Holders of the Tibetan Buddhist Tradition gave Dr. Guenther many of the original copies of the precious ancient yogic texts. These manuscripts contain vivid accounts of research in the spiritual science of pristine awareness, or *Dzog-chen.* Dr. Guenther has long believed this living tradition must be communicated and shared using the language of our times. It is hoped that the ideas gathered together in this book contribute to that process.

I continue to be grateful to my teachers, particularly Peter Melchior and the various pioneers who have gone before to enrich the path. Wayne and Agnes Moore and Alicia Davis kindly provided biographical information on John McDearmon Moore, one of the individuals to whom the book is dedicated. For years, Peter Gascoyne coached me in the implications of the work of Albert Szent-Györgyi and his colleagues that form a major part of the theme of the book.

I thank David P. Knight, Mae-Wan Ho, Deborah Stucker, Jason Mixter, Risa Kaparo, Stanley Rosenberg, Urs Honauer, Ruedi Schaepi, Gerda Braun, Dionysus Skaliotis, Deane Juhan, Randy Mack, Zanna Heighton, Cynthia Hutchison, Alicia Davis, Fritz Smith, Tom Myers, Deborah McSmith, Betty Wall, Tim Wargo, Tessy Brungardt, Jean Luc Connile, Charles Daily, Duane Elgin, Peter Laine, Caryn McHose, Artemis March, Candace Pert, Molly Scott, Benjamin Shield, William Redpath, Beverly Rubik, Mike Weiner, Chloe Wordsworth, Juan Acosta-Urquidi, Mark

Neveu, Timothy Dunphy, Michael Ruff, Kate Leigh, Susannah Hall, Jay Shah, Ray Gottlieb, Eric Jacobson, Ron Mack, Elliot Abhau, Carolyn McMakin, David W. Mitchell, Hank Meldrum, Jason Mixter, Heidi Holt, Karli Beare, Sylvia Meneau, Hendrik Treugut, Erik Rosenberg, Lisa Tully, John Zimmerman, Norm Shealey, Stephen Porges, John Chitty, Tor Philipsen, Deborah Stegmaier, Jim Asher, Vernon Rogers, Mary Lynch, Debra Harrison, and the many others who have continued to provide interest and precious input of new information and insights. Leon Chaitow and Bob Charman were extremely supportive of this project from the beginning and have continued to provide help and resources. I thank those who have invited me to meet with their students and colleagues at various therapeutic schools around the world, and I especially thank my audiences for inspiring comments and questions.

To the various schools, hospitals, and medical centers where I have presented this information, I express my debt for stimulating conversations. These include the Rolf Institute; the Guild of Structural Integration; the Zero Balancing Association; the Polarity Zentrum and the Colorado Cranial Institute in Zürich; Healing Touch International; the Stichting Opleiding Manuele Therapie, Amersfoort, The Netherlands; the Upledger Institute United Kingdom; the Stanley Rosenberg Institute and Nordleys Centret, Silkeborg, Denmark; the Ergoterapeut-og Fysioterapeutskolen, Holstebro, Denmark; the Massage Therapy Research Agenda Workgroup of the American Massage Therapy Foundation; Concord Hospital, NH; University of Vermont Medical School; the Philadelphia College of Osteopathic Medicine; International Sound Symposium, San Jose, California; International Society for the Study of Subtle Energies and Energy Medicine, Boulder, Colorado; the Annual International Weight-Training Injury Symposium in Toronto, Canada; Wentworth-Douglass and Portsmouth Regional Hospitals, New Hampshire; York Hospital, York, ME; University of Wisconsin, Milwaukee School of Nursing; the Holographic Repatterning Association; the Alliance for Integrative Medicine of Northern New England; the American Polarity Therapy Association; John Harvey Gray Center for Reiki Healing; St. Francis Hospital and Medical Center, Hartford, CT; Misty Meadows Herbal Center, Lee, NH; the Annual Conference on Light and Vision, St. Louis, MO; the Connecticut Hospital Association, Wallingford, CT; The Esalen Institute, Big Sur, CA; Continuum, Santa Monica, CA; the University of New Hampshire, Durham, NH; the Bradley Institute, New Bedford, MA; Gaia Healing Center, Frederick, MD; Society for Bioinformation, Germany; University of Westminster, London, UK; Sports Medicine & Rehabilitation International, Parrish, FL; the Osher Institute, Harvard Medical School, Boston, MA; and the National Institutes of Health.

Dr. Philip S. Callahan, Vernon Rogers, Kerry Weinstein, Stephen Birch, and Dr. David P. Knight gave me a number of essential tips and valuable references to the literature. My experience in quantum electronics came from discussions with Drs. Albert Szent-Györgyi, Peter Gascoyne, Ron Pethig, and Herb Pohl, and I thank them for taking the time to educate me. I have been particularly inspired by the writings of Drs. Edward Adolph, Robert O. Becker, Herbert Fröhlich, and Fritz Popp. Finally, I thank Dr. John Zimmerman, Dr. Ray Stevens, Leane E. Roffey, and my students for many valuable discussions.

I express my thanks to the various authors, artists, and publishers who have permitted me to reproduce their material here. And I am extremely grateful to Walter K. Hallett III and Karen Morley for their help with the permissions, and to Larry Azure for supporting the permission process. I especially thank Walter K. Hallett III for many creative cover designs. He had a lot of valuable input into the one that was finally chosen.

Prologue

Round about the accredited and orderly facts of every science there ever floats a sort of dust-cloud of exceptional observations, of occurrences minute and irregular and seldom met with, which it always proves more easy to ignore than to attend too. . . . Anyone will renovate his science who will steadily look after the irregular phenomena, and when science is renewed, its new formulas often have more of the voice of the exceptions in them than of what were supposed to be the rules.
WILLIAM JAMES

This book is dedicated to the pioneers who have made our science and medicine what it is today and who have set the course for the future of biomedicine and human evolution. We depend on them to leap beyond the "accredited and orderly facts of every science," to explore the "irregular and seldom met with" observations that fuel the renewal process upon which our collective future depends.

History shows that the pioneering experience rarely is easy for the pioneer. Here are some examples of pioneers whose ideas were readily accepted and some who had the opposite experience. These examples are related to major themes of this book: *biological communications* and *invisible forces*.

THE PIONEER WITH AN IDEA "WHOSE TIME HAS COME"

It is rare to find pioneers with a new insight that is quickly and enthusiastically recognized and accepted by their contemporaries, and everyone lives happily ever after. These are people who are lucky or clever enough to discover something "whose time has come."

An example is the discovery of the role of nitric oxide (NO) in the regulation of the circulation to muscles. This work, which was first published in 1986, received the Nobel Prize in Physiology or Medicine in 1998. Twelve years is a "fast track" for such recognition. Acknowledgment was swift because the revolutionary discovery solved a huge mystery in regulatory physiology:

> *In severe exercise a given leg muscle can consume oxygen at 40-60 times the resting rate. The heart, lungs, breathing, vasculatures, autonomics, adrenal medullas, spinal synapses, and many more functional units respond to this one demand for action . . . . To this day, no one knows how the demand is accurately conveyed to the heart and to the arterioles that divert the flow hither and thither.* ADOLPH (1979)

NO, a gas, turns out to be the signal molecule that "divert[s] the flow hither and thither."

Signal transmission by a gas produced by one cell and penetrating membranes to regulate the functions of other nearby cells was an entirely new principle in regulatory biology. Muscle cells in need of oxygen produce NO, which quickly migrates to the smooth muscles lining nearby arteries and causes the smooth muscles to relax. The vessels open or dilate, and blood flow increases to the places in need.

This is an example of nature selecting, from all of the possible alternatives, the perfect communication mechanism to regulate a vital function. Using a highly diffusible gas with a short lifetime enables a rapid response to increased demand and an equally rapid reversal when demand returns to normal. Because the signal molecule breaks down on its own in about 10 seconds, no enzyme or other mechanism is needed to switch the process off when it is no longer needed.

Remarkably, this research also led to an explanation of a 100-year-old mystery: How does nitroglycerine relieve angina pain? Nitroglycerine releases NO gas, which in turn dilates coronary arteries, increasing blood flow to the heart muscle and relieving angina.

Ironically, Alfred Nobel, who founded the Nobel Prize, was the inventor of dynamite, a product containing nitroglycerine. When Nobel was taken ill with heart disease, his doctor prescribed nitroglycerine. Nobel refused to take it, knowing that it also caused headaches. At the time he wrote in a letter, "It is ironical that I am now ordered by my physician to eat nitroglycerine."

The discovery of the regulatory role of NO led to many advances in physiology and medicine. The popular prescription drug sildenafil citrate (Viagra) is one application of NO research. As a fast-acting messenger, NO plays a key role in major moments of the cycle of life: fertilization, birth, development of the brain, and each heartbeat. NO acts so quickly that each heartbeat results in a pulse of NO that relaxes the vascular smooth muscles so that the vascular system can accommodate the pressure wave. Hemoglobin transports and releases NO at the same time it transports and delivers oxygen to the tissues. NO plays a key role in the destruction of bacteria in infections and in the stomach. White blood cells produce huge quantities of NO to kill invading bacteria and parasites, hence the use of nitrites to preserve meat from bacterial spoilage. NO triggers the flashes of fireflies. And the list goes on.

Importantly, nature's discovery of NO is not a recent development; it took place early in evolution. Simple animals such as sponges, which lack nerves, use NO as a signal molecule (Giovine et al.).

HOW MANY MORE COMMUNICATION SYSTEMS WILL BE DISCOVERED?

The recent discovery of the role of NO in cell-cell communication raises a fundamental question: How many more communication mechanisms will be discovered in living systems?

We always thought that the nervous system was the prime signaling system in the body. Generations of neuroscientists have gathered remarkable and sophisticated information about the nervous system; however, there are no nerves connecting muscles with the arterioles that supply them with oxygen and nutrients, so there was a mystery that remained unsolved until NO was discovered.

A second well-documented communication system involves the circulatory system, which delivers regulatory hormones to every cell in the body. A third is the immune system, which is capable of mobilizing the body's defenses against "intruders," no matter where they are located, on or in the body. The interaction among all of these systems (nervous, hormonal, and immune) has become a fascinating study unto itself. One discipline exploring this fascinating and important subject is known as *psychoneuroimmunology.*

The bias against exploring the energetic aspects of communication has delayed an important avenue for researching biological regulations. Energy fields are available to nature as a means for providing even faster control than a diffusible gas with a 10-second lifetime. Some energy fields travel at the speed of light; others are instantaneous. For nature to have left these communication modalities out of its "bag of tricks" is highly unlikely. This book explores some communication systems that are relatively new to contemporary biomedical research and for which there is mounting evidence.

THE PIONEER WITH AN IDEA "AHEAD OF ITS TIME"

The second type of pioneering experience is of a monumental struggle to get anyone to listen to the new idea. A classic example is the work of Semmelweis in Vienna in the mid-1800s. Semmelweis introduced a completely radical medical concept, hand washing, to protect patients from an "invisible force," bacteria:

> When Semmelweis proposed that obstetricians wash their hands before delivering babies, the idea was considered preposterous. He was in effect introducing a new and invisible factor at work in healing, which today we call infection. At the time, however, a theory of infectious disease did not exist. So Semmelweis did a simple experiment to prove his point. For a year the midwives on one obstetrical ward washed their hands, and the obstetricians on another ward did not. On the hand washing ward, mortality from childbirth fever declined by 1000%. But, alas, the data made no difference. The skeptical physicians still could not accept the conclusion that there was a lethal factor lurking on the hospital wards that they were helping spread, and which could be controlled by washing one's hands. Semmelweis was regarded as a troublemaker and was vilified. He fled Vienna to Budapest and eventually committed suicide as a result of the emotional strain he experienced. GARRISON (1929)

Most of us grow comfortable and attached to the things we have learned and are not impressed with new and different ideas. If it were not for the pioneers and their struggles at the frontiers of human intellect and accomplishment, most of what we now value in science, medicine, and many other areas would never have become available to us. Science and biomedicine would progress at a much faster pace if people were willing to consider new ideas instead of reacting to them.

ANOTHER EXAMPLE, WITH A HAPPIER ENDING

Magnetic resonance imaging (MRI) is a valuable clinical tool that provides sharper images of tumors and other pathologies than any other method, often enabling diagnosis of serious problems at an early stage. Because of its use of magnetism and electromagnetic fields, which are forms of energy, MRI is an energy medicine diagnostic tool.

Few realize that the inventor of the MRI, Raymond Damadian, once was considered a "mad scientist." His final success was due to his unselfish dedication to his vision of a nuclear MRI system, a system that nobody was willing to support financially but that he knew from the beginning would have a profound impact in diagnostic medicine.

Moreover, few realize that the principle by which the MRI scanner discriminates cancerous from normal tissues is not accepted in mainstream biomedical science. To be specific, we see in Figure 1 that there are two views or models of cell structure, one (A) in which the cell is surrounded by a semipermeable membrane, and a second (B) in which the membrane is either discontinuous or does not exist at all. This is a topic that will be discussed in more detail later in the book.

In the standard model (A) a voltage across the cell membrane arises because of the uneven concentration of ions inside and outside the cell. The concentration difference is due to the permeability barrier at the cell surface and to ion and solute pumps that transfer specific ions and molecules into and out of the cell.

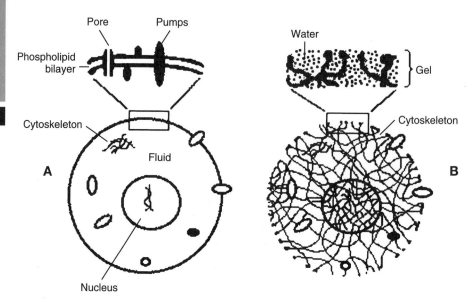

Figure 1 Models of cell structure. **A**, The widely accepted "cell as a bag of solution" model in which the cell is surrounded by a continuous semipermeable membrane composed of a double layer of lipid molecules. A voltage arises across the membrane because of the uneven distribution of ions inside and outside of the cell. The concentration difference is caused by the permeability barrier at the cell surface, diffusion through pores in the membrane, and ion and solute pumps that transfer specific ions and molecules into and out of the cell. This model downplays the presence of structure within the cell (the cytoskeleton). **B**, A different model proposed by Ling, Cope, Damadian, Pollack, and others. Some ions and molecules enter and are retained within the cell because those molecules fit nicely into the organized layers of structured water that virtually fill the spaces between fibers of the cytoskeleton. The water and ions, in turn, are organized or "structured" because of the charges along the gel-like material forming the cytoskeleton. This model takes into consideration the fact that the cytoskeleton is so pervasive that there is little space within a cell for liquid water as such. Instead, virtually all of the water is organized in layers that are held in place by the charges along the cytoskeletal fabric. These charges are primarily negative. In this model the membrane potential actually arises because of the charged nature of the gel within the cell, which establishes a well known phenomenon called the Donnan equilibrium.

In the model proposed by Ling, Cope, Damadian, Pollack, and others (Figure 1, *B*), some ions and molecules enter and are retained within the cell because the molecules and the water that clusters around them fit nicely into the layers of structured water that virtually fill the spaces within the cell. The first model downplays the role of structure within the cell (the cytoskeleton). The second model recognizes that there is so much cytoskeletal material that little if any of the water within the cell is free or is liquid water; instead the water forms multiple layers around the cytoskeletal fabric *(B)*. In this model the "membrane potential" arises because of the charged nature of the gel within the cell, a well-known phenomenon called the *Donnan equilibrium.*

No serious discussion of cell and tissue physiology, or biological communications, can exclude consideration of the merits of these two differing models of the cell surface and interior. These are, after all, only models or guesses about the structure of something that is far too small to be studied directly. In relation to the apparent conflict between the two models shown in

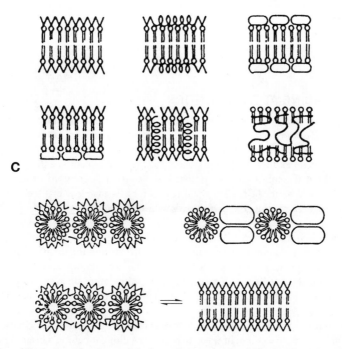

C

Figure 1 (cont'd) C, Consider the models shown in *A* and *B* in relation to the situation in 1970 when texts presented several alternative models for the cell membrane, reflecting that the evidence at the time was insufficient to decide the matter. By an interesting process, the various alternatives were dropped in favor of the model shown in *A,* which became "the structure of the cell membrane" without the important modifying word, "model." (**B**, Modified from Pollack, G.H. 2001, *Cells, Gels and the Engines of Life,* Ebner & Sons, Seattle, Fig. 8.8, p. 122. C, From DeRobertis, E.D.P., Saez, F.A., & DeRobertis, E.M.F. 1975, *Cell Biology,* 6th edn. WB Saunders, Philadelphia, Fig. 8-2, p. 149.

Figure 1, *A* and *B,* it is important to realize that only a few years ago, most biology texts presented several alternative models for the cell membrane. For example, Figure 1, *C,* shows a set of models published in a cell biology text in 1970 (de Robertis, Nowinski & Saez 1970). At that time, there were good reasons to suspect that there was *not* a continuous lipid bilayer membrane encircling the cell. Gradually the alternative models slipped into obscurity, and the model shown in Figure 1, *A,* took precedence to the point that modern texts describe it as "the structure of the cell membrane," dropping the important word "model." The minority perspective continues to be discussed by Ling (2001) and Pollack (2001). We shall return to this subject later in the book (Chapter 8).

Models are just models, and it is always valuable to keep the possible alternatives open. A simple intellectual tool to carry this out is the method of multiple working hypotheses, to be described shortly. It is always important to remember:

Science has no eternal theories. ALBERT EINSTEIN

In his book entitled *A Machine Called Indomitable,* Kleinfield (1986) documents the story of the inventor of the MRI, a commercial success in spite of its completely unacceptable theoretical base. MRI devices are available in virtually all major medical facilities, although the majority of the instruments are manufactured by General Electric rather than by Damadian's company

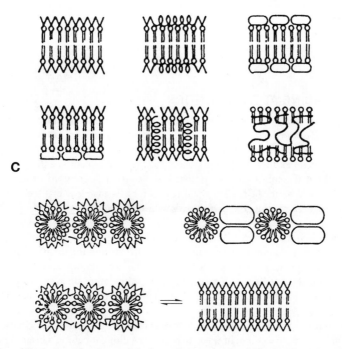

(Fonar Corporation). General Electric actually infringed on Damadian's original patents, leading to a legal action and a settlement to Damadian in excess of $100 million.

Asked about how to spot a pioneer, Damadian replied, "He's the one with all the arrows in his back."

ENERGY MEDICINE AT THE FRONTIER

The whole subject of energy medicine has long been a pioneering and frontier area. The predecessor to this book, *Energy Medicine: The Scientific Basis* (Oschman 2000), summarized the controversies, confusions, and bitter debates that have long dominated the subject. In essence, *Energy Medicine* documented that there no longer are any good reasons to continue the contention and befuddlement about the subject of energy that has distracted generations of academics and slowed the progress of biomedicine. The main reason for this change is that technologies have been developed that can measure the energy fields within and around living systems. These *fields of life* have the very properties that sensitive individuals have been describing and using for healing for thousands of years, properties that science had long thought did not exist.

A vast amount of useful information can be assembled around the theme of energy. There are many energetic approaches to life, health, and disease; much is known about the subject; and energy is just as worthy a topic for biomedical science as any other. To argue against this is to waste precious time and effort that could be better spent solving the problems that face us in so many directions.

The most exciting comment about *Energy Medicine* came from publisher George Quasha: "This book enables people to think in ways they have not thought before." Given the variety of serious issues surrounding us, being able to think in new ways is essential. Problems obviously are not solvable by following the same logic that created them.

If we do have new ways of thinking about energy, where can these new thoughts take us? This book begins to answer that question by looking at what William James referred to as the "dust-cloud of exceptional observations, occurrences minute and irregular and seldom met with," the "irregular phenomena" that are much easier "to ignore than to attend to." The purpose is to intentionally bring out and attend to these phenomena and thereby renovate science. We will do this with the scientific language of energy medicine described in the previous book.

The renovation process is done with humble awe and appreciation of the science that has gone before. It is not in any way intended to prove that past discoveries are wrong; instead the purpose is to open up new vistas for exploration and new possibilities for experiencing life. This is an entirely appropriate scientific endeavor, and anyone who argues that science should not go to these interesting places does not appreciate the scientific process. New vistas are available to us because a vast group of sophisticated and sensitive individuals have been exploring *life in action* and have made fantastic discoveries in the process, but the historic confusion and bias surrounding the word *energy* has prevented scientists from exploring this rich treasure trove of observations.

FINDING THE MISSING PIECES OF THE PUZZLE

We have been in a situation similar to assembling a jigsaw puzzle with half of the pieces missing. Yes, eventually you will have some success, you will fit some of the pieces together, and you will begin to see parts of the image; however, the whole picture will never emerge until you find the remaining pieces and fit them in. In the past, the *energy* pieces have been left out of the puzzle, and now we can fit them into place.

One group that has been virtually ignored in the past consists of the various bodywork, energetic, and movement therapists who are developing increasingly sophisticated, successful, and popular approaches to health, disease, and disorder. Another group is composed of individuals who have explored the energetic properties of living systems using tools that measure electricity, magnetism, light, and sound. Each approach has profound insights into the living organism, but their voices have been muted because their work is branded with such words as *alternative, illogical,* or, even worse, *energetic.*

Complementing this precious information are insights from performers, martial artists, athletes, and the healers, trainers, and teachers who encourage and maintain them. Their goal is to nurture and explore peak performance at the edge of what is humanly possible. This includes restoring healthy function as quickly as possible after an injury.

> *Our senses*
> *our instincts*
> *our imagination*
> *are always a step ahead*
> *of our reason.*   OCTAVIO PAZ

Bringing all of this emerging information into the scientific inquiry into life has been an extremely rewarding and exciting endeavor.

Of course, scientists are trained doubters, and I expect readers of this book will be skeptical about every aspect of it; however, there is always a need to see beneath one's skepticism about a subject like this. Is my skepticism based on some fact of nature or personal observation, or has it been passed on to me from someone who learned it from someone else? Is it genuine and appropriate, or is it a relic from some other time or some other person's bias or misconception?

WAYS OF DOING SCIENCE

There are widespread misconceptions about how science is done and how scientific progress takes place. For example, there is supposed to be a scientific method that all scientists know and follow in order to discern the truth, but the reality is that there are about as many scientific methods as there are scientists, although not everyone will admit this. The entire scientific enterprise—what is accepted as "the one and only correct way to do science" and which subjects are appropriate and which are "off limits" to investigation—is as subject to whims and fashions as are the designs of clothing or the styles of automobiles. Each generation of scientists has had its ideas about the "one and only" way of finding the truth, and the fact that these ideas change periodically points toward the fact that "truth" is actually a relative matter.

I have had many teachers and have learned many approaches to doing research. My favorite approach, advocated by a person who you will learn more about if you read on, is to *enjoy* science. Certainly having fun never distracted Albert Szent-Györgyi from making some of the most profound scientific discoveries of the previous century. The tale of discovery you will find in this book continues to be a fun journey. The implications of the story are serious, important, and profound, but that has never taken away from the pleasure of working with the people and the ideas that gave rise to the book that is in your hands.

THE METHOD OF MULTIPLE WORKING HYPOTHESES

There is a productive approach to the information and ideas presented in this book. It is a mode of thought that was recommended many years ago in a perceptive essay by T.C. Chamberlin (1897) entitled *Studies for Students. The Method of Multiple Working Hypotheses.*

The broad goal of this book, and of science in general, is to stimulate the thinking that will lead to solutions to problems that have been resistant to conventional ways of thinking, to stimulate independent and creative insight, to discover new truths, and to make new and useful combinations of existing information. Chamberlin's essay, which is just as relevant now as it was a century ago, documents how scientists become attached to their explanations and provides a way of disciplining one's mind to avoid this trap:

> *The moment one has offered an original explanation for a phenomenon which seems satisfactory, that moment affection for his intellectual child springs into existence, and as the explanation grows into a definite theory his parental affections cluster about his offspring and it grows more and more dear to him.* CHAMBERLIN (1897)

To avoid the unconscious *pressing of the theory to make it fit the facts and a pressing of the facts to make them fit the theory,* Chamberlin urges the method of multiple working hypotheses. This method distributes the effort and divides the affections. The method involves bringing up every rational explanation of a phenomenon and developing every tenable hypothesis about it, as impartially as possible. Figure 1 on page xx gives an example.

The investigator thus becomes the parent of a family of hypotheses, and by his parental relations to all is morally forbidden to fasten his affections unduly upon any one.

Following Chamberlin's approach, the purpose of this book is to present new members to an interesting and growing family of ideas about the nature of life, health, and disease. I am aware that some will object strenuously to some or all of the ideas presented here. Perhaps time will show that all of these perspectives hold a clue or insight into the puzzle we are working on.

HOW TO READ THIS BOOK

The reader who has reached this point in the book will realize that I am passionate about this subject and want people to think about and discuss it. An attempt is being made to develop a style of writing about scientific material that makes it accessible and useful to a variety of individuals, ranging from those with little or no science background to those who are professionals in science or biomedicine. This is a challenge when discussing a multidisciplinary investigation such as this because each scientific field has its own language system, so the reader must have the patience to shift from one to another. Hence there are various ways of using this book, depending upon the depth to which the reader wishes to investigate the subject. This can range from simply looking at the illustrations, to looking at the parts that are set off in bold type, to reading the entire text and the technical articles that are cited.

Because of the subject matter, there are places in the text where the science must be described with sufficient detail to allow an expert to examine and evaluate my ideas and conclusions. The reader lacking a PhD in biophysics or other scientific discipline will recognize these sections and is encouraged to bypass them.

In general, the book describes certain biological principles that should be apparent to the reader both by the content and by the way the book is organized. For example, whole systems perspectives are described, and if these perspectives are accurate they should be applicable to any endeavor, including the writing of a book. Integration is a whole systems perspective, so the information and ideas in the book should be integrated. Another whole systems perspective is holography. Followed to its logical conclusion, the reader should be able to get a picture of the entire message of the book from reading any part or by looking at a few of the illustrations. Finally, the reader is encouraged to jump to the Afterword, which describes the picture that has emerged from the process of writing the book.

REFERENCES

Adolph, E.A. 1979, 'Look at physiological integration', *American Journal of Physiology*, vol. 237, pp. R255-R259.

Chamberlin, T.C. 1897, 'Studies for students. The method of multiple working hypotheses', *Journal of Geology*, vol. 5, pp. 837-848.

de Robertis, E.D.P., Nowinski, W.W. & Saez, F.A. 1970, *Cell Biology*, 5th edn, WB Saunders, Philadelphia.

Garrison, F.H. 1929, *History of Medicine*, 4th edn, WB Saunders, Philadelphia, pp. 435-437, cited from Dossey, L. 1997, 'The forces of healing: Reflections on energy, consciousness, and the beef stroganoff principle', *Alternative Therapies*, vol. 3, p. 9.

Giovine, M., Pozzolini, M., Favre, A., Bavestrello, G., Cerrano, C., Ottaviani, F., Chiarantini, L., Cerasi, A., Cangliotti, M., Zocchi, E., Scarfi, S., Sara, M., Benatti, U. 2001, 'Heat stress-activated, calcium-dependent nitric oxide synthase in sponges', *Nitric Oxide*, vol 5, pp. 427-431.

Kleinfield, S. 1986, *A Machine Called Indomitable*, Times Books, Random House, New York.

Ling, G.N. 2001, *Life at the Cell and Below-Cell Level: The Hidden History of a Fundamental Revolution in Biology*, Pacific Press, New York.

Oschman, J.L. 2000, *Energy Medicine: The Scientific Basis*, Churchill Livingstone/Harcourt Brace, Edinburgh.

Pollack, G.H. 2001, *Cells, Gels and the Engines of Life*, Ebner & Sons, Seattle, Washington.

Introduction

To reach the source, one has to swim against the current. MARTIN LUTHER

ENERGY MEDICINE EMERGES

After decades of being "off limits" to academic science and medicine, we are witnessing a veritable explosion of research and exploration into energetic approaches to life and health. There is a growing appreciation that all medicine is energy medicine and that the energetic perspective holds the key to the future of the entire medical enterprise.

Several major trends have brought this about, and they are intersecting. One trend is increasing awareness in the biomedical community that electrical and magnetic fields, as well as light and sound, affect cellular processes and can be used to stimulate healing in various tissues. After a period I refer to as "the dark ages," from around 1910 to 1950, when even mention of "energy medicine" was politically incorrect in academic circles, clinical researchers rediscovered the application of different forms of energy for "jump-starting" the healing process.

A second shift is the rapidly growing popularity of complementary and alternative therapies, which are being integrated into standard medical care throughout the United States and elsewhere around the world. In some geographical regions the process of integrating energy therapies (such as those listed in Box 1) in the health care system is proceeding astonishingly rapidly but in others it is laboriously slow; however, the trend is obvious. Many people *like* complementary medicine, they *like* energy therapies, and they will pay for them out of their own pockets if they must (Eisenberg et al. 1998).

A third trend involves the area of human performance. It involves those who are responsible for the training, health, and accomplishments of high-level athletes and other kinds of performers. The demands and stresses of peak cutting-edge human activity have motivated these individuals to explore all practical and theoretical approaches that are available. They not only seek new insights and ideas, they take immediate practical advantage of them.

Box 1

Acupuncture	Cranial-sacral therapy	Radionics
Alexander	Healing touch	Reiki
Aromatherapy	Holographic repatterning	Rolfing
Aura balancing	Homeopathy	Structural integration
Bach flower essences	Massage	Therapeutic touch
Bowen	Polarity therapy	Zero balancing
Consegrity		

CLUES ABOUT LIFE IN GENERAL

As a consequence of these changes in the way health care and biology in general are being viewed, the extraordinary observations of therapists from many traditions finally are being examined from a biomedical perspective and seen for what they really are: vital clues about what is *possible* in the realms of healing and human performance in general. As science, medicine, and complementary medicine converge, entirely new vistas are opening up, revealing what has been hidden about our potentials to heal ourselves and others and about how we can accomplish more with less effort.

In thinking about where all of this is headed, I have concluded that some of the most dramatic consequences will be in the realm of the therapeutic encounter that takes place in all medical traditions. A favorite quotation:

> There is this medicine and that medicine, and this method and that method, and then there is the way the body really is.   KERRY WEINSTEIN

INTEGRATIVE MEDICINE

As complementary and conventional medicine come together as *integrative medicine,* we are discerning a clearer picture of the way the body really is and the meaning of this picture for medicine and for life in general.

New medical technologies will continue to be developed, new drugs will be designed, and new breakthroughs will take place in basic science; however, what the whole enterprise comes down to is human beings helping other human beings. If I have learned anything from watching the various therapeutic approaches in action, it is that there are countless ways of perceiving what is going on with the person who is in front of you and who needs your help and that there are many ways of stimulating healing responses. As we begin to understand the energetic exchanges taking place between individuals who are in physical contact or who are in proximity, we are seeing new possibilities for enhancing the sensitivity, subtlety, and effectiveness of the therapeutic encounter in all branches of medicine.

HUMAN POTENTIAL

These new possibilities extend far beyond medicine. Those who dedicate their lives to achievement in any endeavor are looking for ways to explore their inner resources to maximize their accomplishments. An intention of this book is to open up new possibilities, new ways of thinking, in this direction.

This book will not teach new therapies or provide new instructions for performers. There are many books, schools, and trainings that are excellent sources for such information; instead this book is a description of the communication systems that participate in any kind of human endeavor. It is written for therapists of every tradition and for their patients who wonder deeply about what is happening within themselves and within those around them in a therapeutic encounter. It is for performers of any kind who have experienced remarkable and inexplicable phenomena taking place from time to time and are curious to know if there will ever be a science to help them understand, describe, and discuss what is going on and to extend their limits. It is for educators, trainers, and thinkers from every tradition who are looking for new insights into getting the most enjoyment and success from the resources we all have within us.

NEW WAYS OF THINKING

The book is being written now because times have changed. In the past, energy medicine has been a very controversial and confusing topic, to the point that important findings have been virtually ignored, mainly because they simply were not fashionable or did not seem to fit with the dominant paradigms at the time. Scientists are human, and what they regard as the most interesting direction does not always work out. Science has a long history of blind alleys that have preoccupied generations of researchers but led nowhere. In the meantime, other concepts that could be of great value are left behind to gather dust in the libraries. We need to go through our archives and bring these precious pieces of the puzzle of life back out onto to the table.

This is more than an academic exercise. Our health care system and many of the other systems that sustain us are in crisis.

> *The significant problems we have cannot be solved at the same level of thinking with which we created them.* ALBERT EINSTEIN

Some refer to this as "thinking out of the box," evoking an image that our problems are sustained by the ways we have enclosed our thinking processes.

I have been exposed to many different kinds of thinkers and have learned to value the insights of those who are able to think creatively, "out of the box" so to speak. What is it that enables these individuals to look at the world through new eyes? It is obvious that real human progress actually depends on these people. As long as we look at the world as it is and accept things as they are, comfortable though they may seem to be, evolution ceases.

> *A paradox is not a conflict with reality. It is a conflict between reality and your feelings of what reality should be like.* RICHARD FEYNMAN

The great discoveries of medicine and of any other endeavor have come from individuals who have seen beyond the usual and accepted ways of looking at things. How do they think, how do they view the world around them, and what is the essence of creative thought? Some of the most important ideas have been so violently resisted by the community that their potential contribution to human progress has been hidden from those who could benefit from them. What are the qualities that a pioneer needs to bring his or her ideas to fruition?

EVOLUTIONARY THINKING

I believe that our future lies in the hands of those who have the capacity for clear, constructive, and innovative thought. I further believe that this kind of thinking is both exciting and contagious. It is the kind of thinking upon which our evolutionary future depends—we almost have a biological *urge* for it. Hence the book in your hands chronicles the work of some of the extraordinary thinkers of our times.

For me, the Nobel laureate Albert Szent-Györgyi (1893-1986) was a major influence. His research and writings are featured here not only because of what they have taught us about life, but also because of the infectious and creative spirit they can instill into the research and thought processes.

Szent-Györgyi made a diagram of his own scientific progress (Figure 1). While he made major contributions to science, he described his scientific journey in a way that invites us to go out on the limb:

> *There is but one safe way to avoid mistakes: to do nothing or, at least, to avoid doing something new. This, however, in itself, may be the greatest mistake of all. The selected, who are able to open*

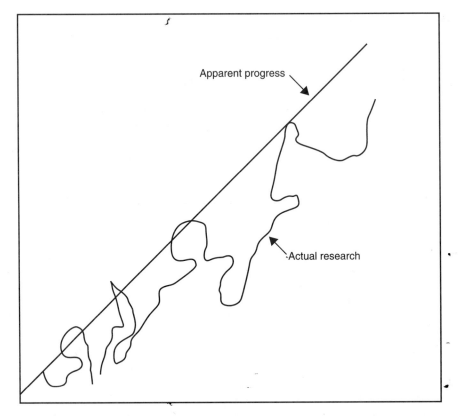

Figure 1 "The basic texture of research consists of dreams into which the threads of reasoning, measurement, and calculation are woven. Once we get somewhere, and present our results, we like to present them as a logical sequence. If the course of our work had to be plotted, we would plot it as the straight line, where A follows B, B follows C, etc. The real curve . . . would be more irregular." (From Szent-Györgyi, A. 1960, *Introduction to a Submolecular Biology,* Academic Press, New York.)

new roads to science without erring, are very few and the author, certainly, does not belong to them. The unknown lends an insecure foothold and venturing out into it, one can hope for no more than that the possible failure will be an honorable one.   SZENT-GYÖRGYI (1957)

Another modest and honest message is that mistakes are made, blind trails are followed, and one needs to admit them.

I have had to eat many of my words, and found the diet nourishing.   WINSTON CHURCHILL

DEMYSTIFICATION

This book brings the emerging information on energy medicine into focus by discussing some common denominators for a wide range of profound experiences that have seemed mysterious and virtually inexplicable in the past. Examples of these experiences are drawn from the realms of "spontaneous healing," cutting-edge athletic and artistic performance, the martial arts, and various contemplative and spiritual practices, as well as from science itself. I believe that these phenomena not only are worthy of study, but that they are vital clues that must be pursued.

The book also discusses some common experiences that have been too difficult or fuzzy for scientists to come to grips with. Just exactly what are intuition and the subconscious? What is consciousness? How are traumas and traumatic memories stored in cells and tissues? In the realms of human performance, precisely what is involved in *practice, recovery from injury, coordination,* and *awareness*?

To begin to answer these questions, we will look at current research on biological communications and energetics. The human body is composed of thousands of billions of cells that must cooperate to enable us to carry out our daily activities. A major part of the success of living systems arises from the ability of cells to communicate with each other through various signaling tactics. In the past, most of the emphasis has been on chemical messengers such as hormones, neurotransmitters, and growth factors. Electrical communication is well documented in the nervous system, but there has been far less interest in exploring the roles of other forms of signaling, such as light, sound, magnetism, and electronic conduction, even though there is documentation that all of these mechanisms exist and are important. Because these other kinds of communication have been relatively unappreciated by academic science, the use of various kinds of energy as therapies has been slow to catch on in conventional biomedicine.

We will see that there is a system or substrate within the human body that transfers and processes energy and information in ways that are different from the nervous, hormonal, and biochemical systems as they usually are portrayed. This substrate consists of the well-known materials of which the body is composed: the connective tissues and the fabrics within all of the cells throughout the body; the genetic material; and the atoms, subatomic particles, and "empty" space that is actually the body's most pervasive component. Water is an intimate and functional part of this matrix. My appreciation of the properties of this substrate has come from three primary sources: the scientists I have met, the scientific literature, and the experiences of therapists and performers of all kinds.

SYSTEMIC INTERCONNECTEDNESS

The exhilarating scientific enterprise of dissecting living systems into increasingly smaller pieces has led to spectacular advances, but in the process the essence of life and health has nearly slipped through our fingers. We have separated out and nearly discarded the single most important attribute of the organism: its *systemic interconnectedness.* We now can put this vital component back into our picture of life and thereby join together the enhanced understandings that modern science has achieved.

We are learning that the *fabric of life* consists of a physical matrix and the communications and energies that are conveyed from a place within it. This energetic and informational fabric has a physical reality and a set of properties that can be described and measured. Viewing the nervous and circulatory systems as the primary communication pathways in the body has hindered us from considering that there might be others. The *structures* of the anatomist, histologist, and cell biologist now are being looked at as the interconnected *circuitry* of the body.

In the past, science has focused mainly on the chemistry of energy and information—on molecular interactions. What we are learning now is that there is vastly more to the story.

Molecules do not have to touch each other to interact. Energy can flow through . . . the electromagnetic field. . . . The electromagnetic field along with water forms the matrix of life.

SZENT-GYÖRGYI (1988)

There is nothing wrong with biochemistry or with its medical application as pharmacology. It is just that by focusing most of our attention on molecular reactions we have missed the rest of the story: the roles of electrons, electromagnetic fields, and related energetic and quantum processes; the properties of space; and consciousness itself. This is the story that will be documented in this book.

INTUITION AND INSIGHT

Complementary therapists have been interacting with these energetic and communication systems and processes for a long time, but they have had difficulty discussing some of their most significant and exciting observations with the scientific community. Because of the illusion of a lack of a sound scientific and logical foundation for the concepts they wish to discuss, many therapists have felt they had to work on the basis of insight and intuition and that science would never be able to appreciate their most exciting observations and discoveries.

There is nothing wrong with intuition or with observations made through intuitive processes. Huge progress has been made this way. Indeed, mature scientists recognize that virtually all research springs from insights about mysteries; however, the therapists who work intuitively often have lacked a scientific language for expressing themselves, and science simply has not had a logical and testable framework to even begin to consider some of their most important observations. This situation has changed, for reasons you will learn about in this book.

THE BODY'S OPERATING SYSTEM

The story that is emerging is that the various entities comprising the body and their interactions, taken together, constitute a sort of "operating system" of the body, which, like the operating system in a computer, works silently in the background, coordinating and regulating all living processes at all levels. These processes include sensation, movement, the formation and reformation of body structures, consciousness, and physiological functioning, as well as the ways these processes come together in the perfect performance or in perfect health. Complementary therapists have invented various successful ways of interacting with the mysterious aspects of the living system, but, as just mentioned, they have had difficulty articulating what it is they are doing.

The iceberg (Figure 2) is an excellent metaphor for this inquiry because, like the patient before you, it is only the visible tip that can be seen and touched. What lies hidden below the surface is most of the mass of the iceberg. What lies hidden within the patient before you are many qualities and systems for which our metaphors are weak and fuzzy approximations and for which our most significant observations are rare and fleeting, that defy our usual logical explanations. In this book we open this entire area, below the surface, for logical analysis, discussion, and investigation.

For the most part, we are not aware of our internal operating system because ordinary consciousness enables us to perceive but a narrow slice of the world picture. It is only during the extraordinary events, when we have inexplicable and untreatable diseases or disorders, when we are involved in traumatic emergencies, or when we connect with others in a deeply nourishing and inspiring manner, that we get glimpses of the remarkable possibilities, capabilities, and vulnerabilities of life's operating system. Any medicine that denies these realities is, by its choice of

Figure 2 The iceberg is a metaphor for our inquiry into the human body because it is only the visible tip that can be seen and touched. What lies below the surface are the systems and processes that give rise to the transcendent qualities for which our metaphors are weak and fuzzy approximations. (From a composite of photographs by Ralph Clevenger, reproduced by permission of Corbis, London.)

paradigms, unable to embrace the organism as a whole and unable to address a number of major medical issues.

A DOCUMENTARY

This book has to begin with a personal story, a sort of documentary. In part it tells of distinguished and accomplished scientists and therapists whose concepts were far ahead of their time and whose work is just now coming into fruition. It is a story of laboratories, clinics, and hospitals and of changing ideas about fundamental aspects of biology and clinical practice. It is a story about scientific insight itself.

In telling this story, it is essential to reference some of the older literature. I make no apology for this. Some have the arrogant attitude that any reference more than 10 years old is useless because so much more has been learned since then. However, good research is timeless, and some of the jewels of scientific literature have been passed over and need to be taken from the shelves, dusted off, and looked at again to see what they can teach us in these times.

Although the theoretical implications of the emerging energetic concepts are staggering, the goal is to find ways that all therapists, including medical doctors, can use this information to enhance their therapeutic effectiveness. As mentioned earlier, the emerging story has implications for all who are exploring the limits of what is possible in the realms of medicine, human performance, and human potential in general. There also are implications for those who wish to open up for scientific inquiry a number of phenomena that have, in the past, seemed fuzzy, that is, too odd or vague to be approached by scientific method and logic.

REFERENCES

Eisenberg, D.M., Davis, R.B., Ettner, S.L., Appel, S., Wilkey, S., Van Rompey, M. & Kessler, R.C. 1998, 'Trends in alternative medicine use in the United States, 1990-1997', *Journal of the American Medical Association*, vol. 280, pp. 1569-1575.

Szent-Györgyi, A. 1957, *Bioenergetics*, Academic Press, New York.

Szent-Györgyi, A. 1960, *Introduction to a Submolecular Biology*, Academic Press, New York.

Szent-Györgyi, A. 1988, 'To see what everyone has seen, to think what no one has thought', *Biological Bulletin*, vol. 175, pp. 191-240.

Contents

A brief history of energy medicine

The cell is a machine driven by energy. It can thus be approached by studying matter, or by studying energy. In every culture and in every medical tradition before ours, healing was accomplished by moving energy.   Szent-Györgyi (1967)

It often is stated that the human energy field, if it does in fact exist, is theoretical. *Healing energy, energy medicine,* and *life force* all are concepts that properly belong to science fiction. However, there is actually a substantial logical and experimental basis for the existence of energy fields within and around the human body, and these fields are vitally important to the health of the organism. Indeed, these energy fields, which give rise to real *forces,* as the word is used in physics, go to the core of life. Physicians and researchers actually know this, but many simply do not realize that they know it.

It is useful to summarize the history of this subject, mainly to remind people that this information has been accumulating for a long time.

THE ELECTROCARDIOGRAM

We have known about the electrocardiogram for about 100 years. The beating of the heart produces a huge electric current that flows throughout the body, mainly because blood and extracellular fluids, with their high salt content, are extremely good conductors of electricity—and electricity is obviously a form of energy.

When Einthoven first recorded the electrocardiogram (Einthoven 1906), the field he was trying to measure was so weak that he had to use the most sensitive galvanometer available (Figure 1-1). We now know that the field of the heart is the

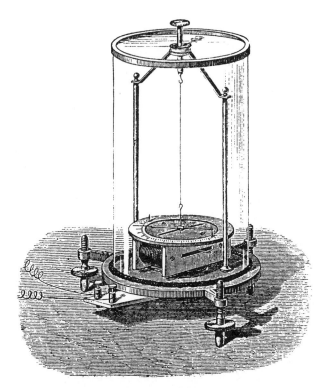

Figure 1-1 Galvanometer, constructed by Nobili, which provides "the most delicate apparatus for determining the existence, strength, and direction of weak electrical currents." (From Guillemin, A. 1872, *The Forces of Nature*, Scribner, Welford & Armstrong, New York, Fig. 415, p. 609.)

strongest field in the body. Hence our description of *strong* or *weak* has more to do with the sensitivity of our measuring instruments than with the actual strength and biological importance of the field.

Einthoven's accomplishments brought him the Nobel Prize in 1924, and electrocardiographic equipment now can be found in all hospital emergency rooms and in many physicians' offices. The electrocardiogram is a diagnostic tool rooted in energy medicine.

THE ELECTROENCEPHALOGRAM

Five years after Einthoven received his Nobel Prize for the discovery of the electricity from the heart, Hans Berger (1929) showed that much smaller electric fields could be recorded from the brain, using electrodes attached to the scalp. With some refinements the recordings, which came to be called *electroencephalograms,* became a standard diagnostic method in neurology.

MAGNETISM FROM ELECTRICITY

In 1820, Hans Christian Oersted noticed that a current flowing through a wire would cause a compass needle to move (Figure 1-2). Between 1820 and 1825, André Marie Ampère (1775-1836) quantified the phenomenon. *Ampère's Law* (see Figure 1-2) is a fundamental law of electromagnetism.

BIOMAGNETISM

Ampère's Law *requires* that electrical currents, such as those produced within the body by the activities of the heart, brain, muscles, and other organs, *must* produce magnetic fields in the space around the body. Researchers actually started measuring the magnetic field of the heart in 1963 (Baule & McFee 1963). But huge advances in this area of research began upon the discovery of a quantum phenomenon called *electron tunneling*, which provided the basis for the development of *magnetometers* of unprecedented sensitivity (Figure 1-3). These magnetometers, called *SQUIDs* (*S*uperconduct-ing *Q*uantum *I*nterference *D*evices), now are being used in university and medical research laboratories around the world to document and evaluate the biomagnetic fields in the space around the human body.

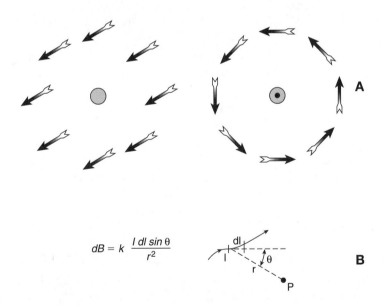

$$dB = k \; \frac{I \, dl \, \sin\theta}{r^2}$$

Figure 1-2 A, Hans Christian Oersted's accidental discovery, made in Copenhagen in 1820, was that current passing through a wire would cause compass needles to orient in a circular manner around the wire. **B,** The equation devised by Ampère, expressing the magnetic density *B* at a point *P* at a distance *r* from a current flow *I* of a particular length *l*, and at a particular angle θ between the current element and the line joining the element to the point. This is known as Ampère's Law or Laplace's Law and is a fundamental law of electromagnetism. (**A** is from Halliday, D. and Resnick, R. 1970, *Fundamentals of Physics.* Reprinted by permission of John Wiley & Sons, Inc. **B** is from Parker, S.P. [editor]. 1993, *McGraw-Hill Encyclopedia of Physics,* 2nd ed, McGraw-Hill Inc., New York.)

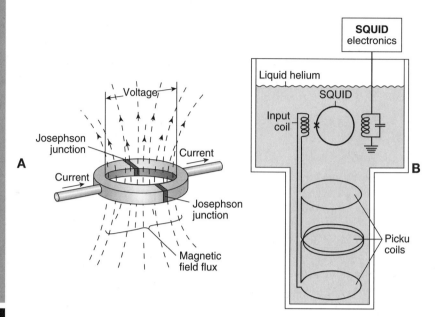

Figure 1-3 Josephson junction (**A**) and its application in the superconducting quantum interference device (SQUID) (**B**), an extremely sensitive magnetometer used to map the biomagnetic fields surrounding the human body. (**A** is from *Modern Physics* by Bernstein/Fishbane/ Gasiorowicz, © Reprinted by permission of Pearson Education, Inc., Upper Saddle River, NJ. **B** is from *Energy Medicine: the scientific basis*, Oschman, Figure 2.2, page 31, Churchill Livingstone, Edinburgh, 2000, with permission from Elsevier Science.)

The first SQUID recordings of the field of the heart were made in 1967. The heart's magnetic field is about one millionth as strong as the Earth's field. The first recordings of magnetic brain waves were reported in 1972 (Cohen 1972). The field of the brain is 100 or more times weaker than the heart's field (Figure 1-4).

Magnetism and biomagnetism are obviously forms of energy, and their application in clinical medicine can certainly be termed *energy medicine.*

A magnetic sense?

There are good reasons to believe that humans can sense biomagnetic fields:

- Biomagnetic fields are present around the organism, as we have just seen. Hence the fields are there to be detected.
- The biomagnetic fields produced by one person's body should induce tiny currents in the tissues of a nearby person (Faraday's Law of Induction, see p. 6).
- Considerable experimental evidence suggests that humans and other organisms possess a compass-like orientation or navigational sense.
- Many therapists report a magnetic-like sensation when they place their hands near an injured or diseased tissue.

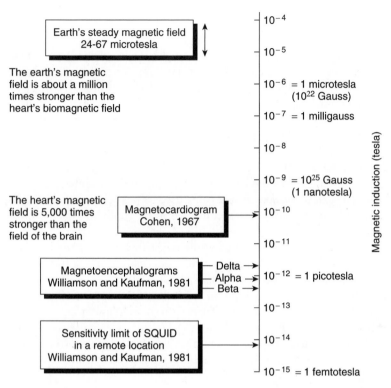

Figure 1-4 Comparison of the strengths of various magnetic and biomagnetic fields. (From *Energy Medicine: the scientific basis,* Oschman, Figure 15.1, page 220, Churchill Livingstone, Edinburgh, 2000, with permission from Elsevier Science.)

- Josephson tunneling, the basis for the SQUID magnetometer (see previous section), takes place in living systems (del Guidice et al. 1989).

Of course, the classic five senses do not include magnetism. But the points listed above, taken together, indicate that a human biomagnetic sense is possible, plausible, and worthy of investigation.

Of particular interest is the research of del Guidice et al. (1989), who found that yeast cells produced signals of the strength and frequency predicted for the presence of a Josephson tunneling effect. These signals were produced during specific stages of cell division, whereas killed cells did not produce them. "[A] pair of nearby correlated domains, namely a pair of living systems or subsystems, may be considered as a Josephson junction." We shall look into this in more detail later in the book.

There is also a large and growing literature on the ways various animals sense and use the Earth's magnetic field for navigation. This is of interest to those who study the possible biomagnetic interactions between organisms, such as those that seem to take place during energetic "diagnosis" and healing. Appendix B lists some of the literature on this subject.

Electricity from magnetism

In 1831, Michael Faraday in England demonstrated that moving a magnet near a coil of wire *induces* a measurable current flow through the wire (Figure 1-5). *Faraday's Law of Induction* is another basic law of electromagnetism.

The biological and medical significance of Faraday's Law of Induction is that moving or time-varying magnetic fields in the space around the body *must* induce current flows within the tissues. This provides a physical basis for a number of medical devices and for various energy therapies.

Clinical biomagnetism. Energetic approaches have been developed and are being applied in conventional clinical practice. In major medical centers, *magnetocardiograms* and *magnetoencephalograms* are being used by physicians for making diagnoses and clinical decisions. This fact is documented in the thousands of sites on biomagnetism on the *World Wide Web*.

The magnetic measurements, such as the *magnetocardiogram* and the *magnetoencephalogram,* are more accurate indicators of events taking place within the body than the traditional electrical measurements, called the *electrocardiogram* and *electroencephalogram,* respectively. The reason for this is that tissues are virtually transparent to magnetic fields, whereas the electrical currents have to follow complex paths of least resistance to reach recording electrodes on the skin surface. Hence electrical signals are weakened and distorted by the time they reach the skin and, therefore, are more difficult to interpret. In technical terms, these concepts can be explained as follows:

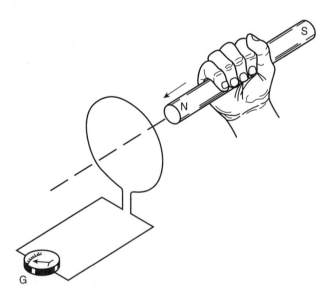

Figure 1-5 Faraday's experiment showing an effect opposite to that of Oersted's (Figure 11-3, *A*). Moving a magnet near a wire causes a current to flow through the wire. (From Halliday, D. and Resnick, R. 1970, *Fundamentals of Physics.* Reprinted by permission of John Wiley & Sons, Inc.)

Biomagnetic fields: The magnetic permeabilities of the various tissues are all about the same, approximately 1, as in a vacuum.

Bioelectric fields: The electrical resistances of different tissues vary by a factor of about 30. Bioelectric fields generated within the body take the paths of least electrical resistance, so the complex patterns measured at the body surface are intricate, diminished in strength, and much more difficult to interpret.

Clinical magnetobiology. *Biomagnetism* is the name given to the study of fields emitted by living systems, and *magnetobiology* is the study of the effects of magnetic fields on the body.

As an example of magnetobiology, medical researchers have found that pulsing electromagnetic fields (PEMFs) can "jump start" the healing process in a variety of tissues. The most widely used example is the application of PEMFs to stimulate the repair of fracture "nonunions." Most orthopedic surgeons have prescribed this method at one time or another (Bassett 1995). This is obviously an example of the clinical application of energy medicine.

PROMISING DIRECTIONS

Success with PEMFs for bone healing led to research on other tissues. It has been discovered that each tissue responds to a particular frequency. Clinical methods are being developed to use PEMFs to stimulate repair of ligaments, nerves, capillaries, and skin.

Particularly promising is the use of pulsing fields in association with silver-coated fabrics that can be placed on a wound or infected tissue. The method was pioneered by Berger et al. (1976) and Spadaro et al. (1974), who developed a method for treating complex bone infections in the early 1970s. Broader clinical applications have been advanced by Becker, Flick, and Becker (1998) and Flick (2000).

Meanwhile, there has been a surge in interest in the use of millimeter waves in the countries of the former Soviet Union. Millimeter waves are in the extremely high-frequency range of 30 to 300 GHz. Extensive research has shown that very low intensity signals (10 mW/cm^2 and less) affect cell growth and proliferation, the activities of enzymes, the state of the cell genetic apparatus, and the function of excitable membranes, receptors, and other biological systems. In animals and humans, local exposure to these fields stimulated tissue repair and regeneration, alleviated stress and pain, and accelerated recovery from a wide range of diseases. Over 50 diseases and conditions have been claimed to be treated successfully with millimeter wave technology alone or in combination with other means. There are more than 1,000 millimeter wave therapy centers in the former Soviet Union, and by 1995 more than three million people had received this type of therapy (reviewed by Pakhomov et al. 1998).

In the United States, research on the use of this noninvasive therapy for analgesic, anti-inflammatory, and immune-stimulating uses is progressing at the Richard J. Fox Center for Biomedical Physics, Temple University School of Medicine, Philadelphia, Pennsylvania. Primary investigators involved are Mikhail A. Rojavin, Ph.D., and Marvin C. Ziskin, M.D. (Rojavin & Ziskin 1997).

ENERGY THERAPIES

As long ago as 1984, Janet Quinn was researching scientific evidence that energy therapies involve an energy exchange (Quinn 1984). A number of studies have shown that practitioners of Healing Touch, Therapeutic Touch, Qi Gong, and other energy therapies are capable of projecting pulsing electromagnetic fields from their hands. Fascinating research by Dr. John Zimmerman, at the University of Colorado Health Sciences Center, showed that the signals practitioners project from their hands correspond in strength and frequency to those produced by various clinical devices that are being developed to stimulate the repair of different kinds of tissues (Zimmerman 1990). Dr. Zimmerman's important findings have been confirmed by researchers in China and Japan studying practitioners of Qi Gong, meditators, and other practitioners (Seto et al. 1992).

Mechanism

Not only is the evidence for energy fields in and around the body compelling, but careful research has given us a logical explanation of *how* these fields affect cells. Healing, like other biological processes, is essentially a cellular process. PEMFs initiate a cascade of reactions, leading from the cell membrane to the cytoplasm to the cell nucleus and the DNA, activating cellular processes (Figure 1-6). The great sensitivity cells have, which enables them to pick up the signals from an energy medicine practitioner, is accounted for by research that led to the 1994 Nobel Prize. *Cellular amplification* enables a single photon of electromagnetic energy to initiate a massive influx of calcium into the cell, triggering cellular activities such as immune surveillance, regeneration, tumor invasion, and injury repair (Gilman 1997). Intimately involved in the amplification process are the G-proteins (so named because they bind guanosine triphosphate). Gilman and Rodbell received the 1994 Nobel Prize in Physiology or Medicine, in part for demonstrating that G-proteins integrate multiple signals from outside the cell and activate the various cellular amplifier systems.

The role of the heart

From what we have learned from energy medicine research, we are beginning to discern a much clearer picture of the nature of the healing response. It is known that the heart generates the largest electrical and magnetic field of the body. The fields of both the heart and the brain contain signals in the biologically important part of the energy spectrum known as the *ELF (extremely low frequency)*.

Important work on this subject was reported in *The Heart's Code* (Pearsall 1998). Research at the Institute for Heart Math (McCraty, Atkinson & Tomasino 2001; McCraty et al. 1998) has shown a relationship between emotional state and the frequency spectrum of the electrical signals from the heart. Feelings of love, caring, and compassion, or of frustration and anger, will affect the signals produced by the heart. These signals are conducted to every cell in the body and are radiated into the space around the body.

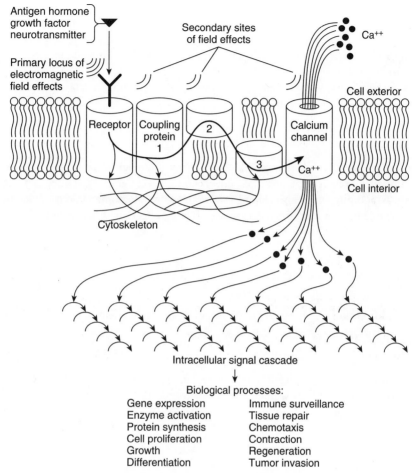

Figure 1-6 Cellular cascade and amplification, including parts of the system that are sensitive to magnetic fields. A single antigen, hormone, pheromone, growth factor, or smell or taste or neurotransmitter molecule, or a single photon of electromagnetic energy emitted from such a molecule as a result of the vibration of its electrically charged components, can activate a cascade of intracellular signals that initiate, accelerate, or inhibit biological processes. This is possible because of enormous amplification—a single molecular event at the cell surface can trigger a huge influx of calcium ions, each of which can activate an enzyme. The enzymes, in turn, act as catalysts, greatly accelerating biochemical processes. The enzymes are not consumed by these reactions and therefore can act repeatedly. Some of the reactions inside the cell also are sensitive to electromagnetic fields, some are not, and a few have not yet been tested. Some frequencies enhance calcium entry, others diminish it. Steps in the cascade involving free radical formation are likely targets of magnetic fields. Some of the products of the cascade are returned back to the cell surface and into the surrounding extracellular space. Molecular events within cells set up electronic, photonic, and electromechanical waves (phonons) that propagate as solitons (see Chapter 19) through the cellular and extracellular matrix. These feedbacks enable cells and tissues to form a functionally organized society. The cells "whisper" to each other in a faint and private language. They can literally "tune into" each other over long distances. (From Oschman, J.L. 2000, *Energy Medicine: The Scientific Basis*, Churchill Livingstone, Edinburgh, Fig. A-1, p. 253, with permission from Elsevier Science.)

Because cellular regulations can be influenced by pulsing electromagnetic fields in this same frequency range, the possibility arises that pulsations originating from the heart may be involved in the effects of various complementary therapies such as Healing Touch, Therapeutic Touch, Reiki, and the other methods listed in the table in the Introduction.

Many therapeutic schools teach the importance of attaining a relaxed and caring state on the part of the practitioner:

> *... focus completely on the well-being of the recipient in an act of unconditional love and compassion.* QUINN AND STRELKAUSKAS (1993)

Freud (1856-1939) was convinced that an energy exchange of some kind between practitioner and patient operates at an unconscious level to affect the patient's mental, emotional, and physical health (Freud 1962). That such phenomena do take place has been documented in a number of studies showing effects with or without physical touch:

- Wound healing (Wirth 1990)
- Pain (Keller 1986; Redner, Briner & Snellman 1991)
- Hemoglobin levels (Krieger 1974)
- Conformational changes in DNA and water structure (Rein & McCraty 1994)
- Tumor growth (Bengston & Krinsley 2000)
- Emotional state (Quinn 1984)

In spite of this research, western biomedicine has continued to be justifiably skeptical because of the lack of a plausible mechanism to explain the nature of the proposed energy exchange or how it could have any physiological or emotional effects. Again, times have changed. There are now plausible and testable scientific explanations that are leading to research on these ideas.

MEASURING ENERGY EXCHANGES

Given the Quinn hypothesis, there is great significance to the research from laboratories that have documented energy exchanges between people who are touching or are in proximity. These findings are summarized in a remarkable paper entitled *The Electricity of Touch* (McCraty et al. 1998). Specifically, one's electrocardiogram signal can be registered in another nearby person's electroencephalogram and elsewhere on his or her body. The signal is strongest when people are in contact, but it is still present when subjects are in proximity without contact. As expected for energy field interactions, the strength of the effect diminishes with distance (McCraty, Atkinson & Tomasino 2001).

All in all, this information is leading to a deeper appreciation of the energetic interactions taking place between therapists and their patients. Physicians palpate for diagnostic purposes, but we can now see that palpation may have a deeper significance as a healing endeavor.

We now have a set of logical, testable, and refutable scientific hypotheses that can account for the effects of the various energetic therapies. Such energy exchanges have been talked about in one way or another for a long time, including in the spiritual context, so it is not surprising that the development of sensitive measuring instruments is enabling us to measure these effects.

OTHER KINDS OF ENERGY

We focus on electrical and magnetic energies because these are the easiest to measure and we know more about their effects. But the body also produces other kinds of energy, including light, sound, heat, chemical energy, gravity, and elastic energy. Research has shown that both light and sound signals can be emitted by the hands of healers. Scientific study of the biological and clinical significance of these different forms of energy is in progress and can appropriately be called *energy medicine.*

There may be additional kinds of energy, not yet measured, and these often are called *subtle energies.* Although such mysterious energies often are referred to in the healing literature, it is the author's opinion that we can build up a good picture of living energetics without referencing mysterious and unknown forces. This certainly should not be taken to mean that there are no such things as subtle energies or that we have discovered all there is to learned about the energetic phenomena in nature.

REFERENCES

Ampère, A.M. *Mémoires sur l électromagnétisme et l électrodynamique,* Paris, Gauther-Villars et cie, 1921.

Bassett, C.A.L. 1995, 'Bioelectromagnetics in the service of medicine' in ed. M. Blank, *Electromagnetic Fields: Biological Interactions and Mechanisms,* Advances in Chemistry Series 250. American Chemical Society Washington, DC, pp. 261-275.

Baule, G.M. & McFee, R. 1963, 'Detection of the magnetic field of the heart', *American Heart Journal,* vol. 66, pp. 95-96.

Becker, R.O., Flick, A.B. & Becker, A.J. 1998, *Iontophoresis system for stimulation of tissue healing and regeneration,* United States patent 5,814,094.

Bengston, W.F. & Krinsley, D. 2000, 'The effect of the "laying-on of hands" on transplanted breast cancer in mice', *Journal of Scientific Exploration,* vol. 14, no. 3, pp. 353-364.

Berger, H. 1929, 'Uber das Elektrenkephalogramm des Menschen', *Archiv fur Psyckchiatriaca,* vol. 87, pp. 527- 570.

Berger, T.J., Spadaro, J.S., Bierman, R., Chapin, S.E. & Becker, R.O. 1976, 'Antifungal properties of electrically generated metallic ions', *Antimicrobial Agents and Chemotherapy,* Nov., pp. 856-860.

Cohen, D. 1972, 'Magnetoencephalography: Detection of the brain's electrical activity with a superconducting magnetometer', *Science,* vol. 175, pp. 664-666.

Del Guidice, E.S., Doglia, S., Milani, M., Smith, J.M. & Vitello, G. 1989, 'Magnetic flux quantization and Josephson behavior in living systems', *Physica Scripta,* vol. 40, pp. 786-791.

Einthoven, W. 1906, 'Le télécardiogramme', *Archives Internationales de Physiologie,* vol. 4, pp. 132-164.

Flick, A.B. 2000, *Multilayer laminate wound dressing,* United States patent 6,087,549.

Freud, S. 1962, 'The anxiety neuroses', in *The Standard Edition of the Psychological Works of Sigmund Freud*, ed. J. Strachey, Hogarth Press, London, UK, pp. 107-111.

Gilman, A.G. 1997, 'G proteins and regulation of adenylyl cyclase. Nobel Lecture presented December 8 1994', in *Nobel Lectures Physiology or Medicine, 1991-1995*, ed. N. Ringertz, World Scientific, Singapore, pp. 182-212.

Kobayashi, A. & Kirschvink, J.L. 1995, 'Magnetoreception and electromagnetic field effects: Sensory perception of the geomagnetic field in animals and humans', in *Electromagnetic Fields: Biological Interactions and Mechanisms*, Advances in Chemistry Series 250, ed. M. Blank, American Chemical Society, Washington, DC, pp. 367-394.

Krieger, D. 1974, 'Healing by the laying on of hands as a facilitator of bio-energetic change: The response of in vivo human hemoglobin', *Psychoenergetic Systems*, vol. 1, pp. 121-129.

McCraty, R., Atkinson, M., Tomasino, D. & Tiller, W.A. 1998, 'The electricity of touch: Detection and measurement of cardiac energy exchange between people', in *Brain and Values: Is a Biological Science of Values Possible*, ed. K.H. Pribram, Lawrence Erlbaum Associates, Mahwah, NJ, pp. 359-379 (also available from the web site of the Institute for Heart Math, Boulder Creek, CA [www.heartmath.org/ResearchPapers/Touch/Touchsum.html]).

McCraty, R., Atkinson, M. & Tomasino, D. 2001, *Science of the Heart. Exploring the Role of the Heart in Human Performance*, Publication No. 01-001, Institute of Heart Math, Boulder Creek, CA.

Pakhomov, A.G., Akyel, Y., Pakhomova, O.N., Stuck, B.E., Murphy, M.R. 1998, 'Current state and implications of research on biological effects of millimeter waves: A review of literature', *Bioelectromagnetics*, vol. 19, no. 7, pp. 393-413.

Pearsall, P. 1998, *The Heart's Code. Tapping the Wisdom and Power of Our Heart Energy. The New Findings About Cellular Memories and Their Role in the Mind/Body/Spirit Connection*. Broadway Books, New York.

Quinn, J. 1984, 'Therapeutic touch as an energy exchange: Testing the theory', *Advances in Nursing Science*, Jan., pp. 42-49.

Quinn, J.F. & Strelkauskas, A.J. 1993, 'Psychoimmunologic effects of therapeutic touch on practitioners and recently bereaved recipients: A pilot study', *Advances in Nursing Science*, vol. 15, no. 4, pp. 13-26.

Redner, R., Briner, B. & Snellman L. 1991, 'Effects of a bioenergy healing technique on chronic pain', *Subtle Energies*, vol. 2, no. 3, pp. 43-68.

Rein, G. & McCraty, R. 1994, 'Structural changes in water and DNA associated with new physiologically measurable states', *Journal of Scientific Exploration*, vol. 8, no. 3, pp. 438- 439.

Rojavin, M.A. & Ziskin, M.C. 1997, 'Therapy with millimeter radiation in Eastern Europe: Treatments unknown to western doctors', *EMF Health Report*, vol. 5, no. 4.

Seto, A., Kusaka, C., Nakazato, S., Huang, W., Sato, T., Hisamitsu, T, Takeshige, C. 1992, 'Detection of extraordinary large biomagnetic field strength from human hand', *Acupuncture and Electro-Therapeutics Research International Journal*, vol. 17, pp. 75-94.

Spadaro, J.A., Berger, T.J., Barranco, S.D., Chapin, S.E. & Becker, R.O. 1974, Antibacterial effects of silver electrodes with weak direct current', *Antimicrobial Agents and Chemotherapy*, Nov., pp. 637-642.

Wirth, D.P. 1990, 'The effect of non-contact therapeutic touch on the healing rate of full thickness dermal wounds', *Subtle Energies*, vol. 1, no. 1, pp. 1-20.

Zimmerman, J. 1990, Laying-on-of-hands healing and therapeutic touch: A testable theory', *BEMI Currents, Journal of the Bio-Electro-Magnetics Institute*, vol. 2, pp. 8.17 (Available from Dr. John Zimmerman, 2490 West Moana Lane, Reno, NV 89509-3936, USA; also see 1985, 'New technologies detect effects of healing hands', *Brain/Mind Bulletin*, Sept. 30, p. 10).

Energy medicine today

Scientists are trained doubters, but the traditional bias against energy medicine has gone deeper than skepticism. Fortunately this situation is changing rapidly. There is a growing appreciation that energetic approaches, released from a historic academic hostility, will play a key role in the future of medicine.

Certainly a large amount of valuable information can be gathered together by looking at life from an energetic perspective. That this information has not been very well assembled in the past, has not been looked upon as a whole, and has been rarely discussed in biomedical research circles is due primarily to a sort of academic myopia that finally has been diagnosed and treated.

For rather arcane historical reasons, those engaged in the pursuit of the knowledge and wisdom of life often have chosen to argue about various issues that, although interesting and even fascinating, now can be seen as distractions from the development of the larger picture. Debates about *wholism* versus *reductionism*, *vitalism* versus *mechanism*, and *quantitative* versus *qualitative* approaches have preoccupied thinkers for centuries and have hindered them from observing the center, the core, the kernel of life, in all of its intricate wonder. In terms of the pace of medical progress, the possibilities for understanding disease and disorder, and reducing human suffering and loss of life, this has been a costly distraction.

A previous book (Oschman 2000) began to open up the subject of energy medicine to cooperative and creative exploration by physicians, researchers, and energetic therapists. For the energy therapists, the book provided scientific information that began the overdue logical validation of their work. For the physicians, there was a new appreciation that the energetic therapies their patients seem to like so much may not be so strange, illogical, and untestable as the medical profession had thought. Indeed, physicians have been using energetic diagnostic and treatment tools for a long time. For biomedical researchers, there were new hypotheses to be examined, validated, or refuted.

As we shall soon see, physicists are joining the inquiry. It is for this reason that the previous book, and this one, contains numerous references to scientific articles. If scientists and physicians wish to question the author's conclusions about any subject discussed here, and I certainly hope they will, they can delve into the published literature and create their own interpretations, hypotheses, and conclusions.

ALL MEDICINE IS ENERGY MEDICINE

The academic myopia mentioned earlier has created a situation in which a valuable perspective on life has simply not been explored as well as it could be. The author sees and reads about all kinds of sophisticated activities taking place in laboratories and research centers, involving multitudes of serious, dedicated, highly trained professionals working in many different areas of inquiry. Meetings with many of these individuals have led to an appreciation of the remarkable work they do. All of their research relates to energy in one way or another, but the academic myopia has prevented many of them from looking at the medical significance of their efforts through an energetic lens. Using energy of any kind in therapy has long been frowned upon. This is, of course, an absurd situation.

The statement that *all medicine is energy medicine* obligates and challenges the author to be specific about what energy medicine really is and what it means for all of medicine. There is another important obligation. When integrating new and old concepts it is important to state the old concepts in a clear and logical manner so that any reader can contemplate both what is already known and what the new synthesis might mean in terms of his or her personal experience or thinking process. The overall goal of this book is not to convince, it is to inquire, learn, and stimulate creative thought.

AN EVOLUTIONARY PERSPECTIVE

There are several premises this book is based upon, and there are good reasons to state them at the beginning. As a biologist, and someone in complete awe of life's amazing accomplishments and possibilities, the author has developed some impressions, hypotheses, and perspectives that can be specifically stated and opened up for discussion, debate, testing, and refutation.

The first of these is an evolutionary perspective. It is through the study of evolution that living structure and function and behavior begin to make sense. If darwinian evolutionary theory is correct, the living being responsible for assembling the words you are now reading is the culmination of uncountable successful and unsuccessful experiments with the laws of physics as they pertain to survival and to the continuation of our species. Indeed, if we are to really understand that area we refer to as *physics,* the author believes that the best place to look is within ourselves, for it is here that we will find the "laws of physics" expressed in their most sophisticated form. We are the culmination of millions of years of evolutionary survival testing; by comparison, our current physical and biological sciences are the culmination of a few centuries of intellectual activity that has been compromised and distracted by issues far less potent than survival, such as egos, vested interests, bickering, and a certain amount of arrogance.

Hence the author has in mind both the application of the laws of physics to living processes and the application of the phenomena of biology to the laws of physics.

The true physics is that which will, one day achieve the inclusion of humanity in a coherent picture of the world.   PIERRE TEILHARD DE CHARDIN

PHYSICAL BIOLOGY AND BIOLOGICAL PHYSICS

A corollary to this idea is that real progress can take place in both biology and physics when we bounce back and forth between the two domains. What has happened, though, is that this creative interactive process has been stifled by the academic myopia mentioned earlier. Energy is a major aspect of physics, and to exclude from biology and medicine the powerful tools physics has developed for the careful and logical study of energetic interactions has short-circuited the whole intellectual enterprise.

Likewise, physics has had its own deep confusions about biology. One of the most obvious of these has been the willingness of some physicists to make pronouncements about what is possible and what is not possible in the realm of biology. These suggestions from experts are appreciated, but they often are wrong. For example, the following statement by a physicist applies equally to biological systems:

A philosopher once said 'It is necessary for the very existence of science that the same conditions always produce the same results.' Well, they do not.   RICHARD FEYNMAN

And the following conclusions, by a distinguished regulatory physiologist, can be applied to physics as well:

The mature scientist knows that cause and effect are elusive because of the presence of multiple correlations. No properties are uncorrelated—all are demonstrably inter-linked. And the links are not single chains, but a great number of criss-crossed pathways. Hypertension provides an example. The volume of the blood is regulated. The large veins and the chambers of the heart have mechanoreceptors that enable both instantaneous and long-term control of blood volume. These stretch receptors report to an elaborate network that has so many feedback loops that the kind of response at any particular instant cannot be predicted. The wealth of sensitivities, pathways, and effectors that have been demonstrated allows many possible modes of regulation, and makes it virtually impossible to identify any one as being predominant.   ADOLPH (1979)

A century of biological research has confirmed that living systems have some capabilities that physics has not yet fathomed. The hallmark of these phenomena, mentioned in Chapter 1, is *amplification,* in which a tiny stimulus yields a huge effect.

To summarize, it is true that accepted physical theory summarized in Chapter 1 allows one to calculate the currents that will be *induced* within an organism by *magnetic fields* in the environment. These fields come from other organisms or from household electrical appliances and the wires going to them, cell phones, radar, and other technologies. Organisms do, indeed, have to obey physical laws. But there is a deep problem. Biologists look at plant and animal behavior and see evidence that organisms have sensitivities far beyond those physicists are able to account for by applying physical principles. What is going on here? Are the biologists hallucinating, or are the physicists missing something?

What is going on is that what we have said earlier is correct, that living systems have "learned" far more about the laws of physics than most physicists can possibly imagine. So when a biologist (or a therapist or an athlete) notices a behavior or a sensitivity that seems to violate what seems to be possible, this author takes notice. Living systems have a long evolutionary history of testing, learning, and devising survival strategies. These strategies have to be quite sophisticated. Living systems are reliable and can be believed, whereas physical theory is, by comparison with life, in its infancy.

LESSONS FROM COMPUTER AND COGNITIVE SCIENCE

There is an exciting place in modern research and engineering where a dynamic interplay between biology and physics can be seen in operation. It is in the multidisciplinary conversations between neurophysiologists and computer designers, symbolized in Figure 2-1. Each advance in neurophysiology leads to improvements in the design of sophisticated high-speed computers; each advance in computer theory leads to a search for its possible application in studies of the neurophysiological basis of behavior and consciousness. This back-and-forth interplay has gone on for years and has led to astonishing advances in two fields that are working with each other cooperatively and synergistically. It is an exciting adventure and one that has not yet begun to approach completion. The information in this book directly relates to the future of that creative process.

BIOMIMICRY

The synergy between cognitive and computer science is a part of a new and revolutionary and extremely promising field of endeavor called *biomimicry* (Benyus 1998).

> *Biomimicry is a new science that studies nature's models and then imitates or takes inspiration from these designs and processes to solve human problems, e.g., a solar cell inspired by a leaf. Biomimicry uses an ecological standard to judge the "rightness" of our innovations. After 3.8 billion years of evolution, nature has learned: What works. What is appropriate. What lasts. Biomimicry is a new way of viewing and valuing nature. It introduces an era based not on what we can extract from the natural world, but on what we can learn from it.* BENYUS (1998)

SURGERY IS ENERGY MEDICINE

What can a surgeon or a psychotherapist or an internist or a family practice physician learn from the energetic perspective? When we say that all medicine is energy medicine, what exactly does this mean?

Look closely at the surgical process as an example. The scalpel can be viewed as a profoundly important energy medicine device that is capable of opening up the body's interior for therapeutic manipulations. We would like to know precisely what is hap-

Figure 2-1 The interface between neurons and an integrated circuit. (The logo of the International Neural Network Society, INNS, Mt. Royal, New Jersey, used with permission of the Society.)

pening, at the cellular, molecular, and atomic levels, when a sharp edge passes through tissues.

In spite of its great medical significance, there is much to be learned about the physics, energetics, and cell biology of cutting (for a recent technical look at this subject, see Forest Van Der Giessen & Kubin 2001). Applications of this knowledge are leading to improvements in the design of scalpels. Surgeons have been using electrical currents to cut and coagulate human tissues since the early 1900s. The method is called *electrocautery*. New surgical devices have been developed recently that provide far more precise coagulation and vessel shrinking that rapidly stops bleeding in highly vascular tissue and eliminates air leaks in lung surgery (TissueLink Floating Ball and Harmonic Scalpel, TissueLink Medical, Dover, NH). The new devices combine a conductive fluid with radiofrequency energy to seal tissue.

Clinical medicine takes place within a larger *energetic context* that we are learning a lot about. In the author's local hospital, surgeons are finding that preoperative and postoperative Reiki sessions seem to enable patients to relax and thereby have a less traumatic experience.

Naturally, physicians want to know what, if anything, is going on in a treatment by a Reiki or other energetic practitioner. Is it simply the presence of a relaxed and caring person, spending time focused on the patient, with soothing music in the background?

Or is some more profound interaction taking place that can directly affect the physiology of the patient? At first glance, the second possibility seems incomprehensible, but let us take a closer look.

ENERGETIC SYMBOLS

One method used in various energy therapies is the visualization of specific symbols. An example, used in Reiki therapy, is shown in Figure 2-2. Known as *Dai-Ko-Myo*, this symbol is alleged to stimulate the immune system and to have other beneficial effects (Stein 1995). It is one of many symbols that have been handed down from teacher to disciple for over 1,000 years.

The logical mind rejects this whole concept, for how could looking at a symbol on a piece of paper conceivably have any physiological effect? But the answer is extremely simple. The hallmark of the process is amplification, in which a tiny stimulus yields a huge effect. The possible physiological significance of symbols, remarkable as it may seem, is worthy of detailed description because it provides the basis for a number of the phenomena we will discuss in this book.

We know an image on the retina, of a symbol or of any object in the visual field, results in a pattern of electrical activity that travels through the optic nerve to the optic lobes of the brain (Figure 2-3). The mechanisms involved have been carefully researched and much is known. The pattern of light on the retina is translated into a pattern of impulses on the occipital cortex. The retina is projected point to point onto the cortex (Polyak 1934; Talbot & Marshall 1941).

We also know that an amplification or magnification takes place. The visual cortex has a topographic map of the retina, but the cortical map has 10,000 times the area of the corresponding retinal area. There are about 100 cortical cells representing each retinal cone cell (Talbot & Marshall 1941).

Figure 2-2 Visualizing Dai-Ko-Myo, a Reiki symbol alleged to stimulate the immune system and to have other beneficial effects. (From Stein, D. 1995, *Essential Reiki. A Complete Guide to an Ancient Healing Art*, The Crossing Press Inc., Freedom CA, p. 99. Used by the permission of the artist, Ian Everard.)

It has been known for a long time that a few photons of light can trigger a nerve impulse that travels into the brain via the optic nerve. There is a huge amplification that takes place because a few photons can trigger the entry of hundreds or thousands of calcium ions into the retinal cell, leading to a depolarization of the membrane and a nerve impulse (Stryer 1985).

Nerves from the retina contact or synapse on many other nerves, so the energy (electrical energy) tends to spread out over a broad area as the signal propagates through the circuits shown in Figure 2-3, *A*.

When a neuron conducts an impulse, a magnetic field *must* be created in the surrounding space (Oersted and Ampère again) (Figure 2-3, *B*). Tissues are virtually transparent to magnetic fields, so the fields extend through the tissues and into the space around the head. We have seen in the previous chapter that modern magnetometers called *SQUIDs (Superconducting Quantum Interference Devices)* permit the mapping of neuromagnetic fields set up by neural activities within the brain (Figure 2-4). This is called *magnetic source imaging* (Cuffin & Cohen 1979; Ko et al. 1998).

CONTINUITY AND INJURY REPAIR

The electrical and magnetic fields produced by neural activity in the brain are not confined to the head and its surroundings—they spread throughout the body via the

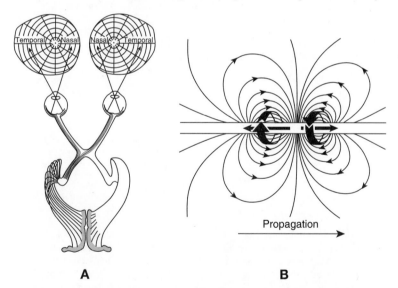

A **B**

Figure 2-3 A, The central visual pathways to the left hemisphere. (Originally from Homans, J., 1941, *A Text-book of Surgery,* 5th edn, Charles C Thomas, Springfield, IL, and reproduced from Ruch, T.C., & Patton, H.D. 1965, *Howell & Fulton's Physiology and Biophysics,* 19th edn, WB Saunders, Philadelphia, p. 444. Reproduced with kind permission of Elsevier Science.) **B,** The first measurements of the magnetic field produced by a nerve impulse. The recordings were made using a SQUID magnetometer (see Figure 1-3). (Reprinted with permission from Wikswo J.P., Barach, J.P. & Freeman, J.A. 1980, 'Magnetic field of a nerve impulse: First measurements', *Science,* vol. 208, pp. 53–55, Copyright 1980, American Association for the Advancement of Science.)

nervous system, its connective tissue sheaths (perineurium, also shown in Figure 2-4), and the circulatory system, which extends into every part of the brain and into every part of the body and is a good conductor of electricity. This kind of whole-body *continuity* has been beautifully expressed by Deane Juhan:

> *The skin is no more separated from the brain than the surface of a lake is separate from its depths; the two are different locations in a continuous medium . . .*
> *The brain is a single functional unit, from cortex to fingertips to toes. To touch the surface is to stir the depths.* JUHAN (1987)

In spite of its importance and the research that has been done, the electrical properties of the perineural system have not been well appreciated. Low-frequency electrical oscillations set up in the perineural tissues have a key role in injury repair and regulate consciousness, as discussed in a series of important papers (Becker 1990, 1991). In terms of the healing response, there is no research that could be more important than the study of the electrical properties of the perineural system, and recent discoveries are encouraging this kind of research. The reason for this is the new discovery that the connective tissue cells in the brain, called *glia* and *astrocytes,* do far more than function as support elements: they form a dynamic communication system of their own

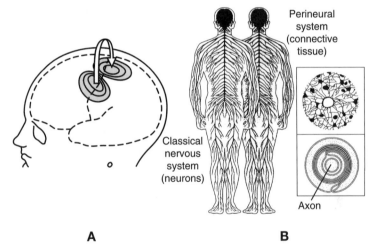

A   **B**

Figure 2-4 **A,** Magnetic source imaging. Current flows set up by neural activities within the brain produce magnetic fields that pass through the various tissues and that can be measured with a magnetometer a distance away from the scalp. (From Okada, Y. 1983, in *Biomagnetism: An Interdisciplinary Approach,* NATO Advanced Study Institute Series, ed. S.J. Williamson, G.L. Romani, L. Kaufman, I. Modena, Plenum Press, New York, p. 409, Fig. 12.6.1.) **B,** The classic nervous system composed of neurons *(left)* and the perineural nervous system composed of connective tissue cells called the *perineurium (right).* Insets show examples of perineural cells. Upper inset shows fibrous astrocytes with end feet around a small blood vessel. Lower inset shows a Schwann cell surrounding an axon in the peripheral nervous system. (From Oschman, J.L. 2000, *Energy Medicine: The Scientific Basis,* Churchill Livingstone, Edinburgh, Fig. 15.2, p. 225, with permission from Elsevier Science. Upper insert is originally from Glees 1955 *Neuroglia: Morphology and Function,* with permission from Blackwell Science Ltd. Lower inset is adapted from Bloom & Fawcett's *Textbook of Histology,* 12th edition, 1994, Fig. 11-21, p. 335, with permission from Hodder Arnold.)

(reviewed by Bezzi et al. 2001; Kirchoff, Dringen & Giaume 2001; Mazzanti, Sul & Haydon 2001; Newman 2001). The initial discovery was that chemical transmitters released from neurons can create calcium waves in glial cells. A new way of thinking about the nervous system is emerging, in which glial and other connective tissue cells have previously unsuspected communication and regulatory roles.

> *For decades, scientists thought that all of the missing secrets of brain function resided in neurons. However, a wave of new findings indicates that glial cells, formerly considered mere supporters and subordinate to neurons, participate actively in synaptic integration and processing of information in the brain.*

VESCE, BEZZI, AND VOLTERRA (2001)

We shall return to this topic in Chapter 15.

That the electrical circuits within the body involved in injury repair and other activities have corresponding patterns in the space around the body has been known for thousands of years (Figure 2-5). As we have seen, modern science only recently has developed the instrumentation to study the magnetic aspects of these circuits. Modern science is beginning to catch up with ancient wisdom.

Hence looking at a symbol, or even thinking about (visualizing) a symbol, results in considerable neural activity involving measurable electrical and magnetic fields. Writing the symbol would activate sensory and motor pathways and therefore bring in many more electrical and magnetic fields. The concept that viewing, visualizing, or writing a symbol can trigger specific patterns of electrical and magnetic fields in and around the body is certainly reasonable, logical, and testable. The idea that some of these fields might be beneficial to a nearby person also is reasonable and testable. A thousand or so years of experience with these symbols may actually connect with modern science.

Many other examples could be provided to show that the energy perspective is useful and viable and a way of looking at all medical interventions.

ENERGY MEDICINE IN HEALTH CARE

The introduction of Reiki and other complementary therapies into the hospitals in the New England region is viewed by many as a huge advance in the health care system. A consequence of this development is that the author has been repeatedly invited to present Grand Rounds and continuing medical education courses in hospitals and other facilities around the world. A goal of these presentations is to show that the energetic approach to health and disease is a productive, profitable, and useful line of inquiry. Energy medicine has meaning for all therapies. In spite of the academic myopia cited earlier, much more progress has actually been made that most people realize.

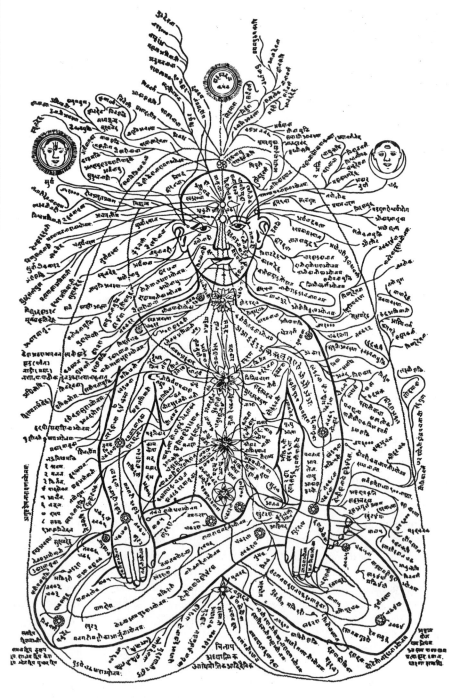

Figure 2-5 Ancient diagram of the energetic pathways in and around the human body. (Attributed to the prophet Ratnasara, Tibet, ca 19th century.)

REFERENCES

Becker, R.O. 1990, 'The machine brain and properties of the mind', *Subtle Energies*, vol. 1, no. 3, pp. 79-97.

Becker, R.O. 1991, 'Evidence for a primitive DC electrical analog system controlling brain function', *Subtle Energies*, vol. 2, no. 1, pp. 71-88.

Benyus, J.M. 1998, *Biomimicry. Innovation Inspired by Nature. Inside the Revolutionary New Science That is Rediscovering Life's Best Ideas—and Changing the World*, William Morrow & Co., New York.

Bezzi, P., Domereq, M., Vesce, S. & Volterra, A. 2001, 'Neuron-astrocyte cross-talk during synaptic transmission: Physiological and neuropathological implications', *Progress in Brain Research*, vol. 132, pp. 255-265.

Cuffin, B.N. & Cohen, D. 1979. 'Comparison of the magnetoencephalogram and electroencephalogram', *Electroencephalography in Clinical Neurophysiology*, vol. 47, pp. 132-146.

Forest Van Der Giessen, S.E. & Kubin, L., 2001, Fifth European Mechanics of Materials Conference on Scale Transition from Atomistics to Continuum Plasticity: Proceedings, Delft, The Netherlands, 5-8 March 2001, *Journal de Physique IV*, vol. 11, no. 5, pp. 1-337.

Juhan, D. 1987, *Job's Body. A Handbook for Bodywork*, Station Hill Press, Barrytown, NY.

Kirchoff, F., Dringen, R. & Giaume, C. 2001, 'Pathways of neuron-astrocyte interactions and their possible role in neuroprotection', *European Archives of Psychiatry and Clinical Neuroscience*, vol. 251, no. 4, pp. 159-169.

Ko, D.Y., Kufta, C., Scaffidi, D. & Sato S. 1998, 'Source localization determined by magnetoencephalography and electroencephalography in temporal lobe epilepsy: Comparison with electrocorticography: Technical case report', *Neurosurgery*, vol. 42, no. 2, pp. 414-422.

Mazzanti, M., Sul, J.Y. & Haydon, P.G. 2001, 'Glutamate on demand: Astrocytes as a ready source', *Neuroscientist*, vol. 7, no. 5, pp. 396-405.

Newman, E.A. 2001, 'Calcium signaling in retinal glial cells and its effect on neuronal activity', *Progress in Brain Research*, vol. 132, pp. 241-254.

Oschman, J.L. 2000, *Energy Medicine: The Scientific Basis*, Harcourt Brace/Churchill Livingstone, Edinburgh.

Polyak, S. 1934, 'Projection of the retina upon the cerebral cortex, based upon experiments with monkeys', *Research Publication of the Association for Research in Nervous & Mental Disease*, vol. 13, pp. 535-557.

Stein, D. 1995, *Essential Reiki. A Complete Guide to an Ancient Healing Art*, The Crossing Press Inc., Freedom, CA.

Stryer, L. 1985, 'Molecular design of an amplification cascade in vision', *Biopolymers*, vol. 24, no. 1, pp. 29-47.

Talbot, S.A. & Marshall, W.H. 1941, 'Physiological studies on neural mechanisms of visual location and discrimination'. *American Journal of Ophthalmology*, vol. 24, pp. 1255-1264.

Vesce, S., Bezzi, P. & Volterra, A. 2001, 'Synaptic transmission with the glia', *News Physiological Sciences*, vol. 16, pp. 178-184.

3 A place for science

CHAPTER OUTLINE

There is a spectacular place where biological research is carried out with an intensity and enthusiasm that is virtually unparalleled for educational and research institutions. It is located on scenic Cape Cod, Massachusetts, in a small fishing village called *Woods Hole* (Figure 3-1). My experiences as a scientist at Woods Hole are a major component of the story in this book.

THE MARINE BIOLOGICAL LABORATORY

In his award-winning book, *The Lives of a Cell: Notes of a Biology Watcher,* Lewis Thomas (1974) devotes a chapter to science at the Marine Biological Laboratory (MBL) in Woods Hole:

Figure 3-1 Aerial view of Woods Hole, Massachusetts, and the campus of the Marine Biological Laboratory. (Photo courtesy Doug Weisman and the Woods Hole Oceanographic Institution.)

MBL's influence on the growth and development of biologic science has been equivalent to that of many of the country's universities combined, for it has had its pick of the world's scientific talent for each summer's research and teaching. If you ask around, you will find that any number of today's leading figures in biology and medicine were informally ushered into their careers by the summer course in physiology; a still greater number picked up this or that idea for their key experiments while spending time as summer visitors in the laboratories, and others simply came for a holiday and got enough good notions to keep their laboratories back home busy for a full year. Someone has counted thirty Nobel Laureates who have worked at the MBL at one time or another. THOMAS (1974)

During a summer in Woods Hole, one can meet and talk with famous and accomplished scientists from around the world, who come to share ideas and to experiment with marine organisms. Professors, physicians, heads of laboratories, and directors of scientific institutions who are difficult to even get a short appointment with at their home institutions are ready to chat for hours on MBL Street, on the front steps of the laboratory, in the library, or at the beach.

A typical Woods Hole story: a friend and colleague, a scientist from Germany, won a prize for "best costume" at an MBL party. The judges were five Nobel Laureates.

Again, from *The Lives of a Cell: Notes of a Biology Watcher:*

If you can think of good questions to ask about the life of the earth . . . you might begin at the local MBL beach, which functions as a sort of ganglion . . . biologists seem to prefer standing on beaches, talking at each other, gesturing to indicate the way things are assembled, bending down to draw diagrams in the sand. By the end of the day, the sand is crisscrossed with a mesh of ordinates, abscissas, curves to account for everything in nature. THOMAS (1974)

TWENTY-ONE BUSLOADS OF PHYSIOLOGISTS

The beginnings of this book can be traced to an event that took place at the MBL in 1929, a decade before I was born. Scientists from around the world had gathered in Boston for the 13th International Physiological Congress. As a special treat, the entire Congress was invited to the MBL for a day. The scientists who accepted the invitation filled 21 buses and were taken an hour or so south, to Cape Cod and to Woods Hole, as guests of the MBL. They were treated to a day of science exhibits and a clambake with lobsters.

Among the visitors was a 34-year-old biochemist from Szeged University in Hungary, whose research would soon earn him the Nobel Prize. His name was Albert von Szent-Györgyi Nagyrapolt, M.D. (Budapest), Ph.D. (Cambridge). For the synthesis of vitamin C (ascorbic acid) and his other major findings in biochemistry, he was awarded the Nobel Prize in Physiology or Medicine in 1937.

KEY WORK ON MUSCLE AND PROTEINS

Szent-Györgyi went on to receive more awards for outstanding work on muscle. He made two fundamental discoveries in the early 1940s that laid the basis for all subsequent work on the mechanism of muscle contraction. He discovered that the muscle protein that was then called *myosin* was actually a complex of two proteins, *myosin* and *actin*. He also discovered that threads made of an actin-myosin complex would contract in vitro (in a glass dish) in the presence of both the energy-rich compound adenosine triphosphate (ATP) and calcium. This was the first time a fundamental physiological process, *muscle contraction,* was accomplished with molecules that had been isolated from cells. It was a huge accomplishment that changed the course of physiological research.

In retrospect we can see that this period, during and after the turmoil of World War II, was an incredibly productive time for Szent-Györgyi. Not only was he doing the fundamental work on muscle described above, but he also was thinking about other matters that led directly to the book you have in your hands. Remarkably, he published an extremely important paper in 1941, at a time when his country and all of Europe was falling into turmoil. It was part of his Korani Memorial Lecture given in Budapest. This paper, which is discussed later, has the unusual distinction of being published in the two leading academic periodicals, *Science* and *Nature* (Szent-Györgyi 1941a, 1941b).

This was also a period of great productivity among some of the other scientists who became major figures in the Woods Hole scientific community. For example, Heilbrunn and Wiercinski (1947) achieved great recognition for studies, performed at the University of Pennsylvania and at Woods Hole, in which they injected calcium into muscle fibers and triggered contraction. Heilbrunn went on to become a distinguished figure in physiology, partly because of his recognition that calcium ions play key roles in the regulation of virtually all cellular processes (see Figure 1-6).

A THRILLING MOMENT

Szent-Györgyi's research and his way of going about it have influenced many people. In this book I discuss his contributions and give a number of quotations from his own works for those who have not encountered them before. For example, in relation to his profound discoveries concerning muscle contraction, Szent-Györgyi displayed his typical modesty:

> *To see these little artificial muscles jump for the first time was, perhaps, the most exciting experience of my scientific life, and I felt sure that in a fortnight I would understand everything.*
>
> *Then I worked for twenty more years on muscle and learned not a thing. The more I knew, the less I understood; and I was afraid to finish my life with knowing everything and understanding nothing. Evidently something very basic was missing. I thought that in order to understand I had to go one level lower, to electrons, and—with graying hair—I began to muddle in quantum mechanics. So I finished up with electrons. But electrons are just electrons and have no life at all. Evidently on the way I lost life; it had run out between my fingers.*
>
> *I do not regret this wild-goose chase—because it made me wiser and I know, now, that all levels of organization are equally important and we have to know something about all of them if we want to approach life.* SZENT-GYÖRGYI (1974)

THE INSTITUTE FOR MUSCLE RESEARCH

In 1947, when Szent-Györgyi fled Hungary after the Nazi oppression and Russian occupation, he fondly remembered Woods Hole and the taste of the lobsters, so he returned there to make his home and establish his Institute for Muscle Research. His institute was housed in a modest space on the third floor of the Lillie Building, overlooking the harbor and Vineyard Sound. In 1954, his work on muscle was acknowledged with the prestigious Albert Lasker Award for Basic Medical Research, for "his distinguished research achievements in the field of cardiovascular diseases, including the discovery of actomyosin, the essential contractible element of muscle."

Much of the content of this book has followed from experiences at Woods Hole and from the people and ideas encountered there, including some fantastic ideas from Albert Szent-Györgyi and those around him. It is a story of a remarkable place, remarkable times, and remarkable science, far ahead of its time, which just now is finding its rightful place in physiology and medicine. However, even more it is a story of people, famous and not so famous, who contributed to my evolution as a scientist and as a human being, and of pioneering scientists who have given us some of the key ideas that will continue to shape the future of medicine and the human spirit.

A LITTLE GIANT

For nearly half a century, the Institute for Muscle Research was visited by many of the world's foremost scientists, all of whom experienced warm, affectionate, and stimulating discussions with the little man from Budapest, who was short but sturdy in physical stature and considered by many to be one of the giants of twentieth century science. To his colleagues, he was known affectionately as *Prof* or *Albi*.

For decades, Prof's annual talks in the MBL physiology course were the most popular academic events of the season. He inspired generations of scientists with his unabatable excitement, wisdom, creativity, and humor (Klotz 1988).

An equally stimulating international focus was Prof's home, The Seven Winds, on a spectacular peninsula called *Penzance Point,* which jutted into the channel between Buzzard's Bay and Vineyard Sound. Here Prof and his wife Marta hosted informal seminars and extraordinary parties.

As one example of the typical goings-on around Prof's home, James D. Watson (1988) has described how his famous book, *The Double Helix,* had its origins in the little study off the entrance of Prof's home, while he was Prof's guest. This was what the biology watcher Lewis Thomas was talking about: many of the world's leading scientists had their successful careers "jump started" by a few ideas freely and generously given to them by Prof.

STUDYING NATURE

There is a large banner in the library at Woods Hole with the simple admonishment, *Study nature, not books.* It is signed by Jean Louis Rodolphe Agasiz (1807-1873). The famous naturalist visited Woods Hole in the early part of the nineteenth century. There is no question that Prof was a keen observer of nature and that his home, so close to spectacular natural wonders, was an ideal place to observe sea life. A favorite story comes from Prof's famous and totally revolutionary article, *Drive in living matter to perfect itself*:

> In the winter, at Woods Hole, the sea gulls are my main company. These gulls, the "herring gulls," have a red patch on their beaks. This red patch has an important meaning, for the gull feeds its babies by going out fishing and swallowing the fish it has caught. Then, on coming home, the hungry baby gull knocks at the red spot. This elicits a reflex of regurgitation in mama, and the baby takes the fish from her gullet. All this may sound very simple, but it involves a whole series of most complicated chain reactions with a horribly complex underlying nervous mechanism. How could such a system develop? The red spot would make no sense without the complex nervous mechanism of the knocking baby and that of the regurgitating mother. All this had to be developed simultaneously, which, as a random mutation, has a probability of zero. I am unable to approach this problem without supposing an innate "drive" in living matter to perfect itself.
>
> I know that many of my colleagues, especially the molecular biologists, will be horrified, if not disgusted, to hear me talk about a "drive" and will call me a "vitalist," which is worse than to be called a communist. But I think that the use of such words as "drive" does no harm if we do not imagine we have found an explanation by finding a name. If we look upon such words as simply denoting great unsolved problems of science, they can even lead to useful experimentation.   SZENT-GYÖRGYI (1974)

The pioneering spirit is contagious, for those exposed to it are never the same afterward. It is as though we all possess within us the drive for perfection Prof mentioned in his essay. This is the spirit that moves humanity toward greater accomplishment in all endeavors. The contagious qualities exemplified by Albert Szent-Györgyi and the other pioneers one can think of include humility, awe of nature, insight, kindness toward our fellow beings on the planet, and an aware presence, to name a few.

REFERENCES

Heilbrunn, L.V. & Wiercinski, F.J. 1947, 'The action of various cations on muscle protoplasm', *Journal of Cellular and Comparative Physiology*, vol. 29, pp. 15-32.

Klotz, I.M. 1988, in 'To see what everyone has seen, to think what no one has thought: A symposium in the memory and honor of Albert Szent-Györgyi', *Biological Bulletin*, vol. 174, p. 228.

Szent-Györgyi, A. 1941a, 'Towards a new biochemistry?' *Science*, vol. 3, pp. 609-611.

Szent-Györgyi, A. 1941b, 'The study of energy levels in biochemistry', *Nature*, vol. 148, pp. 157-159.

Szent-Györgyi, A. 1974, 'Drive in living matter to perfect itself', *Synthesis*, vol. 1, pp. 14-26.

Thomas, L. 1974, *The Lives of a Cell: Notes of a Biology Watcher*, Bantam Books, Toronto, Canada.

Watson, J.D. 1988, in 'To see what everyone has seen, to think what no one has thought: A symposium in the memory and honor of Albert Szent-Györgyi', *Biological Bulletin*, vol. 174, p. 229.

Prof takes on cancer

A MEMORABLE LECTURE

In 1972, I spent my first summer at the Marine Biological Laboratory (MBL) as a visiting scientist. Curiosity about the title of an evening lecture led me to Albert Szent-Györgyi's talk on *Electronic Biology and Cancer* on a warm Tuesday evening in July. Prof's book with this same title came out 4 years later (Szent-Györgyi 1976).

The MBL's Lillie Auditorium was filled to capacity with scientists and their families, as well as nonscientists from the surrounding community.

We soon learned why Prof's lectures were so popular. A very vibrant and charming man came out on the stage and energetically, dramatically, and enthusiastically described his experimentation on the role of electrons in life and disease. His melodious Hungarian accent had an engaging warmth and spirit. It was obvious that he was really enjoying his research and the opportunity to talk to the community. The author was enchanted by Prof's charisma and fascinated with the creativity and novelty of his science!

Angered by the tragic loss of his dear wife Marta and his only daughter Nelly to cancer, Prof was determined to find a cure. He turned all of his abundant energy, brilliance, and insight in this direction. The result was an entirely new and revolutionary perspective on biology and disease.

BREAKING NEW GROUND

It has been fascinating to follow the ideas Prof developed from his explorations of biology and cancer. The story involves some remarkable insights he had that were way out of touch with the research going on around him at the MBL and elsewhere at the time. It always seems this way for those who break new ground.

As is usual with new ideas, Prof's insights were greeted with great skepticism. He expected this. He often mentioned that when one of his scientific papers was rejected for publication in a journal, he was encouraged because it meant that he had discovered something new. When a paper was accepted, it meant that nothing new had been discovered.

Writing this book has enabled me to chronicle how prophetic Prof often was, for many of his most remarkable ideas only recently have been shown to be correct. Others were wild goose chases, and he was the first to admit it (see Figure 1, Introduction).

We shall see that Prof's ideas shed light on a number of fundamental mysteries and controversies related to both conventional and complementary medicine.

BIOCHEMISTRY IN TWO VOLUMES

In essence, Prof told us that the story of biochemistry comes in two "volumes." *Volume I* is the well-known biochemistry of the water-soluble molecules. These are the various enzymes and substrates that interact to carry out the chemical reactions vital to life. This is the biochemistry that has given us pharmacology, molecular biology, the human genome project, and countless other dramatic advances.

There is nothing wrong with this biochemistry. Indeed, Prof was a gigantic figure in its development. He had played a huge role in elucidating a fundamental biochemical pathway, the citric acid cycle, which breaks down sugar and releases energy. This sequence, carried out by water-soluble enzymes, was once called the *Szent-Györgyi cycle*, later became known as the *Krebs/Szent-Györgyi cycle*, and then simply the *Krebs citric acid cycle*, after Prof's famous and dear friend Hans Krebs. Indeed, it was through Prof's generosity of spirit and personal connections that Krebs escaped from Germany in time to avoid the Nazi catastrophe. Prof arranged for Krebs to be given a position in the Biochemistry Department in Cambridge, England, where Krebs was able to focus on the research that won him the Nobel Prize in 1953.

HELPING A COLLEAGUE

Sir Hans A. Krebs (1900-1981) came to Boston University School of Medicine in 1975 for a symposium entitled *Search and Discovery*, celebrating Prof's scientific career and his eighty-second birthday (Kaminer 1977). Krebs had met Szent-Györgyi in Boston at the 1929 International Physiological Congress mentioned in Chapter 3. In telling his story of his relationship with Prof, Krebs showed a slide of a remarkable note Prof had written him, in German, in 1933:

> *The Hague, 12 April 1933*
>
> *Dear Colleague:*
> *I am glad to know that you are interested in ascorbic acid. Unfortunately, your letter reached me while traveling and it will be three weeks before I shall be back home again. I will then send you ascorbic acid immediately. Should you no longer require it, please let me know as our supplies are at present limited. If I do not hear from you I will send the substance promptly after my return.*
>
> *I am very sorry to hear that you have personal difficulties in Germany. During the last few days I was in Cambridge where people have in mind helping you somehow. Of course, I have*

encouraged them as much as possible and I hope that my words will have contributed a little toward the realization of the plans.

With kind regards,
A. Szent-Györgyi

P.S. If you really would like to come to Cambridge it would be best if you wrote to Hopkins and told him that you would be content with very modest opportunities. Senior posts are not available and perhaps he might be diffident to offer you a junior position.

Therefore ask Hopkins (if you would like to come to Cambridge) to give you an opportunity there. If it does not embarrass you, do refer to my encouragement.

In June 1933, the National Socialist Government terminated Krebs' appointment at the Medical Clinic of the University of Freiburg-im-Breisgau. Following Prof's advice, Krebs wrote to Sir Frederick Gowland Hopkins at the School of Biochemistry, Cambridge, who offered Krebs a Rockefeller Studentship. This "modest opportunity" led to a distinguished career, including a Nobel Prize shared with Fritz Lipmann. Krebs was rewarded by the Staff of Karolinska Institutet, which was "pleased to reward your achievement when with intuitive perception you were able to see in the chaotic and fragmentary mass of known enzymatic processes the way, the primary pathway, of combustion, and with consummate skill to prove the reality of your vision" (Hammarsten 1953).

At the 1975 symposium in Boston, Krebs commented, "So it was Albert Szent-Györgyi's kindness and considerateness, over 42 years ago, which decisively influenced my life at the most awkward, almost catastrophic, stage and I lived happily in England ever after, needless to say, with everlasting memories of deep-felt gratitude to Albi" (Krebs 1977).

DEMOLISHING CELLS TO LEARN ABOUT THEM

The major technique of research in biochemistry involves the disruption of tissues and cells by homogenization, followed by centrifugation at high speed to force the solid and insoluble materials (the *precipitate*) to the bottom of the tube. Most of modern biochemistry is based on the study of the water-soluble molecules dissolved in the upper layer (called the *supernatant*). The solids at the bottom of the tube are generally discarded down the sink (Figure 4-1). Prof was actually a pioneer in this kind of research, but he did not stop there.

Eventually Prof decided to begin a study of the material the other biochemists were throwing away. These are the insoluble structural proteins, such as collagen, keratin, actin, myosin, and elastin, which form the *fabric of life*, the connective tissue, and cytoskeletons of the various cells in the body. In essence, the other biochemists were virtually ignoring *structure*. They viewed the cell as a bag containing a solution of enzymes and other molecules, salts, and water. Biochemistry as it is usually portrayed is solution biochemistry.

Prof referred to the protein fabric his colleagues were discarding as *the stage on which the drama of life is enacted.* Study of this material was the beginning of Prof's *Biochemistry, Volume II.* In essence, he was laying the foundation for a new branch of biochemistry that now is called *solid-state biochemistry.* In his own words:

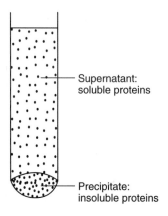

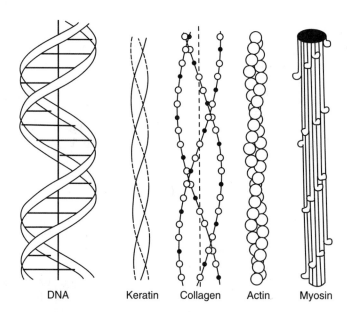

DNA   Keratin   Collagen   Actin   Myosin

Figure 4-1 Classic biochemistry involves homogenizing a tissue, centrifuging it at high speed, and collecting the supernatant with its dissolved proteins and enzymes for further study. Pasteur had discovered that such cell-free extracts could carry out metabolic processes such as fermentation—a gigantic advance that led the way to a century of molecular biology. For a long time, the precipitate at the bottom of the centrifuge tube was discarded down the sink. This material, long considered unworthy of study, consisted of deoxyribonucleic acid (DNA), keratin, collagen, actin, and myosin. (Lower drawing: From *Energy Medicine: the scientific basis*, Oschman, Figure 15.6, page 235, Churchill Livingstone, Edinburgh, 2000, with permission from Elsevier Science.)

It was at an early date that I began to feel that the wonderful subtlety of biological reactions could not be produced solely by molecules, but had to be produced partly by much smaller and more mobile units which could hardly be anything else than electrons. The main actors of life had to be electrons whereas the clumsy and unreactive protein molecules had to be the stage on which the drama of life was enacted. Electrons, to be mobile, need a conductor, which led me to the conclusion that proteins have to be electronic conductors. Toward the end of the 1930s theories began to appear about the submolecular structure of condensed matter. This opened the possibility of electronic mobility in proteins, and thus in 1941 I proposed that proteins may be conductors.

SZENT-GYÖRGYI (1978)

The significance for cancer research:
Do we understand cancer? No, nobody does. There is something missing. Our real problem is not "What is cancer?" but "What is life?" We can't understand cancer until we understand life, because cancer is just distorted life.

Biochemistry, Volume II, Prof told us, is electronic or solid-state biochemistry. He published a diagram of an important biological molecule, carotene, found in plants. Next to the molecule he placed a diagram of the power cord for his toaster (Figure 4-2). The obvious analogy: the cord on the toaster conducts electrons to the heating element; the backbone of the carotene molecule conducts electrons from one end to the other.

If we are to understand cancer, he said, we need to understand the communication system that exists in the very fabric of the body, in the material the other biochemists had been throwing away, and to understand this communication system we need to study subatomic particles. How do we do this?

VENTURING INTO THE QUANTUM REALM

For Prof, the method of study of electronic biology had to be quantum mechanics.
The cell is like a very involved watch. The watch is made up of hundreds of little wheels and parts. First you must know how it is put together, how it works. There are four dimensions with which the biologist must be concerned: macroscopic (anatomy), microscopic (cells), molecular (proteins), and submolecular or electronic. Biology readily followed physics into the first three, but took practically no cognizance of the fourth; it stopped at the molecular level.

In this book we will follow Prof's advice mentioned in the previous chapter, "*all levels of organization are equally important and we have to know something about all of them if we want to approach life.*" We will see the extraordinary value of understanding the phenomena taking place in the quantum realm. We also will see that he was correct—information from the subatomic realm is opening up new avenues for exploration of the greatest mystery of life: consciousness. We will see that many others now have followed Prof's footsteps by venturing into the subatomic world.

At this point we have a hint of a kind of communication in living systems that is rapid and subtle and involves electrons. Where did Prof get this incredible idea? Is there any real evidence indicating that such a system exists?, and, if it does exist, precisely where is it located in the organism? We shall take these questions in order.

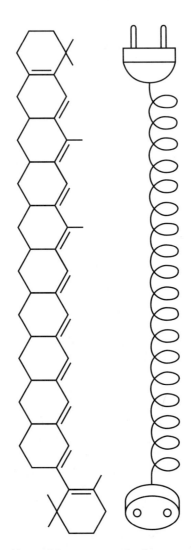

Figure 4-2 The molecular backbone of the carotene molecule compared with the power cord for a toaster. (Reprinted from *Bioelectronics*, Szent-Györgi, p.23, Figure 8, 1968, Academic Press, New York, with permission from Elsevier Science.)

REFERENCES

Hammarsten, E. 1953, 'Presentation speech for the Nobel Prize in Physiology or Medicine 1953', in *Les Prix Nobel 1953*, The Nobel Foundation.

Kaminer, B. (ed.) 1977, *Search and Discovery. A Tribute to Albert Szent-Györgyi*, Academic Press, New York.

Krebs, H. 1977, 'Errors, false trails, and failures in research', in *Search and Discovery. A Tribute to Albert Szent-Györgyi*, ed. B. Kaminer, Academic Press, New York, pp. 3-15.

Szent-Györgyi, A. 1976, *Electronic Biology and Cancer*, Marcel Dekker, New York.

Szent-Györgyi, A. 1978, *The Living State and Cancer*, Marcel Dekker, New York, p. 5.

5 A kitten and a fly

To see what everyone has seen, to think what no one has thought.

Albert Szent-Györgyi (1988)

NEW WAYS OF THINKING ABOUT FAMILIAR THINGS

During his brief lecture in 1972, Albert Szent-Györgyi traced the beginnings of an entirely new science, which he referred to as *electronic biology*. I vividly recall being totally bewildered by this lecture because it all sounded quite logical, Prof was quite eloquent and confident about his subject, yet it was all new. Years of university life and taking courses, listening to lectures, and attending international conferences was no preparation for Prof's talk. The puzzlement was due to the fact that what Prof was saying was totally unfamiliar. We shall follow the ways this story has developed since that memorable lecture in 1972.

One of Prof's experiences that inspired this book concerned a kitten. It was a dramatic example of one of his famous quotations about his philosophy for doing research: to *see what everyone has seen, to think what no one has thought* (Figure 5-1). In his own words:

> *Perhaps I could make this clearer by a little story about a kitten which shared my tent once in Cornwall, England. One day a snake crept into our tent. My kitten stiffened in horror. When I touched its tail, the kitten jumped up vertically about two feet high. This happened because the nerve fibers which ended on the motor nerve cells conveyed the message that there was danger of life and any motion had to be fast and violent. These messages came, as I said, from faraway complex nerve centers which worked up and evaluated the visual impressions of my kitten.* SZENT-GYÖRGYI (1974)

From the biological perspective, jumping high in the air gives the kitten a fraction of a second to decide what to do when it comes back down to the ground, that is, *fight or flight* (see Figure 5-1).

Figure 5-1 A kitten exhibiting a typical "flight-or-fight" response to a dangerous stimulus. (From Fogle, B., *The Encyclopedia of the Cat*, DK Publishing, Inc., New York, p. 65, Copyright Dorling Kindersley and used with permission from Dorling Kindersley Limited, London.)

What changed Prof's life, and the course of science, was what Prof thought about this event. He wondered how the extremely complex networks in the brain could become organized to carry out this exceedingly rapid response. He thought his kitten's reaction was far too fast to be explained by the familiar neuromuscular response system that the neurophysiologists had so painstakingly traced out from decades of research. Something far more rapid and sophisticated was taking place. In essence, he began to suspect that there was another way of sensing and moving that was different from the way we usually pictured neuromuscular responses. This, of course, was a completely radical idea, one of many that Prof put forward during his life.

THE MYSTERIOUS BRAIN OF THE CAT

Figure 5-2 shows a textbook explanation of how a cat's brain is supposed to work. Figure 5-3 shows the basic wiring diagram of the neural systems that are thought to control sensation and movement.

We soon shall see that, although these schemes are valid, there is far more to the story. In the next chapter we will see that the whole subject of neurophysiology, which previously had reached some sophisticated understandings of the neural origin of human behavior, now has gone into a period that can best be described as *uncertainty*. After a period of being off limits to researchers, the key attribute of living matter, consciousness, has become a respectable topic for scientific exploration. However, the closer we have looked at consciousness, the more elusive it has become. Before attempting to shed some new light on consciousness, we need to look at another observation that changed Prof's thinking about life.

A FLY IN THE EYE

In his lecture Prof described another experience that seemed to confirm that the nervous system and nerve signals were too slow to preserve and protect the body in emergency situations. He was riding his motorbike to the laboratory one morning and

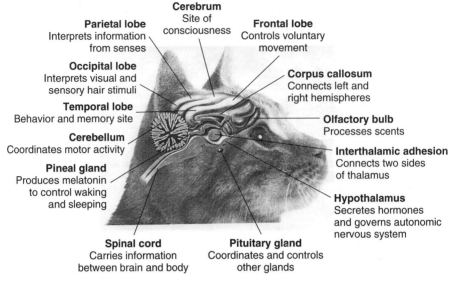

Cerebrum
Site of
consciousness

Parietal lobe
Interprets information
from senses

Frontal lobe
Controls voluntary
movement

Occipital lobe
Interprets visual and
sensory hair stimuli

Corpus callosum
Connects left and
right hemispheres

Temporal lobe
Behavior and memory site

Cerebellum
Coordinates motor activity

Olfactory bulb
Processes scents

Interthalamic adhesion
Connects two sides
of thalamus

Pineal gland
Produces melatonin
to control waking
and sleeping

Hypothalamus
Secretes hormones
and governs autonomic
nervous system

Spinal cord
Carries information
between brain and body

Pituitary gland
Coordinates and controls
other glands

Figure 5-2 The brain of a cat as it is sometimes depicted by those who would determine the function of each part. It is our thesis that this type of description is superficial and masks the deep mysteries of neuronal and brain functioning and consciousness. (From Fogle, B., *The Encyclopedia of the Cat*, DK Publishing, Inc., New York, p. 65, Copyright Dorling Kindersley and used with permission from Dorling Kindersley Limited, London.)

drove his eye straight into a fly. What he noticed was that he had not seen the fly in his path, but when the fly touched the tip of his eyelash his eyelid closed so fast that it prevented the fly from smashing into his cornea.

A simple calculation shows that at a modest velocity of 10 mph the time between the fly's encounter with the tip of the eyelash and the closing of the eyelid has to be about a thousandth of a second (Figure 5-4). Now, one knows the neurological sequence coupling the stimulation of the eyelash with the closing of the eyelid, and it involves a number of nerve messages and synaptic events that move through the pathways shown in Figure 5-4. Although such responses seem to us to be immediate, they are not. Nerves conduct impulses with a certain velocity, and there is a delay at each synapse while one nerve stimulates another. The response is a *reflex*, a very fast kind of reaction that enables the body to protect itself by generating rapid avoidance movements before there is a conscious awareness of the stimulus.

We shall explore the mechanisms involved in reflexes in a later chapter. In order to understand the ideas in this book we need to understand the well-known neuromuscular pathways involved in sensation and action. What we are going to suggest is that these pathways do, indeed, work the way neurophysiologists think they do, but that there also is another kind of coupling between sensation and action. One place this

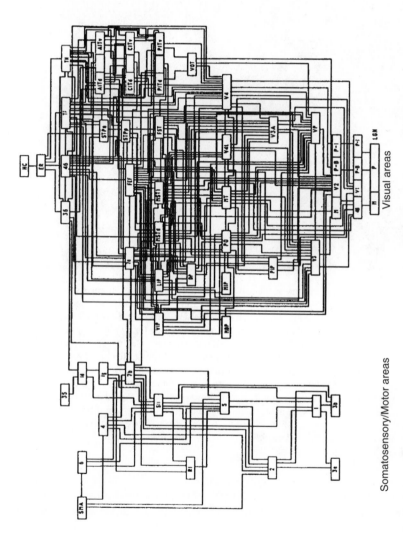

Somatosensory/Motor areas

Visual areas

Figure 5-3 The visual and motor pathways depicted as linear circuits. It is our thesis that this type of mapping may be accurate but may not tell the whole story of the connections between sensation and action. (From Van Essen, D.C., Felleman, D.J., et al. 1990, 'Modular and hierarchical organization of extrastriated visual cortex in the macaque monkey', *Cold Spring Harbor Symposia on Quantitative Biology*, vol. LV, pp. 679–696, Fig. 8.)

Figure 5-4 Prof going to his laboratory at the MBL on his motorbike. It was on one such excursion that he drove his eye straight into a fly. Remarkably, by the time the fly would have smashed into his cornea, his eyelid had closed. Thinking what no one had thought before, Prof concluded that this was another example of a response that was too rapid to be accounted for by the well-understood neuromuscular pathways. A simple calculation shows that at a modest velocity of 10 mph the time between the fly's encounter with the tip of Prof's eyelash and the closing of his eyelid has to be slightly more than a thousandth of a second. The calculation assumes the motorbike is moving at 10 mph, which amounts to 52,800 feet/hr or 14.62 feet/sec or 175.46 inches/sec. If the distance between the tip of the eyelash is one-quarter inch, that distance will be traversed in 0.001435 second. The calculation assumes a stationary fly and would be altered somewhat if the fly were moving toward or away from the eye. (From Szent-Györgyi, A. 1988, 'To see what everyone has seen, to think what no one has thought', *Biological Bulletin*, vol. 175, pp. 191–240. Reproduced with permission from the Biological Bulletin.)

other kind of coupling reveals itself to us is in life-threatening situations, and there are other examples, from observations of human performance, that we shall discuss in later chapters.

We shall see that even the fastest of reflexes, called *monosynaptic reflexes,* cannot accomplish a response, such as Prof's "fly in the eye" response, in such a short time. Prof was correct: living systems have to have mechanisms that can operate far faster than the nervous system when such responses are needed. These rapid response systems are ancient in terms of evolutionary history. They are built into all living matter, even in organisms too "primitive" to have a nervous system.

For the classic neurophysiologist, this is an extraordinary idea and seems to go against decades of careful research that has mapped out the neural pathways in the body involved in sensing the environment, analyzing the information, and generating a response. We shall see, though, that the idea that there might be different ways of sensing and moving is supported by a variety of extraordinary experiences that people have reported from time to time, and from some extraordinary research, and scientists have begun to describe the mechanisms involved. The next two chapters give some examples of such experiences, and then we will come back to Prof's kitten, the fly in the eye, and the scientific basis for the extraordinary.

REFERENCE

Szent-Györgyi, A. 1974, 'Drive in living matter to perfect itself', *Synthesis*, vol. 1, pp. 14-26.
Szent-Györgyi, A. 1988, 'To see what everyone has seen, to think what no one has thought', A symposium in memory and honor of Albert Szent-Györgyi, *Biological Bulletin* vol. 174, pp. 191-240.

6 Sensation and movement at the edge

THE "ZONE"

This chapter gives examples of extraordinary achievements of which the human body is capable under special circumstances. Some of these are familiar to all of us as captivating athletic and artistic performances. Other examples are found among more exotic or esoteric circumstances, although these are becoming more and more popular and familiar: martial arts; contemplative, meditative, and spiritual practices; and the healing encounter itself. Later we discuss traumatic events, for which these considerations are particularly relevant.

Occasionally athletes and performers of all kinds experience astonishing levels of coordination and dexterity during which accomplishments take place that seem to lie far beyond our normal functioning. These are profound experiences of being totally prepared, present, or focused, or reacting to the surrounding circumstances in ways that seem absolutely extraordinary. For example, performers such as jazz musicians, athletes, and dancers often report finding a "zone" in which their actions are effortlessly synchronized and integrated with their fellow performers in rapid and sophisticated ways that defy explanation by a slow-moving neurological consciousness.

Examples from the martial arts

In the martial arts, what seems extraordinary to us is commonplace. Consider the student who sneaks up behind his master and quietly attacks with all the force he can muster. His master cannot see or hear the attacker, yet the master's body responds extremely rapidly in a way that sends the attacker flying across the room. Ask the master about what happened, and he will report that he was not *consciously* aware of the attack until he saw his attacker flying through space. Asked, "How can you do that?"

the master might reply, "You can never touch me. I can always touch you. I have control over *time*."

Two kinds of time. What, exactly, could this mean? Is there something we do not understand about time that the martial arts master does understand?, or are we confronted by a mystery that we shall never resolve?

There is something essential about the Now which is just outside the realm of science.
ALBERT EINSTEIN (1963)

We shall try for a partial explanation for this. We will describe two kinds of time: (1) *neurological time* and (2) *connective tissue* or *quantum time*, a time based on a clock that runs much faster than the nervous system. These two different kinds of time, we shall suggest, are experienced, respectively, by our ordinary *neurological consciousness* and by *connective tissue consciousness*, a kind of consciousness of which we only get rare glimpses. It is a consciousness that uses the entire fabric of the body, the entire living matrix, as a sensory antenna, communication system, and medium for producing responses.

Here is an example of an Aikido student describing his daily experience with his master (Figure 6-1). His remarkable description is as thoughtful and honest as it is worthy of serious consideration:

> If a child who had never tasted sugar should ask, "What does it taste like?," your reply might be, "It is sweet." But what is sweetness? How can a taste be explained in words? Each person tastes differently. Past experience, knowledge, and personality are all determining factors in an individual's reaction to a particular taste. It must be experienced to be fully understood.
>
> I would often receive ukemi from O Sensei at early morning class, sometimes being thrown as far as thirty feet.
>
> Do not see my experience through the clouded eyes of a mystic, but with the clear, crisp vision of a seeker of truth. The experience, and my interpretation of it changing and growing with time, has been so vital and is still so vital to my search that I must try to give it voice.
>
> During the demonstration of a technique at practice, I attacked my master with all my power, and my only thought was to strike him down. The walls of the dojo shook as his kiai (a piercing scream originating in the lower abdomen) shattered the air, and my entire body was imprisoned by the shock. I perceived the power of a hurricane; the violent winds of a typhoon lashed my body. And the force of his gravity, drawn from the energies of the universe into the vacuum of a "black hole" from which there was no escape. Deep within my core a bomb exploded, and the whole universe expanded. There was nothing but light—blinding, searing light and energy. I could not see or feel my body, and the only reality was the enormous energy expanding from it. For those watching, it happened in the split of a second, but for me time had stopped. There was no time, no space, no sound, no color, and the silence was more deafening than his scream. In the light so completely, I was the light, and my mind and spirit were illuminated and completely clear. Then I was unconscious. As my body connected with the tatami mat, I was revived.
> (Saotome 1993)

The dance of coordination. Athletes who work together in team sports report that there are times when their coordination with their teammates achieves an uncanny degree of perfection. A friend who played football as a linebacker recalls times when

Figure 6-1 *"I would often receive* ukemi *from O Sensei at early morning class, sometimes being thrown as far as thirty feet."* (From Aikido and the Harmony of Nature by Mitsugi Saotome. © 1986, 1993 by Mitsugi Saotome. Reprinted by arrangement with Shambhala Publications, Inc., Boston, www.shambhala.com.)

he and his partner had an extraordinary sense of each other's movements and the whereabouts of the opposing ball carrier that enabled them to cooperate perfectly on each tackle. Similar synchronization takes place in ballet and other types of performance where a number of participants are moving together.

Jazz musicians provide a classic example of cooperation when they simply let loose and improvise together, often for long periods of time, playing music they are creating in the moment that is nonetheless in precise synchrony and harmony, as if the group were a single musician. Similar experiences are reported for circle drumming.

The dance of conversation. Although there may be no "science of jazz," there is a science of the coordinated movements that take during conversation. William S. Condon (1975, 1977; Condon & Sander 1974) at Boston University School of Medicine studied a film of two people having a conversation. He studied the movements related to the parts of individual words. For example, it takes about one fifth of a second to say the word "ask." Observed at 48 frames per second, the word can be broken into four parts:

ae lasts 3/48 second

E lasts 2/48 second

S lasts 3/48 second

K lasts 2/48 second

During the first part of the word, "ae," the speaker's head moves left and up slightly, eyes hold still, mouth closes and comes forward, four fingers begin to flex, and the right shoulder rotates slightly inward. Each part of the word has a distinct cluster of movements associated with it. The body dances in precise rhythm with each part of each word.

For a year and a half Condon carefully analyzed a 4.5-second piece of conversation. After 100,000 sequences, the film was worn out, so he made another copy and looked at it. After making millions of observations on 130 copies of the same film, he reached his conclusions:

> *The speaker and the listener are both performing a synchronous "dance" consisting of micromovements of different parts of their bodies. Remarkably, each part of each word has a cluster of movements associated with it, and the listener moves in precise shared synchrony with the speaker's speech. There is no discernible lag between movements of speaker and listener.* CONDON (1975).

Even total strangers display this synchronization. Listeners usually do not move as much as speakers. The speaker's and listeners movements may be different. The listener is not responding or reacting to the speaker. The listener is one with the speaker.

A spectacular example occurred when there was a silence in one of the conversations Condon filmed. At the precise 1/48th of a second when the speaker resumed talking, the listener resumed his series of synchronized movements.

These studies raise many interesting questions. One is, when does the *understanding* of a word begin and end, in relation to where the *sound* of the word begins and ends? Here there is an obvious problem with *time*. We hear a word and think we understand it when we hear it, but Condon's studies indicate that we have a *sense* of what the word will be before we actually hear it.

WHAT IS HAPPENING IN THE BRAIN

Additional examples of extraordinary physical phenomena are discussed in Michael Murphy's book, *The Future of the Body* (Murphy 1992). Murphy makes note of remarkable research showing that there is a marked *decrease* in the brain's metabolic rate during periods of intensive concentration (Haier et al. 1992). In an article about this metabolic disease, Shainberg (1989) proposed that these phenomena might be mediated by the basal ganglia or other subcortical elements of the central nervous system. Meditation, he suggested, helps sensorimotor command centers operate more efficiently, unimpeded by unnecessary mental activity or memories of past performances.

Synesthesia

A related phenomenon is *synesthesia,* in which one sense (such as a sound) conjures up one or more other sensations (such as a taste, color, or smell). The condition is rare, occurring in about ten people in one million. Synesthetes often say that they have had their "gift" for as long as they can remember, and they are surprised to learn that others do not perceive the world this way. As an example, one person explained that the sound of a piano is definitely a sound like any other person would hear and is accompanied by shapes that are not distinct from hearing but are part of what hearing is. "The vibraphone, the musical instrument, makes a round shape. Each note is like a little gold ball falling. That's what the sound is. It couldn't possibly be anything else."

The phenomenon is described in an extraordinary book, *Man Who Tasted Shapes,* by Dr. Richard Cytowic (1993). This book is cited in another remarkable book, *The Man Who Mistook His Wife for a Hat,* by renowned clinical writer Oliver W. Sacks (1998). Research into this topic began when Cytowic, a neurologist, encountered a man who could literally feel shapes (spheres, columns, points) on his hands in response to different food tastes. When this phenomenon is taking place a variety of sensory inputs are being combined and integrated. In researching the phenomenon using positron emission tomography scan techniques, Cytowic discovered that the sensory cortex actually is shut down while so-called *synesthetics* are having synesthetic experiences.

The phenomenon is linked with photographic memory (eidetic memory) or heightened memory (hypermnesis), as described by Luria (1988) in *The Mind of a Mnemonist: A Little Book About a Vast Memory.* Luria tells the life story of a man who had a literally limitless memory accompanied by an especially vivid synesthesia. He converted what he saw or heard into vivid visual imagery, with powerful gustatory and auditory aspects.

Baseball and archery

Murphy quotes the retired baseball catcher Tim McCarver: "The mind's a great thing, as long as you don't have to use it."

Baseball player Ted Williams was the last of the .400 hitters and wrote a book entitled *The Science of Hitting.* He suggests that hitting a baseball is the single most difficult thing to do in sport:

> *I get raised eyebrows and occasional arguments when I say that, but what is there that is harder to do? What is there that requires more natural ability, more physical dexterity, more mental alertness? That requires a greater finesse to go with physical strength, that has as many variables and as few constants, and that carries with it the continuing frustration of knowing that even if you are a .300 hitter—which is a rare item these days—you are going to fail at your job seven out of ten times? If Joe Montana or Dan Marino completed three of every ten passes they attempted, they would be ex-professional quarterbacks. If Larry Bird or Magic Johnson made three of every ten shots they took, their coaches would take the basketball away from them.* WILLIAMS (1986)

His key pieces of advice to other batters are not what you might expect:

- Hitting is 50% above the shoulders.
- Be natural—hit according to your style. Follow the rule he got from Lefty O'Doul: "whatever you do, don't let anybody change you." Your style is your own.
- Proper thinking. In this case, careful study of pitchers.
- Wait for a good ball to hit. Don't swing at a pitch you have not seen before.
- Be quick. From the moment the ball leaves the pitcher's hand, a batter has about two fifths of a second to make up his mind whether to swing at the ball, and, if he does, to complete his swing.
- Get your hips into the action. The way you bring your hips into the swing is directly related to the power that is generated.
- Guess. Think, do your homework, study the pitchers, and, when the time comes, guess or anticipate. Get an idea of what is going to come at you.

Another relevant quote is from Denise Parker, a teenage archery prodigy, describing her mental state when she became the first American female to score 1,300 points in a particular competition:

> *I don't know what happened. I wasn't concentrating on anything. It didn't feel like I was shooting my shots, but like they were shooting themselves. I try to remember what happened so I can get back to that place, but when I try to understand it, I only get confused. It's like thinking how the world began.* QUOTED IN SHAINBERG (1989)

SPIRITUAL AND MEDITATIVE PRACTICES

Extraordinary perception seems to be related to transcendent states of consciousness that are available to all of us under certain circumstances. This is telling us something profound about consciousness, cognition, and life itself.

Extraordinary perception and action arise in the martial arts and other contemplative traditions, in which "extraordinary" and "mindless" activities are cultivated. A martial arts master is capable of extremely rapid and precise responses to "invisible" stimuli, such as a silent attack from behind. These phenomena, in which sensation and action occur ahead of normal neurological cognition, are well documented in what might be called *esoteric* literature (Saotome 1993). To us, they are key clues in the mystery that is the topic of this book.

There exists in the ancient Buddhist literature of the *Dzog-chen* concepts of "pristine awareness," "atemporal cognitiveness," or "originary awareness" in which the organism responds authentically with little or no intervention of "mind." The quoted concepts are English renderings of the Tibetan *ye-shes*, which refer to a temporal dimensionality. The coming together of modern science and these ancient traditions has been referred to as *The Third Enlightenment* (Kimura 2000).

Emerging understandings of the energetics of the connective tissue are beginning to provide a logical scientific basis for the experience of pristine awareness, atemporal cognition, originary awareness, cosmic consciousness, and luminosity described in the ancient texts translated and interpreted by Dr. Herbert V. Guenther (1977). These experiences have deep meaning for all ages, as described by Guenther in the Foreword to Yasuhiko Kimura's book, *Think Kosmically Act Globally* (Kimura 2000).

METANORMAL HEALING

There are occasions in the healing encounter when a patient and a therapist slip into a zone of connection in which all things seem to become possible. Often these temporary moments are turning points for both participants. We would all like to know more about these metanormal or surpassing phenomena, but science has had difficulty recognizing them and finding ways of approaching them.

TRANSFORMATIVE PRACTICES

Again, Michael Murphy provides deep insight into the meaning of these phenomena for our times:

> Through self-observation we can clarify the nature of our shifting or enduring motives, some of them rooted in basic needs, some of them acquired by social conditioning, some healthy, some destructive, many of them unnoticed or repressed. The ongoing self-reflection enjoined by religious leaders and moral authorities of every culture, and by psychotherapists of modern times, can help to illumine secret or half-concealed attitudes that rule our thought and behavior. Recognizing our competing impulses more clearly, we can more freely choose among them, suppressing or sublimating some while acting upon others. To the extent that religious, moral, or therapeutic disciplines succeed, a person's many wills tend to become one will, single but articulated. As they are integrated, once-divergent intentions produce stronger results. Like a body in which each muscle functions in coordination with other body parts while retaining its own integrity, a well-articulated self can harmonize its separate volitions to achieve its deepest ends.
>
> If successful, transformative practices extend the capacity for purposeful action produced by animal evolution. They can strengthen the capacity for one-pointed behaviors evident in earlier forms of life while providing more options for creative behavior. They can make us better animals and better humans at once, as it were, more single-minded when we choose to be, but more various in our realized intentions. In the words of psychiatrist Roberto Assagioli [Assagioli 1973], it is possible to have a strong, a good, and a skillful will. (Murphy 1992)

Taken together, the remarkable phenomena we have described so far raise questions about whether there are ways of sensing the world and responding to our environment that are different from the classic neuromuscular processes so well known to physiologists and neurophysiologists. Emerging understandings of the energetics of living tissues are giving us a logical scientific basis for the experiences described here. This book gathers this fascinating information together for the first time to present a coherent picture.

REFERENCES

Assagioli, R. 1973, *The Act of Will,* Penguin, New York.

Carnap, R. 1963, 'The Philosophy of Rudolph Carnap', in *Library of Living Philosophers,* vol. 11, ed. P.A. Schlipp, Open Court Publishing Company, Chicago.

Condon, W.S. 1975, 'Multiple response to sound in dysfunctional children', *Journal of Autism and Childhood Schizophrenia,* vol. 5, p. 43.

Condon, W.S. 1977, 'A primary phase in the organization of infant responding behavior', in *Studies in Mother-Infant Interaction,* ed. H.R. Schaffer, Academic Press, New York, pp. 153-176.

Condon, W.S. & Sander, L.W. 1974, 'Neonate movement is synchronized with adult speech: Interactional participation and language acquisition', *Science,* vol. 183, pp. 99-101.

Cytowic, R. 1993, *The Man Who Tasted Shapes,* Putnam Publishing Group, New York.

Guenther, H.V. 1977, *Tibetan Buddhism in Western Perspective: Collected Articles of Herbert V. Guenther,* Dharma Publications, Emeryville, CA.

Haier, R.J., Siegel, B.V. Jr., MacLachlan, A., Soderling, E., Lottenberg, S. & Buchsbaum, M.S. 1992, 'Regional glucose metabolic changes after learning a complex visuospatial/motor task: A positron emission tomographic study', *Brain Research,* vol. 570, pp. 134-143.

Kimura. Y, 2000. *Think Kosmically Act Globally. An Anthology of Essays on Ethics, Spirituality, and Metascience,* The University of Science and Philosophy Press, Blacksburg, VA, p. iii.

Luria, A.R. 1988, *The Mind of a Mnemonist: A Little Book About a Vast Memory,* Harvard University Press, Cambridge, MA.

Murphy, M. 1992, *The Future of the Body. Explorations into the Further Evolution of Human Nature,* Jeremy P. Tarcher/Perigee, Los Angeles, CA, pp. 118-120, 443-447.

Sacks, O.W. 1998, *The Man Who Mistook His Wife for a Hat: And Other Clinical Tales,* Simon & Schuster, New York.

Saotome, M. 1993, *Aikido and the Harmony of Nature.* Shambhala, Boston.

Shainberg, L. 1989, 'Finding the zone', *New York Times Magazine,* 9 April, pp. 34-36, 38-39.

Williams, T. & Underwood, J. 1986, *The Science of Hitting,* Simon & Schuster, New York.

Another way of knowing and moving

The whole of science consists of data that, at one time or another, were inexplicable.   Brendon O'Regan (O'Regan and Pole 1993)

In June 1999, Emmett Hutchins asked me a question about some passages in Valerie Hunt's fascinating book, *Infinite Mind. The Science of Human Vibrations* (Hunt 1989). In her book, Dr. Hunt described remarkable findings from her studies of the electromyographic patterns produced by various dancers, including Emilie Conrad. Could these results be explained?

It took over a year to research Emmett's question. Gradually a logical explanation could be pieced together to provide a basis for further exploration of some extraordinary phenomena. The process was accelerated greatly when the author experienced some of Emilie Conrad's evolutionary movement work, which is called *Continuum Movement*.

HOW SCIENCE IS DONE

Before describing the results of this investigation, it is appropriate to express an opinion regarding the scientific thought process itself. Mature scientists agree that *any* explanation for a phenomenon is far better than *no* explanation. An important part of accepted scientific procedures is to develop hypotheses that can be verified or refuted. Without hypotheses, the scientific process cannot even begin.

Albert Einstein, who had much experience in this realm, stated it this way:

If at first the idea is not absurd, then there is no hope for it.   ALBERT EINSTEIN

There is a tendency to become attached to one's hypotheses and to have regrets should one have to admit that the first, second, or third model proved incorrect or inadequate. For science to proceed, these tendencies toward attachments and regrets are to be avoided. The story that emerged from researching Hunt's observations is presented in the scientific spirit of adventure and exploration. Following you will find discussion of phenomena that some consider to be outside of normal scientific understanding. Such phenomena are called *anomalies* (Kuhn 1970). In *real* science, it is always the annoying *anomalies* that give rise to new discoveries. Today's anomalies become tomorrow's theories and laws. Stated differently:

> *The scientist knows that in the history of ideas, magic always precedes science, that the intuition of phenomena anticipates their objective knowledge.*
>
> MICHAEL GAUGUELIN (1974)

SENSATION AND ACTION

The mechanisms by which organisms sense and adapt to their ever-changing environments are central to survival and therefore are of paramount importance in biology. These mechanisms have to be looked upon as ultimate miracles of nature.

Textbook explanations imply that these are solved problems. Maps of the neural pathways in the body (such as those shown in Figure 5-3) have been known about for a long time and have been thoroughly verified by neuroanatomists and neurophysiologists. Prof's observations implied that there is more to the story. This book begins to tell that story. It is a tale that will take us deeply into the nature of life itself.

As a sensory and response system, the nervous system has the advantage that it is relatively easy to study. One can insert microelectrodes into neurons and document the signals passing through them. Sensory fibers can be stimulated, and the resulting nerve impulses can be followed into and through the brain. Stimulation of specific neurons, called *motor neurons,* will cause particular muscles to contract. Much research has been done on the nervous system, and we have sophisticated understanding of how it operates. Of course, no neurophysiologist will proclaim that the process of discovery is finished or even nearly complete.

We do not wish to diminish the remarkable and brilliant research that has been done on nerve cells using microelectrodes, patch clamps, dye injection, electron microscopy, and many related techniques. Instead, we shall see that what has been discovered in this way may not reveal all that is going on. Modern neurophysiologists actually have confirmed the incompleteness of their understandings, as we soon shall see. We now are *certain* that there are other communication systems in the body that are not so easy to study at present but that are nonetheless as important as the nervous system.

A serious misconception is that what we refer to as *sensation* and *cognition, learning* and *memory,* and *consciousness* and *mind* are activities that reside primarily in the brain. When a computer scientist is asked if computers will be able to mimic human consciousness, the answer is often, "Yes, obviously, because the brain is just a sophisticated computer." This kind of statement refers to an outdated "neuron doctrine" that states that "brain = mind = computer." It is a model that perpetuates the mistaken notion that consciousness arises from the massive patterns of interconnectivity between firing neurons in the brain. This is an idea that persists in recent reports from neuroscientists (LeDoux 2002) even though most neuroscientists recognize that although massive interconnectedness definitely exists, there is far more to the story. We will return to this topic later, in Chapter 15.

DR. HUNT'S OBSERVATIONS

Dr. Hunt's book, *Infinite Mind. The Science of Human Vibrations* (Hunt 1989), presents some remarkable findings on certain dancers, including Emilie Conrad, who appeared to be capable under certain conditions of moving vigorously yet virtually effortlessly. Hunt's observations were made using electromyography, a method that records the electrical activity at the myoneural junction, the place where the motor nerve impinges on the muscle. According to classic neuromuscular theory, when a motor nerve fires there is a depolarization at its termination on the muscle that activates the contraction process. It has been known since 1867 that an electrical field applied at these "muscle points" will elicit contraction, and, vice versa, electrical fields are set up at these points when the muscle is stimulated by the nerve (Figure 7-1).

Here are the relevant passages from Dr. Hunt's book, *Infinite Mind: The Science of Human Vibrations.* The first passage describes electromyographic recordings from a

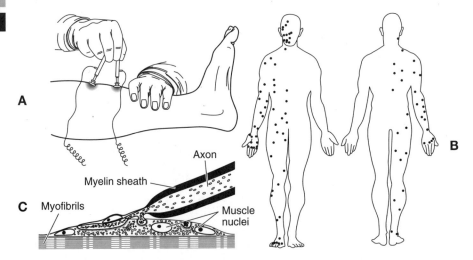

Figure 7-1 Electromyography evolved from observations of Duchenne (1867) showing that electrical stimulation of certain points on the body surface causes specific muscles beneath them to contract (**A**). Since then, an array of "muscle points" corresponding to each muscle in the body have been established (**B**). Electrical stimulation of a point causes the corresponding muscle to contract, and voluntary contraction of the muscle gives rise to a measurable electrical field at the point. The muscle points generally correspond to the locations of the motor end plates (**C**), where motor nerves synapse on the muscle fiber. (**A** is from *Energy Medicine: the scientific basis*, Oschman, Figure 1.6, page 11, Churchill Livingstone, Edinburgh, 2000, with permission from Elsevier Science. **B** is based on Shestaack, R. 1977, *Handbook of physical therapy*, 3rd ed, pp. 36-45, used by permission of Springer Publishing Company, Inc., New York. **C** is from Bloom & Fawcett's *Textbook of Histology*, 12th ed, 1994, Fig. 10-31A, p. 288, with permission from Hodder Arnold.)

dancer who wanted to know what was happening when she danced in an "altered state."

When the electrodes were secured, she started her dance routine. At the beginning nothing unusual happened. . . .Yet in five minutes the recordings remarkably changed. The muscular signal from her lower arm stopped. The baseline activity characteristic of all living tissue was absent on the scopes. Next, the upper arm recording dropped out. The engineer believed there was no equipment failure, although there was no ordinary energy in the arms. Soon she sat down in a "Taylor position" which required back muscle activity to balance her body. Again the spinal muscles showed no recording, no energy expended.

We have known as long as there is life, skeletal muscles give off signals. Next, electromagnetic energy poured from the top of her head with intensity beyond what our equipment could handle. This state lasted for seven minutes, followed by a reverse sequence of reactivating the spine, upper arm, and lower arm muscles. In my years of neuromuscular research I had never witnessed any similar situation, nor had any been described in the literature. I was at a total loss to explain, but I could not forget these happenings.

Emilie Conrad

Later in her book, describing Emilie Conrad's movements, Dr. Hunt reported:

Spontaneously, I checked her heart rate and blood pressure before she started. She then danced strenuously, even acrobatically, for 30 minutes with a perfection and repertoire superior to any I had seen in a single dancer. Her movements took in all directions, large and small, were fast and slow, with complicated neuromuscular rhythms flowing through various parts of her body. She demonstrated amazing flexibility and strength as she effortlessly glided about. Beyond her technical elegance, I sensed that she communicated powerful feelings and ideas.

I was not prepared, however, for the shock that came when I realized that my long study of superior athletes, the handicapped, and native ritualistic movements from all over the world did not help me to understand what I observed with Emilie. Furthermore, my training as a movement observer and neuromuscular physiologist didn't help. I rechecked her heart rate and blood pressure, hoping that her physiological changes would clear up the issue. But to my surprise, neither the heart rate nor the blood pressure had elevated; in fact both had dropped slightly. To make things worse, she was not perspiring nor was she breathing heavily. With total disbelief, I asked for her explanation, not anticipating her simple answer.

She said, "I create a field of energy and ride it." Now it is easy to understand riding the outside force of a wave or wind, or skiing downhill by gravity, but what was this energy that created movement without a physical force or physiological happenings? She seemed to be trying to tell me that there are other ways to move that are beyond the classical neuromuscular contraction that we physiologists accept.

These remarkable and anomalous observations raise fascinating questions. Are there, indeed, ways of moving that are different from the classic neuromuscular processes that are so well known to physiologists and neurophysiologists? Why would the baseline on the electromyogram disappear during the operation of such an alternative movement process, if there really is such a phenomenon?

In one case, the electromyographic recordings decreased to zero during active movement. In another case, vigorous and acrobatic dancing was accomplished with a *decrease* in heart rate and blood pressure.

Until recently these observations seemed to be totally outside of any logical scientific paradigm. They were *anomalies*—scientists had no way of even thinking about them. To our knowledge, Hunt's observations have not been replicated. Why would anyone bother to try to repeat a phenomenon as inexplicable as this?

Continuum movement. In the meantime, Emilie Conrad has continued to evolve her movement work, which now is called *Continuum Movement.* Her process has awakened many people to the remarkable movement possibilities that are not ordinarily experienced. Its dramatic practical application is in enabling individuals with spinal cord injury to regain a part of their sensory and motor function. Because normal neurological pathways have been damaged in these individuals, the possibility arises that alternative pathways for energy and information flow are being activated by the *Continuum* process.

This book summarizes information from physiology and biophysics that provides a possible mechanism for these extraordinary observations. In making these suggestions, it is important to understand a basic fact of science and how it progresses. Some ideas will be presented that are new and untested. Each of these ideas has a background that will be described. Science progresses because people look at mysterious phenomena and try to explain them. As mentioned earlier, it is always better to have an idea than no idea.

We propose that Valerie Hunt's observations on the seemingly effortless movements of dancers, in the absence of electromyographic signals, Emilie Conrad's successes with paralysis, Prof's observations described in Chapter 5, and the other ways of sensing and moving described in this chapter, show us that we have available to us another sensory/movement system that can operate under certain conditions. It is not suggested that the conventional neuromuscular pathways so well studied by physiologists are wrong, but that they can be supplemented or replaced by an alternative parallel system under extraordinary conditions. We have termed this alternative method of sensing and responding the *continuum pathway*. It is illustrated in Figure 7-2.

The continuum pathway

The diagonal line shown in Figure 7-2 summarizes the neural pathway connecting *sensation* and *action* as you will find it in physiology texts. The energetic *stimuli* in the environment, heat, light, sound, smell, and so on, are detected by various *receptors*, which trigger the flow of impulses through *sensory nerves* to the *brain*. If an *action* (movement) is needed, signals are sent via *motor nerves* to *muscles*, which then contract, leading to an appropriate *action*. The brain regulates and coordinates such voluntary muscular activities.

Neurophysiologists have described a shortcut, called the *monosynaptic reflex arc* or *spinal reflex* (Figure 7-2). When you touch something hot, your hand is rapidly and involuntarily withdrawn, even before you are aware of the pain.

Figure 7-2 also shows a much faster way of responding to stimuli that operates under certain circumstances, such as during life-saving reactions that take place in an

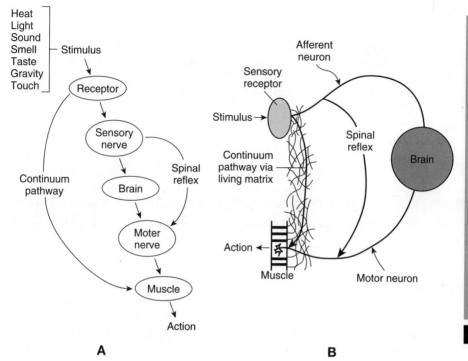

Figure 7-2 Three pathways connecting sensations with actions. **A,** This diagram summarizes the pathways, with the diagonal sequence representing the classical neuromuscular mechanism in which a stimulus activates a receptor, triggering an impulse to flow through an afferent sensory neuron to the brain. If appropriate, the brain initiates a signal that is sent via a motor neuron to the appropriate muscle or muscles to produce an action. A faster shortcut is the spinal reflex that can trigger a muscle contraction before one is consciously aware of the stimulus, such as when the hand is pulled back from a hot surface. The proposed continuum pathway is even faster, involving semiconduction through the excitable medium of the living matrix. The signal is propagated from the base of the sensory cell directly through the connective tissue to the myofascia and muscle. **B,** The anatomical correlates of these schemes.

emergency, in peak athletic or artistic performances, and in the martial arts. We refer to this as the *continuum pathway*. It is a shortcut between the receptor and the muscle, between sensation and action. It conducts both information from the senses and energy to power the muscles.

Here, *continuum* refers to a continuous pathway for sensory energy/information that includes but is not limited to nerves. It is a pathway that is more ancient, in terms of evolution, than the nervous system. In essence, we have come to the main goal of this book: to explain the continuum system and its significance for a wide range of phenomena that in the past have seemed inexplicable.

The operation of the continuum pathway enables one to move faster and in a more coordinated manner than the nervous system allows—to move before one is consciously aware of the reason for moving. It is this system that Albert Szent-Györgyi suspected existed on the basis of his experiences with his kitten and with the fly in his eye (Chapter 5). Most people can recall examples from their lives when they experienced brief moments when they moved in a surprisingly rapid and appropriate manner, and "time has seemed to slow down." It is suggested that this is the continuum pathway in action.

The continuum concept, extraordinary as it may seem, has come from many sources that are detailed in this book.

TWO KINDS OF TIME

We conclude that there are at least two major kinds of consciousness. *Neurological consciousness* arises from the operations of the nerves and brain. A second kind of consciousness we call *connective tissue* or *continuum consciousness.* Its pathway is via the continuum pathway in the living matrix, which will be introduced in Chapter 8 and described in greater detail in subsequent chapters.

Connective tissue consciousness is far more rapid and actually *precedes neurological* consciousness. Connective tissue consciousness arises because the living matrix is an *excitable medium,* just as nerve cells are excitable. We only become aware of this second kind of consciousness during extraordinary moments, such as during an accident, in the martial arts, or in peak athletic or artistic performances, when we become aware of all parts of our body working together and responding with an astonishing speed and coordination, far faster than thought.

Therapists, athletes, and performers may do their best work when they are tapping into the wisdom of this *precognitive consciousness.*

Precisely where is this continuum system located, and how does it work? We shall see next that this system is none other than the system that Albert Szent-Györgyi had come across in his research on electronic biology.

REFERENCES

Duchenne, G.B.A. 1867, *Physiologie des Mouvements,* 1st ed, trans. E.B. Kaplan, 1959, WB Saunders, Philadelphia.

Gauguelin, M. 1974, *The Cosmic Clocks,* Avon Books, New York.

Hunt, V. 1989, *Infinite Mind. The Science of Human Vibrations,* Malibu Publishing Company, Malibu, CA, pp. 10-11.

Kuhn, T. 1970, *The Structure of Scientific Revolutions,* University of Chicago Press, Chicago, IL.

LeDoux, J. 2002, *Synaptic Self.* Viking, New York.

O'Regan, B. & Pole, W. 1993, *The Heart of Healing,* Institute of Noetic Sciences, Sausalito, CA.

8 Introducing the living matrix

Continuum: a continuous extent, succession, or whole, no part of which can be distinguished from neighboring parts except by arbitrary division; an uninterrupted ordered sequence; an identity of substance uniting discrete parts.

Webster's Seventh New Collegiate Dictionary (1965);
The American Heritage Dictionary of the English Language (1978)

BIOLOGICAL COMMUNICATIONS

When we are asked about the communication systems in the human body, we immediately think of the nervous system. We also might add the circulatory system, because it delivers hormonal messages to cells throughout the body. The immune system is another vital whole-body communication network.

These perspectives are correct, but there has to be more to the story of biological communications. This book summarizes the evidence that there must be a high-speed communication system in living systems that is *not* the nervous system; instead it is a body-wide energetic and communication system that *includes* the nervous, circulatory, and immune systems, and it also includes *all* of the other systems in the body. This *system of systems* has come to be called *the living matrix*.

Nobel Laureate Albert Szent-Györgyi became aware of this system in two ways: by contemplating and researching the subatomic properties of the protein fabric of the body, and by observing the natural world in front of him and thinking new thoughts. He watched a kitten react (see Figure 5-1), and he thought about the extremely rapid responses that protected his cornea when he drove his motorbike into a fly (see Figure 5-4).

From her study of the trance dancer, Valerie Hunt concluded that there has to be some form of movement that differs from the usual neuromuscular pathways. Emilie Conrad's work with spinal cord injury provides another hint of an extraordinary system that can be activated when the nervous system is disconnected. Condon's research on the dance taking place during conversation implies that there is more to listening than we realize. Experiences in the martial arts and peak athletic and artistic performances all suggest that there are aspects of biological communications that are not readily explained by the nervous system as we usually think of it.

When we left Prof's story back in Chapter 4, he had concluded that the protein fabric of the body, the material at the bottom of the centrifuge tube that most biochemists were discarding, the molecular scaffold that makes a liver look like a liver and a kidney look like a kidney, was a conductor of electrons. To be more precise, he thought these materials were *semiconductors*. He drew Figure 4-2 to compare the backbone of a molecule with the wire going to his toaster. Life, he thought, is sustained by high-speed interactions taking place within this matrix. Electrons, protons, and perhaps other subatomic entities are the units of communication. These are entities that are well known to physics but that are rarely discussed in biology. They do not belong to the individual proteins in the body but instead belong to the organism as a whole. They are good candidates for a body-wide system that informs, energizes, and integrates all of the diverse processes involved in every moment of life. To Prof, cancer appeared to be characterized by a disconnection in these *circuits of life*. This disconnect prevents individual cells from being well-behaved participants in the community of cells that form the organism.

LIFE VERSUS NONLIFE

We are far more than a living machine consisting of mechanical parts held together with threads and fibers. The rapid and subtle flows, forces, and movements that characterize life and set it apart from nonlife originate within the molecules and the molecular, atomic, and subatomic frameworks of which we are composed. "Empty" space must have an active role as well. At the subatomic level, the distinction among matter, energy, and space becomes hazy, as does the distinction among the different forces such as gravity, elasticity, heat, motion, electricity, magnetism, and chemical bonding. The mechanical properties of a molecule, such as its strength, flexibility, and ability to conduct information, become inseparable from its electronic and other submolecular characteristics. Because enzymes also are proteins, their ability to participate in chemical reactions, to catalyze living processes, likewise depends on their electronic and submolecular features. It is here that biology, *the science of life*, becomes one with physics, *the science of matter and motion*. It is in the living of our lives that we experience this oneness. We, as sentient beings, are the experts on the *real* laws of nature.

ELECTRONIC MOBILITY IN THE FABRIC

We mentioned the two insightful papers Prof published in 1941. The statement he made in those papers continues to be fascinating. It is a statement that has led to many connections between modern science and complementary medicine:

> *If a great number of atoms be arranged with regularity in close proximity, as for example in a crystal lattice, single valency electrons cease to belong to one or two atoms only, and belong instead to the whole system. A great number of molecules may join to form energy continua, along which energy, namely excited electrons, may travel a certain distance.* SZENT-GYÖRGYI (1941A, 1941B)

We have reproduced this statement in countless articles and lectures over the years. Each reading spawns a new appreciation for the depth and significance of Prof's insight made over half a century ago. To appreciate the statement it is important to realize that most of the tissues in the body are composed of molecules that are arranged "with regularity in close proximity, as for example in a crystal lattice." We give many examples of this arrangement later (see Figure 9-2).

To paraphrase the above quotation:

In crystals, a great number of atoms or molecules are packed closely together in a lattice arrangement. The bonding electrons that hold the crystal together, which are called valency electrons, are actually mobile. They are not fixed in place, they do not belong to any particular atom, molecule, or bond, but instead they belong to the whole system. Hence a great number of atoms or molecules can join together to form an energetic continuum. Energy, in the form of excited electrons, can move about within the crystal lattice.

I first became aware of Prof's work by attending a lecture in 1972. Eventually I moved into the laboratory across the hall from Prof's Institute for Muscle Research at the Marine Biological Laboratory in Woods Hole, Massachusetts. Later the name of Prof's lab was changed to The National Foundation for Cancer Research. Much of what you are reading stems from discussions with Prof and his many colleagues and from their published papers.

THE BODY IS MAINLY MADE OF COLLAGEN AND WATER

One of the proteins studied in Prof's laboratory is a fascinating substance called *collagen*. This is the most abundant protein in the animal kingdom. It is the main structural ingredient in the connective tissues forming bones, tendons, ligaments, cartilage, and fascia (Figure 8-1).

Years of looking at various tissues in the electron microscope had led to familiarity and fascination with this ubiquitous material. By the end of the 1970s, I accumulated a large collection of reprints of scientific articles related to connective tissue. The

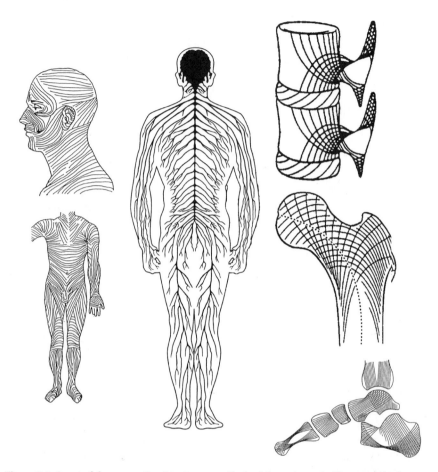

Figure 8-1 Some of the connective tissue systems. To the *left* are the main fiber tracts in the superficial fascia, immediately under the skin. These are the tension lines, also called *Langer's lines* (Hutchinson & Koop 1956; Langer 1861). In the *center* is the important connective tissue system, the perineurium, which surrounds nerves. On the *right* are the fiber tracts running through and between various bones in the skeleton. (Illustrations to the left, Langer's Lines, are from Matsumoto & Birch 1988, *Hara diagnosis: reflections on the sea*, and are used by permission of Paradigm Press, Brookline, Mass. The central illustration is from *Energy Medicine: the scientific basis*, Oschman, Figure 15.2, page 225, Churchill Livingstone, Edinburgh, 2000, with permission from Elsevier Science. The fiber tract system, upper right and lower right, is from *Energy Medicine*, Fig. 12.1, page 168, with permission from Elsevier Science. Originally obtained from Thompson, D.W. 1992, *On growth and form*. Dover Publications, Inc., New York, Fig. 464, page 980.)

purpose was simply to keep learning about this interesting material. At the time the articles were collected, I had no idea of the significance connective tissue would play in my future research.

This research took on new meaning when I experienced *structural integration*, or *Rolfing*, in the late 1970s. Practitioners of this approach to the body have a keen appreciation of human structure in relation to human functioning. Their work is based in part on careful study and experience with the connective tissue as a system. Eventually I acquired a set of reprints on connective tissue that had been accumulated by Dr. Ida P. Rolf, the founder of Rolfing or Structural Integration, and her colleague, George Hall III. The experience of Dr. Rolf's work, and subsequent research on connective tissue, led to a series of publications (Oschman & Oschman 1997) and to this book.

CONNECTIVE TISSUE

To the left in Figure 8-1 we see the main fiber tracts in the superficial fascia, immediately under the skin. These are the tensions lines, also called *Langer's lines* (Hutchinson & Koop 1956; Langer 1861).

In the center in Figure 8-1 is the important form of connective tissue that surrounds nerves and reaches into nearly every part of the body that is innervated. This is called the *perineurium*. On the right we see the fiber tracts running through and between various bones in the skeleton.

We could draw a similar system in relation to the vasculature. Each artery, vein, and capillary has its overall shape, microscopic form, and physical properties because of the ways the connective tissue has shaped it. Likewise, the lymphatic system and the digestive tract are systems of collagenous connective tissue. Every organ and gland has its particular form because of the ways the connective tissue fibers are arranged. Finally, the muscular system is invested with a connective tissue system called the *myofascia*.

The importance of connective tissue

In terms of the goals of this book, there is no perspective that is more important than the systemic properties of the connective tissue and the relation of the connective tissue to the other systems in the body. To summarize:

The connective tissue is the most abundant component of animal matter and forms the bulk of the animal body. The overall form of the body, as well as the architecture and mechanical and functional properties of all of its parts, are largely determined by the local configuration and properties of the connective tissue. All of the so-called great systems of the body—the circulation, nervous system, musculoskeletal system, digestive tract, the various organs—are ensheathed in and partitioned by connective tissue. The connective tissue forms a continuously interconnected system throughout the living body. All movements, of the body as a whole and of its smallest parts, are created by tensions carried through the connective tissue fabric. It is a liquid crystalline material and its components are semiconductors, properties that give rise to many remarkable properties. One of the semiconductor properties of connective tissue is *piezoelectricity*, from the Greek, meaning "pressure electricity." Because of piezoelectricity, every movement of the body, every pressure and every tension anywhere, generates a variety of oscillating bioelectric signals or microcurrents and other kinds of signals that will be described later. These signals are precisely characteristic of those tensions, compressions, and movements. Because of the continuity and conductivity of the connective tissue, these signals spread through the tissues. Because of continuity with cell interiors, to be described next, these microcurrents also are conducted into cells. If the parts of the organism are cooperative and coordinated in their functioning and every cell knows what every other cell is doing, it is due to the continuity and signaling properties of the connective tissue.

Because of a fascination with the electronic properties of collagen and connective tissue, I revised Prof's diagram (Figure 4-2) to compare the collagen molecule with the electronic circuit in an FM tuner (Figure 8-2).

In anticipation of the material we are about to present, it is important to mention that proteins such as collagen and the components of the cytoskeleton are *polymers*. These are materials that are made by attaching units (called *monomers*), such as amino acids, end to end to create a molecular chain of considerable length. We shall return to

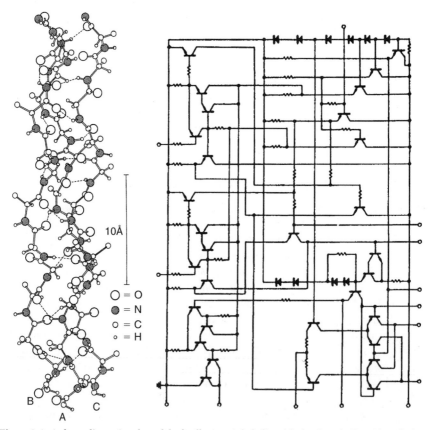

Figure 8-2 A three-dimensional model of collagen triple helix with the three helices identified as A, B, and C *(left)* is analogous with an integrated circuit such as those found in electronic devices *(right)*. (Right illustration is reprinted with permission of National Semiconductor Corporation. Left illustration is Figure 4, page 47, from Ramachandran GN 1963, *The Triple Helical Structure of Collagen*, in Ramachandran GN, ed *Aspects of Protein Structure*, Proceedings of a Symposium held in Madras 14-18 January 1963 and organized by the University of Madras, published by Academic Press, London, and reproduced by permission of Ramesh Narayan and the estate of GN Ramachandran.)

the subject of polymers when we consider the Nobel Prize–winning discovery that polymer plastics can be made that are electrically conductive.

Cells contain a connective tissue of their own

Every cell contains a miniature musculoskeletal system called the cytoskeleton. This substance includes stiff components such as *microtubules,* the "bones" of the cell; connective elements called *microtrabeculae,* the "tendons" of the cell; and contractile units called *microfilaments,* made of actin and myosin, the "muscles" of the cell (Hepler & Palevitz 1974; Pollard & Weihing 1974; Porter 1984; Porter & Tucker 1981; Soifer 1975; Wessels et al. 1971).

Studies of this dynamic cytoskeletal system have revealed how cells move from place to place and how the organelles move about within the cytoplasm.

We cannot place too much emphasis on the importance of cell movement and cell migration, because these processes are vital to injury repair and defense against disease. Research into the ways cells move has been given a high priority for funding by the National Institutes for Health, for example, and there are good reasons for this.

Interface between the two domains

In 1971, Dr. Mark S. Bretscher, in Cambridge, England, made a fascinating discovery during his studies of the proteins in red blood cell membranes. Some proteins are on the inside of the membrane, others are on the outside, and there was one important protein, called *glycophorin,* that seemed to be on both sides (Figure 8-3, *A*). A number of scientists objected to Bretscher's observations, but further investigation revealed that he was correct. Glycophorin actually extends across the cell membrane, from the inside to the outside (Marchesi et al. 1973). This was a surprise to many scientists. Following on this work, others began to discover many different proteins connecting the cell interior with the exterior. Some of these actually go back and forth across the membrane as many as seven times (Figure 8-3, *B*). These molecules are called *glycoproteins* because they contain sugars (Gottschalk 1960).

We now know that there is a whole class of molecules, called *integrins,* that play key roles in communicating information back and forth between the cell interior and the connective tissue system in which the cells are embedded (Figure 8-4, *A*) (Horwitz 1997). These transmembrane structures anchor the cell membrane and the cytoskeleton to the extracellular matrix or connective tissue. Hence the cytoskeleton and the extracellular connective tissue matrix are structurally continuous. This connective tissue/cytoskeleton system, or living matrix, is functionally and energetically continuous. The significance of this statement will begin to emerge in the following chapters.

Soon after the discovery of the transmembrane glycoproteins, it was recognized that the sialic acid–rich extensions of the cytoskeleton and cell surface are the sites where cells acknowledge extracellular proteins, whether they be antigens, hormones, cell-to-cell linkers, or anchors to the substrate (Mannery 1968; Oschman 1978; Steinberg 1958). The substrate, in turn, consists of the ground substance and collagen fibrils of basement membranes and connective tissue. Because all of these materials

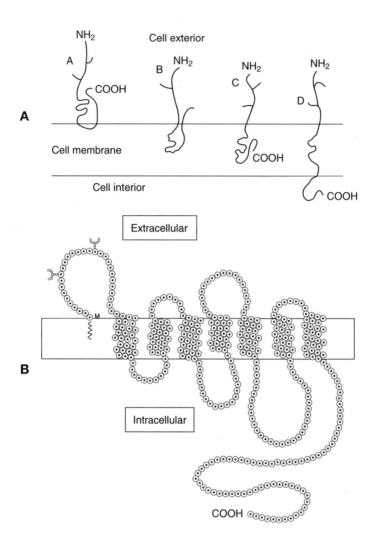

Figure 8-3 A, Four possible ways the glycoprotein glycophorin might be oriented in the intact erythrocyte membrane. On the basis of their labeling studies and other considerations, Bretscher (1971) and Marchesi et al. (1973) concluded that model D is the most likely orientation. Such transmembrane molecules now have been described in many different kinds of cells, and some of them are referred to as *integrins.* **B,** One example of a transmembrane protein that traverses the cell membrane seven times. Comparable serpentine transmembrane proteins are found in many different kinds of cells. (**A** Reprinted from *Journal of Molecular Biology,* 59: 351-357, Bretscher MS, A major protein which spans the human erythrocyte membrane, 1971, reprinted with permission from Elsevier Science.)

possess many negatively charged groups, the position of a cell in relation to neighboring cells and to the substrate is maintained by divalent calcium bridges. Because these bridges are labile rather than permanent, connections can form, break, and reform as cells change shape and/or crawl through the interstitial material. Inside the cell, the

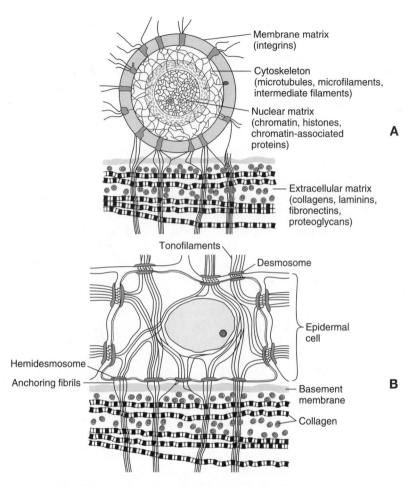

Membrane matrix
(integrins)

Cytoskeleton
(microtubules, microfilaments,
intermediate filaments)

Nuclear matrix
(chromatin, histones,
chromatin-associated
proteins)

A

Extracellular matrix
(collagens, laminins,
fibronectins,
proteoglycans)

Tonofilaments

Desmosome

Epidermal
cell

Hemidesmosome

Anchoring fibrils

Basement
membrane

Collagen

B

Figure 8-4 A, *The living matrix* consists of the connective tissue system, the integrins, the cytoskeleton, and the nuclear matrix. **B,** The living matrix continuum as described in 1984 by Ellison and Garrod. This is the epidermal-dermal junction. Adjacent epidermal cells are attached to each other by desmosomes and are anchored to the dermal connective tissue by hemidesmosomes. All of the anchors are traversed by tonofilaments, which form a continuous fibrous matrix joining together all epidermal cells throughout the skin. Anchoring fibrils link the cellular matrix with the connective tissue. The dermal connective tissue is part of a continuously integrated and interconnected system extending throughout the body. The cytoskeletons of all other cells in the body are similarly linked to the connective tissue system. (Figure A is from *Energy Medicine: the scientific basis*, Oschman, Figure 4.6, page 66, Churchill Livingstone, Edinburgh, 2000, with permission from Elsevier Science. Figure B is from *Energy Medicine: the scientific basis*, Oschman, Figure 3.3, page 47, Churchill Livingstone, Edinburgh, 2000, with permission from Elsevier Science. After Ellison & Garrod 1984, Fig, 10, p. 1790, *Journal of Cell Science* and the Company of Biologists, Ltd.)

connections between glycoproteins and cytoskeleton are labile, allowing for the positioning and repositioning of cell surface receptors (Albertini & Clark 1975; Chen 1982; Gabbiani et al. 1977; Koch & Smith 1978; Rapraeger & Bernfield 1982).

These molecules serve as recognition sites involved in a wide variety of vital processes. They are glycolipids or glycoprotein molecules consisting of a region that is embedded in the membrane and at least one carbohydrate chain extending from the cell surface into the region surrounding the cell. The carbohydrate groups are of seven different types. The carbohydrate portions form the active sites for recognition, reception, and adhesion.

MHC (major histocompatibility complex) molecules are responsible for enabling mammalian cells to recognize each other as self or nonself, and they are the reason for blood type matching (ABO blood groups) in transfusions. These molecules also are referred to as *transplantation antigens* because they are involved in transplant rejection. To be specific, individuals with type A blood have a glycolipid containing an amino sugar called *N*-acetylgalactosamine. People with type B blood have a galactose molecule at that position. People with AB blood have both of these sugars, whereas in people with type O blood the corresponding positions on the glycolipid are empty.

There are several types of cell surface molecules that have different functional roles. Receptors recognize molecules suspended in solution in the extracellular medium, molecules that form parts of the extracellular matrix, or molecules that are attached to the surfaces of other cells. Thus one type of receptor binds peptide hormones, growth factors, and neurotransmitters diffusing within the extracellular medium. These so-called *first messengers* activate processes within the cell. A second type of receptor binds molecules in solution or linked to cell surfaces and takes them into the cell by a process known as *receptor-mediated endocytosis*. A third type of receptor links the cell matrix to the extracellular matrix or to other cells, to hold the cell in its place.

An individual cell may contain hundred or thousands of receptor molecules, depending on the type of cell. Receptor combinations can change during development, and cells can modify their cell surface as part of a response to stimulation.

Examples of important receptors are those that bind insulin or epidermal growth factors, and those that activate processes such as hormone secretion, cell division, fertilization, cell migration, contraction of smooth muscles, and metastasis.

Examples of different classes of cell surface molecules include the cadherins, fusin, integrins, the immunoglobulin (Ig) superfamily, and selectins. Some receptors involve calcium ions in the binding process, others do not.

A review of the literature described the various materials, called *ground substances*, that form the extracellular matrix, the cytoplasmic matrix, the nuclear matrix, and the matrixes within various cell organelles (Oschman 1984). On the basis of the various studies mentioned earlier, as well as the work of Berezney et al. (1982), I suggested that all of the ground substances form a continuous supramolecular network extending throughout the animal body and into every nook and cranny. This network has no fundamental unit or central aspect, no part that is primary or most basic. The functioning of the whole network depends upon the integrated activities of all of the components. It is a cooperative, collective, or synergistic system.

Hence we now have a detailed "road map" of all of the living material forming an organism. This map reveals an interconnected webwork extending everywhere in the body. Since the discoveries listed above, additional research has filled in remaining gaps in the network by the characterization of a variety of linkers, including fibronectin, chrondronectin, laminin, and a whole class of molecules known as *integrins*. A worldwide computer database has been established to keep pace with the rapid proliferation of research papers, books, and guidebooks on cytoskeletal and contractile proteins (Kreis & Vale 1993). A listing of the major extracellular matrix proteins has been published by Ayad et al. (1994).

That the cytoskeleton forms a structural and functional linkage between the cell exterior and the cell nucleus has been affirmed in a variety of studies (Bennett 1982; Nicolini 1980; Otteskog, Ege & Sundqvist 1981; Scott 1984; Staufenbiel & Deppert 1982).

> *The cytoskeleton appears to function as a linkage between the nuclear and the plasma membranes, permitting environmental stimuli acting at the cell membrane to be transmitted to the nucleus for the initiation of an appropriate response.* (Scott 1984)

The concept of a direct and continuous structural and functional linkage between the extracellular matrix and the genetic material has been supported from a variety of perspectives for a long time.

> *The wisdom of the matrix, to borrow from Cannon's wisdom of the body, is more likely to be expressed in a language apart from the hereditary one, though interlinked with it.* (Grobstein 1975)

A functional reciprocity via these linkages is a key to understanding living structure:

> *Based on the existing literature, a model is presented that postulates a "dynamic reciprocity" between the extracellular matrix (ECM) on the one hand and the cytoskeleton and the nuclear matrix on the other hand. The ECM is postulated to exert physical and chemical influences on the geometry and the biochemistry of the cell via transmembrane receptors so as to alter the pattern of gene expression by changing the association of the cytoskeleton with the mRNA and the interaction of the chromatin with the nuclear matrix. This, in turn, would affect the ECM, which would affect the cell, which . . .* (Bissell, Glenn Hall & Parry 1982)

A MAJOR CAVEAT

As mentioned in the Prologue, an entirely different perspective arises from the work of Gerald F. Pollack and is described in his book, *Cells, Gels and the Engines of Life* (Pollack 2001).

Pollack presents an important and unorthodox view of cell structure and function. It is a view that has been pioneered by Gilbert Ling, a scientist whose work has been derided frequently because it seems extreme to many of his colleagues. In our opinion, Ling's pioneering work and Pollack's expansion upon it are vitally important and much worthy of consideration.

Pollack launches his book by describing an extraordinary phenomenon cell biologists have noticed again and again. It is the observation that a cell crawling along a surface can have a portion of its membrane stick to the substrate so strongly that the leading edge of the cell tears itself loose, continues on its journey, and leaves the stuck part of its membrane behind. The cell moves and functions normally, even though it has been "effectively decapitated" to use Pollack's words.

Pollack goes on to describe a variety of experimental situations in which cell membranes have been intentionally torn open or punctured full of large holes without compromising cellular functions. In some cases observable punctures or openings persist for periods up to hours, but the contents do not leak out. Large molecules added to the cell's environment penetrate into the cytoplasm, confirming that there is no barrier.

The problem, and the hostility engendered by even considering the problem, is with a fundamental belief that all cells are surrounded by a continuous and impenetrable membrane. This is hallowed ground, akin to motherhood and apple pie, and so deeply ingrained in modern thinking that it is heretical to even consider it as a possibly refutable part of the logic of modern cell science.

There is a serious problem if the membrane can be torn off and the cell survives. According to conventional wisdom, all of the ions, solutes, and enzymes within it should immediately leak out, and this should shortly bring about a severe case of cell death. However, it has repeatedly been observed that this does not happen. The cells truly do not seem to care that their precious membrane has been torn off or poked full of holes.

The observations summarized by Pollack cause us to question the concept of the cell membrane as a continuous barrier. Where did this barrier concept originally come from? Pollack reviews the history of the subject and reaches the remarkable conclusion that the cell membrane concept arose from studies with the light microscope. There seemed to be a barrier at the cell surface, but the light microscope did not have sufficient resolution to determine its composition. Pollack believes the "barrier" the light microscopists were seeing was a layer of cytoskeletal protein material at the cell surface. This layer is some 100 times thicker than the supposed phospholipid cell membrane.

Cell membranes and membranes within cells have a huge protein content, ranging from 50% to much higher levels on a by-weight basis. The idea of a phospholipid bilayer "studded" with occasional proteins is misleading. Pollack suggests that if the protein had been discovered before the lipid, the concept of a protein membrane might have been adopted instead of the lipid membrane. Indeed, it was not long ago that membrane models in which protein formed the major component were still being actively considered (DeRobertis et al. 1975) (see Figure 1 in the Prologue).

Pollack goes on to describe other experiments that do not fit with the concept that the cell membrane is a phospholipid bilayer structure. The crux of the matter is that the membrane barrier was required, historically, because it was assumed that cell is a bag, and its contents were free to diffuse about. If this is the case, the contents of a cell without a membrane would soon leak out into the surroundings. In addition, the electrical measurements of the potential across the cell surface suggested that there was an electrical barrier. However, if the cell is not a bag (and there is a lot of evidence that it is not, as we shall soon see), if the matter inside the cell is not free to diffuse, and if the electrical potential across the cell surface and unequal distribution of ions arise as a consequence of the charges on the cytoplasmic gel, a continuous membranous barrier is not necessary.

Pollack then explores the water inside cells. One feature stands out immediately: when cells are frozen the water inside of them does not crystallize as it should if it were an aqueous solution of molecules and ions. Cell water is in a different state than ordinary water.

Taken together, the evidence Pollack summarizes must be considered if one wishes to understand the nature of cells and their contents. Many of the concepts discussed in this and other books must be reconsidered.

BIOCHEMISTRY ON THE LIVING FABRIC

Generations of physiologists and biochemists have operated on a simplifying assumption that every cell is merely a "bag" filled with a solution of enzymes and substrates that randomly diffuse from place to place. The approach to studying biochemical reactions has been to burst open the bag to let the enzymes and substrates float out into a test tube where they can be studied. In the "bag" model, biochemical reactions take place when reactants and catalysts accidentally collide, each in the correct orientation, bringing the active sites together.

We now know that this "bag" image is totally inadequate. We know that life is too rapid, subtle, and sophisticated to wait for accidental or chance interactions of enzymes and metabolites randomly bumping about within the cell. Studies of biochemical and transport phenomena based on such a model must be reevaluated. Not only is the cell interior highly ordered, but it appears that the so-called *soluble* enzymes, upon which much of modern biochemistry is based, are, in the intact cell, attached in a highly regular manner to the cytoplasmic framework or cytoskeleton (Arnold & Pette 1968; Masters 1978, 1979, 1981; Mowbray & Moses 1976; Opperdoes & Borst 1977; Walsh, Clarke & Masters 1977). The picture is nicely summarized by Adolph:

> *Metabolic processes are like assembly lines in that materials are handed from one enzyme to the next. Each enzyme is a specialized machine that forms a specific operation on the substance. Biological oxidation is one of the longer assembly lines with sugar molecules being taken apart piece by piece by a sequence of 17 enzymes. Hence enzymes typically work in teams. During development the enzymes usually appear in clusters rather than individually. Individual enzymes only stand out when they are absent, as in inborn errors of metabolism. Team-work means integrative action, and an absentee may interrupt the entire task. (Adolph 1979)*

The reader may notice that the references just listed are from the period 1968 to 1981. These are not outdated references. The passage of time has not led to refutation of these findings; instead the important information in those articles has simply not been fully appreciated because of the dominance of the inaccurate "bag" model of the cell. It is only recently that a few serious scientists have begun to bring cell architecture back into biochemistry (Ingber 1993). Ingber's paper contains more recent studies showing that biochemical processes take place *on* cytoskeletal surfaces rather than in fluids in the spaces between the fibers.

Classic biochemists continue to be convinced that the best way to study cells and tissues is by homogenizing them, putting them in the centrifuge, and studying the dissolved materials in the supernatant (see Figure 4-1). Hence the persistence in textbooks of images of the cell as a fluid-filled bag, with a few organelles floating about within it (see Figure 1, *A,* in the Prologue).

Although this image of the cell and the pharmacology based on it continue to be productive, an entirely new and extremely valuable biochemistry is gradually making its way into the academic scene, and it is none other than Prof's *Volume II,* solid-state biochemistry. A detailed picture of this approach will emerge in the chapters that follow.

The systems the biochemists have studied so diligently actually take place in the context of a three-dimensional architecture that is vitally important and that participates in all living activities. Leaving out the structural context, the molecular scaffolds that tissues and cells are made of created a disconnect between body mechanics and the chemistry of life. Ingber (1993) has shown that cellular architecture actually regulates metabolic processes and not the other way around. This is a profoundly important concept with many implications for the manual therapies. The emerging study of living architecture is in its infancy and will occupy the attention of scientists for many years to come.

For example, solid-state biochemistry is providing a scientific basis for the emerging field of *biomimicry,* mentioned in Chapter 2. It is also leading to a scientific basis for many of the more extraordinary observations that therapists, performers, and masters of the martial arts encounter from time to time, and for Valerie Hunt's observations on the trance dancer summarized in Chapter 7.

In essence, we are finding that life's reactions, elaborated in such great detail in the biochemistry texts and research literature, take place in and around a continuous living fabric that I have termed the *living matrix.*

Physical and conceptual boundaries

The discovery of the continuity of the living matrix system has broken down the artificial conceptual boundary separating the study of the connective tissue outside of cells, the cell surface, and the matrix within cells, for they now are seen to form a continuum (Oschman 1984). The clearly delineated boundary between what is within a cell and what is outside of the cell has become hazy (Figure 8-4, *A* and *B*). These drawings provide an overview of the cytoskeleton and its connections, across the cell surface, to the extracellular connective tissue fabric. The continuous system has led to a new concept of how the various parts of the organism, including those within cells, are joined together and their functions coordinated. The nervous and circulatory systems, the focus for study of whole-body coordination, are but a part of this larger fabric.

Do not forget our caveat

If validated, the discoveries of Gilbert Ling, so well documented and articulated by. Pollack (2001), will simplify the picture even more. If the membrane is not actually composed of a continuous phospholipid bilayer, the cell membrane concept will have to be redefined along lines that were being considered some 50 years ago, when the protein component seemed to be dominant (Danielli 1954; Lehninger 1971). Regardless of how the arguments about the fine details of the architecture of the cell surface turn out, whether membrane proteins extend across a phospholipid bilayer or extend out from a blob of gel, as Pollack proposes, the continuity of the fibrous systems seems assured. The living matrix is, in fact, a living matrix.

An old idea

The concept of a molecular fabric of the animal body, extending into all of its parts, is not new. Over 200 hundred years ago, Haller described an extracellular fibrous matrix that now is known as the connective tissue:

> *This web-like substance in the human body is found throughout the whole, namely, wherever any vessel or moving muscular fiber can be traced, without exception. The principal use of the fabric is to bind together contiguous membranes, vessels, and fibers . . . .* HALLER (1779)

In addition, Lamark (1809) stated that all of the organs in animals without exception are enveloped in this material.

The well-accepted concept of the connective tissue as a continuous system throughout the animal body applies equally to the ground substance matrix, which forms a finer reticulum within the intervals between connective tissue fibers and forms the links across the cell surface to the cytoplasmic and nuclear matrices (Burridge et al. 1987; Hynes & Yamada 1982; Oschman 1984).

THE NERVOUS SYSTEM OF THE CELL

The cytoskeletal network now is considered to represent the "nervous system" of the cell, with the ability to communicate and process electromechanical, electrochemical, and electrooptical signals. It has been realized that components of the cytoskeleton can vibrate with complex but measurable harmonics. Every time a cell moves or changes its shape, specific sets of lattice vibrations (phonons) travel like waves throughout the nuclear and cytoplasmic matrix and then into the extracellular matrix (Pienta & Coffey 1991). Likewise, muscle contractions produce specific sound vibrations that travel through the tissues (Oster 1983; Oster & Jaffe 1980). Long considered by physiologists and physicians to be a nuisance or useless by-product, muscle sounds now are being recognized for their potential role in communication. Because components of the cytoskeleton are piezoelectric, cell and muscular movements and vibrations will produce corresponding electrical oscillations, which also may be conducted for a distance throughout the cellular matrix and beyond, into the connective tissue matrix.

Hence the idea of a "nervous system of the cell" must be placed in the larger context of the larger matrix in which the cell is embedded. Although some cells secrete molecules that diffuse to other cells and act upon them, there is a much faster means of communication provided by the continuity and vibratory character of the living matrix.

In the past it was thought that electrical activity in cells and tissues primarily involved the movement of ions such as sodium, potassium, calcium, and chloride. Although a sophisticated electrophysiology has been developed around study of the fluxes of these ions across membranes, we now know that there are more subtle and far more rapid flows of charged particles in the form of electrons, protons, and other subatomic particles/waves. These forms of energy/information do not diffuse slowly from place to place like the large bulky ions, but instead travel extremely rapidly within the living matrix.

Expanding the concept of "connective tissue"

From the research being done by Prof and his colleagues and from contemporary work in the field of cell biology, it became apparent that the collagen fibers of the connective tissue must form a continuously interconnected electronic and protonic communication system that also is connected to the cytoskeletons within all cells, throughout the body. Within the cytoskeleton is the nucleus, with its own matrix and the genetic material. The "connective tissue" concept has to be expanded to encompass *all* of the continuously interconnected fibrous systems in the body: those forming the classic connective tissues, as well as those forming the matrices of the cytoskeletons and nuclei.

Many therapists appreciate that touching the body in one place is, in reality, touching the entire organism. The *living matrix* provides a material and conceptual basis for the effects of the various holistic or systemwide therapies. Prof's ideas about semiconduction and electronic conduction in the matrix provide a physical basis for the interconnection of the energy/information systems of one person with those of another, through touch or through energy field interactions.

To summarize:

The connective tissue and cytoskeletons together form a structural, functional, and energetic continuum extending into every nook and cranny of the body, even into the cell nucleus and the genetic material. All forms of energy are rapidly generated, conducted, interpreted, and converted from one to another in sophisticated ways within the living matrix. No part of the organism is separate from this matrix.

SEMICONDUCTORS OR INSULATORS

When Prof suggested in 1941 that the proteins in the body might be semiconductors, the scientific community reacted as it often reacts to new ideas: the idea is absurd, and Prof obviously had become confused. However, Prof was definitely an important scientist, and others wanted credit for *proving* that he was wrong, so many experiments were done to disprove his thesis.

There were two fundamental objections. First, many different proteins had been isolated and purified, and all of them appeared to be insulators rather than conductors. Second, semiconduction seemed impossible for proteins. For a substance to be a semiconductor, it was essential for its electrons to have a certain energy structure, and this was not present in proteins (Tomita & Riggs 1970).

To be more specific, solid matter can exhibit a wide range of electronic conductivity. Conductivity depends on two factors, the number of charge carriers and their mobility. Some substances, like diamond, have a much higher electron mobility than the copper used to make wires. However, diamond is an insulator because it has virtually no free electrons available to carry a charge. In contrast to the situation Prof was proposing for proteins, all of the bonding electrons in diamond are tightly held in a rigid bonding arrangement between adjacent atoms. Mobility of electrons is much lower in metallic copper, which is used to make electrical wires, but there are abundant electrons available to conduct charge from place to place (Pethig 1979).

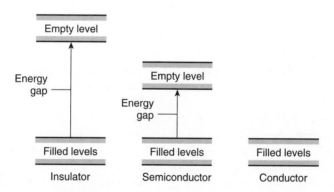

When an electron is excited to a higher energy level, called the *conduction band*, the electron becomes mobile (we shall see examples of this later on in the book). As stated by Szent-Györgyi (1941a), these electrons "cease to belong to one or two atoms only, and belong instead to the whole system." The energy band structure of insulators, semiconductors, and conductors are shown in Figure 8-5.

As most scientists expected, extensive research soon showed that Prof was absolutely wrong. The biochemists seeking to discredit Prof did what they knew how to do so well: they homogenized cells, centrifuged the resulting cell soup, dried out the material at the bottom of the tube, compressed it into pellet form, and inserted the dehydrated remains of the tissue between two electrodes and attempted to pass a current through it. Figure 8-6 shows some of the arrangements the different researchers used to do these kinds of experiments to determine how well electrons could flow through proteins. Electrons did not flow. Proteins are neither conductors nor semiconductors. They are good insulators. Prof was mistaken. Prof was reaching an advanced age, and wouldn't it be more dignified if he simply retired?

The problem with these experiments was that the protein preparations had been dehydrated. Prof and other scientists around the world soon demonstrated that the removal of water converted the proteins from conductors to insulators. Water is essential for life—take water away, and you are studying nonlife. Keep the proteins hydrated, and they are semiconductors.

Electronic engineers and materials scientists took up this line of research and demonstrated that virtually all of the substances in the body are semiconductors:

Figure 8-5 Schematic representation of the energy band structure in an insulator, semiconductor, and conductor. In the insulator there is a large energy gap between the occupied orbitals and the first empty level. Should an electron be capable of making this jump, it will be in what is termed the *conduction band* and will be able to move from place to place. However, at normal temperatures this amount of energy is not available, and virtually no electrons can make the transition. In a semiconductor the energy gap is small enough that thermal excitation can force an electron into the conduction band. In metals, such as sodium, the conduction band is always partially occupied with electrons. (From Pethig, R. 1979, *Dielectric and Electronic Properties of Biological Materials*, Reproduced with permission of John Wiley & Sons Limited.)

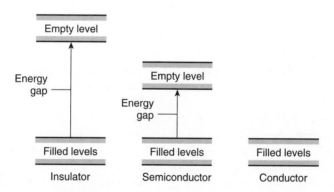

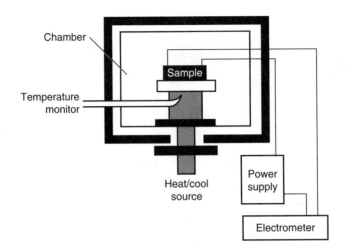

Figure 8-6 Device that was used to demonstrate that proteins are not conductors or semiconductors. These studies were rendered meaningless because the proteins being studied were dehydrated. If the proteins had been kept moist, they would have remained semiconductors. (Based on Figure 5, page 171, in Rosenberg B, Postow E. Semiconduction in proteins and lipids: Its possible biological import. Copyright 1969, *Annals of the New York Academy of Sciences*, Volume 258.)

> *We can now, a quarter of a century later, argue from both an experimental and a theoretical viewpoint that Szent-Györgyi's curious (for 1941) suggestion of conduction bands (i.e., energy continua) in protein systems was quite correct.*
>
> ROSENBERG AND POSTOW (1969)

Water is absolutely essential for electronic and protonic mobility. In fact, as we soon shall see, water is an absolutely essential ingredient for the formation and functioning of *all* living structures. The conductivity of any tissue proved to be hugely dependent on its water content. A study done in Prof's lab (Gascoyne, Pethig & Szent-Györgyi 1981) revealed how much his postulated communication system relies on water. A 10% change in water content can trigger a million-fold change in charge transport along a protein.

Many more studies have shown that organic systems have semiconductor properties and elucidated the mechanisms by which these properties are acquired and regulated in living systems (Boguslavskii & Vannikov 1970; Eley 1968; Forster & Minton 1972; Gutmann & Lyons 1967). This work is being continued and extended by a growing number of scientists around the world. Because of their technological and commercial importance, we now know far more about the physics of semiconductors than of conductors. However, a search of the literature will still reveal a number of articles, even recent ones, that reject the idea of semiconduction in proteins (Compton 1989).

ARCHITECTURE AND VIBRATIONS IN THE LIVING MATRIX

Appreciation of the architecture and vibratory properties of living matrix has profound implications for biology and medicine. The nervous and circulatory systems extend throughout the body, but there are cells that are not in direct contact with either of them, and the nervous and circulatory systems are relatively slow-speed communication systems.

For the body to function as a whole, there must be a system that reaches to and into every cell. For the body to function at its absolute peak of performance, all parts and processes must be interconnected by a system that delivers energy and information at the fastest possible means that nature has available. The living matrix is the system of systems that accomplishes these integrative activities.

The mechanical properties of the connective tissue have long been a topic of research, but scientists now have begun to study the entire living matrix as a structural, energetic, and informational system.

Robert O. Becker has contributed a significant model for regulatory interactions between the nervous system and the extracellular matrix. Becker's research over a long period of time has shown that the cells surrounding the nerves, called *perineural cells,* form an interconnected continuum that generates direct current fields. These fields play key roles in regulating both the nervous system and wound healing (summarized in Becker 1991). Whereas many researchers have confirmed Szent-Györgyi's ideas about the semiconductor nature of molecules, Becker is one of the few physiologists to confirm these concepts at the tissue level.

Relationship to trauma

From Dr. Ida P. Rolf we learned that the connective tissue is also the medium in which the body records the history of traumas and adaptations to them. Her work focused on the physiological and clinical importance of *gravity,* which is arguably the most potent yet largely unappreciated physical influence in any human life. She explained that the ways the body responds to physical trauma also apply to the responses to emotions and to chronic psychological states (Rolf 1962).

To Dr. Rolf's perceptive eye, the shape of a person's body told a history of that individual's encounters with gravity. With her skilled and sensitive hands, this history could be systematically resolved. Connective tissues can be restored to the balance and flexibility that they were originally designed to have. For many, this led to a renewed love of life and energetic participation in it. Her work came to be known as *structural integration,* or *Rolfing.*

INTEGRATING STRUCTURES

The very concept that one's physical structure can be integrated is fascinating. Watching the effects of the Structural Integration process on athletes and performers led to an appreciation of the fact that an outstanding feat of any kind involves the participation of the entire body. The performer at the peak of his or her "game" is an individual who is able to achieve total cooperation, coordination, and participation of every tissue, cell, molecule, and atom within his or her body to produce every aspect of the desired performance. Although this may be obvious, the research on subatomic particles in Prof's laboratory was giving a clearer explanation of precisely what is going on.

Table 8-1 Description of the meridian system and classical and modern acupuncture text references

Fat, greasy membranes; fasciae; systems of connecting membranes that through which the yang qi streams

Han Shu, 16 AD
Nan Jing, 100 AD
Wan Shu He, third century AD
Chen Yan, 1174 AD
Hua Shou, 1361 AD
Yu Bo, 1577 AD
Ippo Okamoto, 1703 AD
Royoan Terashima, 1712 AD
Xie Lin, 1895 AD
Wu Kao Pan, 1980 AD

From Matsumoto, K. & Birch, S. 1988, *Hara Diagnosis: Reflections on the Sea,* Paradigm, Brookline, MA.

The studies on the relationship between water content and conduction fitted in with the experiences of bodyworkers. Often painful or inflexible tissue feels palpably dehydrated, like parchment paper. Massage, Structural Integration, and other hands-on techniques can soften such tissues, and they seem to hydrate. One bodyworker even has her clients drink water during her manipulations because this seems to make pain go away much faster (R. Margoulis, personal communication, 2000).

Relationship to acupuncture

Some of Dr. Rolf's colleagues dissected a cadaver to see what connective tissue looks like. They noticed tiny fibers coursing through the connective tissue and fascia. Some of these fibers seemed to correspond to the locations of the meridians described in Oriental Medicine. The author wondered if the movements of electrons and protons Prof was describing in collagen molecules might be related to the *Ch'i* or *Qi* energies described so long ago in the theory of acupuncture.

The author mentioned this in a presentation at the New England School of Acupuncture. One of the teachers explained that the ancient acupuncture texts were completely congruent with this idea. Steve Birch and Kiko Matsumoto had translated ten classic acupuncture texts, published between 16 AD and 1980, that contained Chinese characters (Figure 8-7, *A*) interpreted as:

> *Fat, greasy membranes, fasciae, systems of connecting membranes, that through which the yang qi streams.*   MATSUMOTO AND BIRCH (1988)

"Greasy" membranes are referred to this way because they are shiny. To see this, look at the tendons in a fresh uncooked chicken leg, for example, or the teased muscle fibers shown in Figure 8-8 that expose the surrounding and investing endomysial fascia. Light reflection like this is a characteristic of materials such as metals, with free electrons in their outermost orbitals. These electrons are responsible for absorbing and emitting photons (reflection). They are the "conduction electrons" responsible for the conduction of electricity. These are the same electrons Prof was talking about in his famous 1941 quotation given earlier.

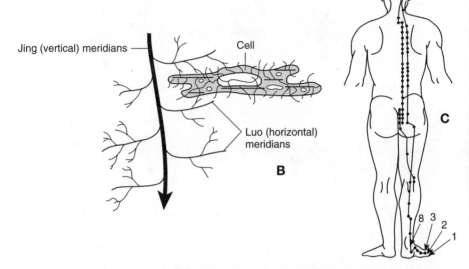

A 膜 膏 膜肓 絡 膜膏

Jing (vertical) meridians

Cell

Luo (horizontal) meridians

B

C

8 3 2 1

Figure 8-7 A, Chinese characters interpreted by Matsumoto and Birch as referring to *"Fat, greasy membranes, fasciae, systems of connecting membranes, that through which the yang qi streams."* **B,** The acupuncture meridian system with its horizontal *luo* meridians that branch and rebranch and reach into every part of the organism. The meridian system with all of its branches corresponds to *the living matrix system.* **C,** Stimulation of vision-related acupoints on the lateral aspect of the foot rapidly activates neural circuits in the occipital lobes of the brain. The activation, which is measured with functional magnetic resonance imaging (fMRI), is far too rapid to be explained by any form of nerve conduction. This approach is opening up study of the nature and velocity of the signals moving through the living matrix system and acupuncture meridians. (A and B are from Matsumoto & Birch 1988. Hara diagnosis: reflections on the sea, reproduced by permission of Steve Birch and Paradigm Press, Brookline, MA. From Cho, Z.H., Chung, S.C., Jones, J.P., Park, J.B., Park, H.J., Lee, H.J., Wong, E.K. & Min, B.I. 1998, *Proceedings of the National Academy of Sciences of the USA,* vol. 95, pp. 2670-2673.)

The *yang qi* is an ancient Oriental designation for energy, and the acupuncture literature shows clearly that this energy was thought to flow through the fascia. To be more specific, Matsumoto and Birch found that this description of the meridian system is repeated again and again in both the ancient and modern literature. Table 8-1 lists some of the references they uncovered in their search.

Prof and his colleagues were developing a scientific basis for the conduction of energy and information that might relate to phenomena acupuncturists had been talking about for millennia. This probably was the same energy conduction Dr. Rolf was talking about, and that Polarity, Healing Touch, Therapeutic Touch, Reiki, Zero Balancing, and other energy therapists tap into during their work. It is an energetic communication system that energizes, integrates, organizes, and informs the body as a whole. It is also the main system responsible for responses to injuries and traumas and for protection against diseases.

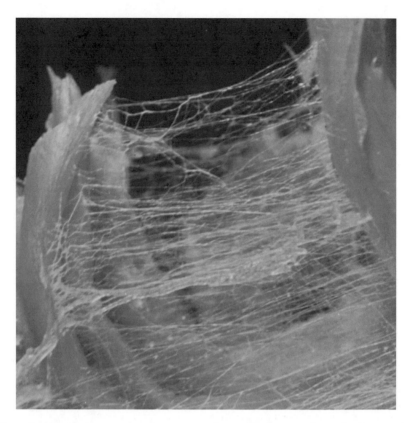

Figure 8-8 In numerous acupuncture texts over the past 2000 years, reference has been made to *"fat, greasy membranes, fasciae, systems of connecting membranes, that through which the yang qi streams."* Here we see endomysial fascia from a teased muscle fiber. Note the shiny aspect of the fibers, a characteristic of metals that contain free electrons that absorb and reemit light (reflective quality) and conduct electricity. (Courtesy Ronald Thompson. From *Anatomy Trains*, Myers, pages 8 and 9, Churchill Livingstone, Edinburgh, 2001, with permission from Elsevier Science.)

The acupuncture teachers explained that there are both vertical and horizontal acupuncture meridians. The horizontal meridians, called *luo*, are thought to branch and rebranch, reaching every cell in the body (see Figure 8-7, *B*). Hence the living matrix concept provides a connection between the latest understandings from cell biology (see Figure 8-4) with the wisdom of the ancients.

Super highways and local roads. Most acupuncture is done to points on the primary channels, also known as the 12 organ meridians. These channels are named Stomach, Spleen, Lung, Bladder, etc., after the organs that they reach internally. However, an important point is that these channels represent only the *superhighways* of the body's energy system, carrying a lot of traffic and causing huge traffic jams when they become blocked (R. Margoulis, personal communication, 2000).

That the skin is connected to everything else in the body via the living matrix system was recognized by 1984, when Ellison and Garrod described the cellular system of connecting filaments of the epidermis and dermis that are directly linked or continuous with the connective tissue fibers of the superficial fascia (see Figure 8-4, *B*). Hence an acupuncture needle inserted in the skin is in contact with a system that extends throughout the body. It is therefore not surprising that the functioning of organs, including the brain, can be influenced via the meridian system.

AN IMPORTANT CONFIRMATION

Validation of the living matrix and continuum pathway concepts has come from research of Dr. Joie Jones and his colleagues at Irvine, California. They have discovered that stimulation of vision-related acupoints on the lateral aspect of the foot rapidly activates neural circuits in the occipital lobes of the brain (see Figure 8-7, *C*). This activation, which is measured with functional magnetic resonance imaging (fMRI), is far too rapid for the signal to be conducted by nerve impulses (Cho et al. 1998; Jones 2001). This approach is opening up studies of the nature and velocity of the signals moving through the living matrix.

SUMMARIZING

We have reached the point in the book where it is worthwhile to examine where we are and where we are headed. The expanded concept of connective tissue, the continuity of connective tissue, and its remarkable energetic properties are the "launching pad" for a deeper understanding of life, disease, disorder, healing, and human performance. We shall return to the acupuncture system for a closer look at how these ideas may be applied. We conclude with a list of submolecular vibratory entities that are candidates for moving energy and information from place to place within the living matrix (Table 8-2).

Aspects of connective tissue that we have discussed so far give rise to *emergent properties* such as consciousness. These aspects include semiconduction, piezoelectricity, crystallinity, hydration, continuity, and information processing. When an injury, attitude, memory, stress, or other factor limits human performance, the problem is to be found within the living matrix system. Any therapeutic approach that focuses on one feature of this global system, such as the nervous system or the joints, for example, will have limited success with addressing the organism as a whole.

Table 8-2 Subatomic and molecular candidates for units of energy and information flow through the living matrix

Electrons
Holes
Phonons
Photons
Excitons
Protons
Solitons
Polarons
Conformons

REFERENCES

Albertini, D.F. & Clark, J.I. 1975, 'Membrane-microtubule interactions: Concanavalin A capping induced redistribution of cytoplasmic microtubules and colchicine binding proteins', *Proceedings of the National Academy of Sciences USA,* vol. 72 pp. 4976-4980.

American Heritage Dictionary of the English Language, 1978, Houghton Mifflin Company, Boston.

Arnold, H. & Pette, D. 1968, 'Binding of glycolytic enzymes to structure proteins of the muscle', *European Journal of Biochemistry,* vol. 6, pp. 163-171.

Ayad, S., Boot-Handford, R.P., Humphries, M.J., Kadler, K.E. & Shuttleworth, C.A. 1994, *The Extracellular Matrix Facts Book,* Academic Press, London.

Becker, R.O. 1991, 'Evidence for a primitive DC electrical analog system controlling brain function', *Subtle Energies,* vol. 2, pp. 71-88.

Bennett, V. 1982, 'The molecular basis for membrane-cytoskeleton association in human erythrocytes', *Journal of Cellular Biochemistry,* vol. 18, pp. 49-65.

Berezney, R., Basler, J., Bucholtz, L.A., Smith, H.C. & Siegel, A.J. 1982, 'Nuclear matrix and DNA replication', in *The Nuclear Envelope and the Nuclear Matrix,* ed. G.G. Maul, Alan R. Liss, New York, pp. 183-197.

Bissell, M.J., Glenn Hall, H. & Parry, G. 1982, 'How does the extracellular matrix direct gene expression?' *Journal of Theoretical Biology,* vol. 99, pp. 31-68.

Boguslavskii, L.I. & Vannikov, A.V. 1970, *Organic Semiconductors and Biopolymers,* Plenum Press, New York.

Bretscher, M.S. 1971, 'A major protein which spans the human erythrocyte membrane', *Journal of Molecular Biology,* vol. 59, pp. 351-357.

Burridge, K., Beckerle, M., Croall, D. & Horwitz, A. 1987, 'A transmembrane link between the extracellular matrix and the cytoskeleton', in Molecular Mechanisms in the Regulation of Cell Behavior, ed. C. Waymouth, *Modern Cell Biology,* vol. 5, pp. 147-149

Chen, W.T. 1982, 'Development of the attachment sites between the cell surface and the extracellular matrix in cultured fibroblasts', *Journal of Cell Biology,* vol. 95, p. 100a.

Cho, Z.H., Chung, S.C., Jones, J.P., Park, J.B., Park, H.J., Lee, H.J., Wong, E.K. & Min, B.I. 1998, 'New findings of the correlation between acupoints and corresponding brain cortices using functional MR', *Proceedings of the National Academy of Sciences of the USA,* vol. 95, pp. 2670-2673.

Compton, R.G. (ed.) 1989, 'Electron tunneling in biological systems', in *Comprehensive Chemical Kinetics,* vol. 30, p. 273.

Danielli, J.F. 1952, 'Structural factors in cell permeability and secretion', *Symposia of the Society for Experimental Biology,* vol. 6, pp. 1-15.

DeRobertis, E.D.P., Saez, F.A., & DeRobertis Jr., E.M.F. 1975, *Cell Biology,* 6th edn, WB Saunders Company, Philadelphia, p. 159.

Eley, D.D. 1968, *Organic Semiconducting Polymers,* ed. J.E. Katon, Edward Arnold, London, pp. 259-295.

Ellison, J. & Garrod, D.R. 1984, 'Anchoring filaments of the amphibian epidermal junction traverse the basal lamina entirely from the plasma membrane of hemidesmosomes to the dermis', *Journal of Cell Science,* vol. 72, pp. 163-172.

Forster, E.O. & Minton, A.P. 1972, 'Electric properties of biopolymers; Proteins', in *Physical Methods in Macromolecular Chemistry,* vol. 2, ed. B. Carroll, Marcel Dekker, New York, pp. 185-344.

Gabbiani, G., Chaponnier, C., Zumbe, A. & Vassall, P. 1977, 'Actin and tubulin co-cap with surface immunoglobulins in mouse B lymphocytes', *Nature,* vol. 269, pp. 697-698.

Gascoyne, P.R.C., Pethig, R. & Szent-Györgyi, A. 1981, 'Water structure-dependent charge transport in proteins', *Proceedings of the National Academy of Sciences of the USA*, vol. 78, pp. 261-265.

Gottschalk, A. 1960, *The Chemistry and Biology of Sialic Acids and Related Substances*, Cambridge University Press, Cambridge, UK.

Grobstein, C. 1975, 'Developmental role of intercellular matrix: Retrospective and prospective', in *Extracellular matrix influences on gene expression*, eds H. Slavkin & R. Greulich R, Academic Press, New York, pp. 9-16, 809-814.

Gutmann, F. & Lyons, L.E. 1967. *Organic Semiconductors*, John Wiley and Sons, New York, pp. 492-504.

Haller, A. 1779, 'First lines of physiology', translated from the correct Latin edition. Edinburgh, E., cited from Baker, J.R. 1948, *Quarterly Journal of Microscopical Science*, vol. 89, p. 113.

Hepler, P.K. & Palevitz, B.A. 1974, 'Microtubules and microfilaments', *Annual Review of Plant Physiology*, vol. 25, pp. 309-362.

Horwitz, A.F. 1997, 'Integrins and health. Discovered only recently, these adhesive cell surface molecules have quickly revealed themselves to be critical to proper functioning of the body and to life itself', *Scientific American*, vol. 276, pp. 68-75.

Hutchinson, C. & Koop, C.E. 1956, 'Lines of cleavage in the skin of the newborn infant', *Anatomical Record*, vol. 126, pp. 299-310.

Hynes, R.O. & Yamada, K.M. 1982, 'Fibronectins: Multifunctional modular glycoproteins', *Journal of Cell Biology*, vol. 95, pp. 369-377.

Ingber, D.E. 1993, 'The riddle of morphogenesis: A question of solution chemistry or molecular cell engineering?' *Cell*, vol. 75, pp. 1249-1252.

Jones, J.P. 2001, *Bridging Worlds and Filling Gaps in the Science of Spiritual Healing*, Kona, Hawaii, 29 November-3 December.

Koch, G.L.E. & Smith, M.J. 1978, 'An association between actin and the major histocompatibility antigen H-2', *Nature*, vol. 273, pp. 274-278.

Kreis, T. & Vale, R. (eds.) *1993 Guidebook to the Cytoskeletal and Motor Proteins*, Oxford University Press, Oxford, UK.

Lamark, J.B.P.A. 1809, 'Philosophie zoologique, ou exposition des considerations relatives a l'histoire naturelle des animaux. Paris (Dentu)', cited from Baker, J.R. 1948, *Quarterly Journal of Microscopical Science*, vol. 89, p. 113.

Langer, A.K. 1861, 'Zur Anatomie und Physiologie der Haut. I. Über die Spaltbarkeit der Cutis', *Sitz. d. K. akad. d. Wiss. Wein*, vol. 44, pp. 19-46.

Lehninger, A.L. 1971, *Bioenergetics*, 2nd ed, A. Benjamin, New York.

Mannery, J.F. 1968, 'Ca^{++} ions and nucleotides at cell surfaces', *Experimental Biology and Medicine*, vol. 3, pp. 24-39.

Marchesi, V.T., Jackson, R.L., Segrest, J.P. & Kahane, I. 1973, 'Molecular features of the major glycoprotein of the human erythrocyte membrane', *Federation Proceedings*, vol. 32, pp. 1833-1837.

Masters, C.J. 1978, 'Interactions between soluble enzymes and subcellular structure', *Trends in Biochemical Sciences*, vol. 3, p. 206.

Masters, C.J. 1979, 'Assemblies, interactions and ambiguities', *Proceedings of the Australian Biochemical Society*, vol. 12, p. Q17.

Masters, C.J. 1981, 'Interactions between soluble enzymes and subcellular structure', *CRC Critical Reviews of Biochemistry*, vol. 11, pp. 105-143.

Matsumoto, K. & Birch, S. 1988, *Hara Diagnosis: Reflections on the Sea*, Paradigm, Brookline, MA.

Mowbray, J. & Moses, V. 1976, 'The tentative identification in *E. coli* of a multienzyme complex with glycolytic activity', *European Journal of Biochemistry*, vol. 66, pp. 25-36.

Nicolini, C. 1980, 'Normal versus abnormal cell proliferation: A unitary and analytical overview', *Cellular Biophysics*, vol. 2, pp. 271-290.

Opperdoes, F.R. & Borst, P. 1977, 'Localization of nine glycolytic enzymes in a microbody-like organelle in *Trypanosoma brucei*', *FEBS Letters*, vol. 80, pp. 360-364.

Oschman, J.L. 1978, 'Morphological correlates of transport', in *Membrane Transport in Biology*, vol. III, eds G. Giebisch, D. Tosteson, & H.H. Ussing, Springer-Verlag, Berlin, pp. 55-93.

Oschman, J.L. 1984, 'Structure and properties of ground substances', *American Zoologist*, vol. 24, pp. 199-215.

Oschman, J.L. & Oschman, N.H. 1997, *Readings on the Scientific Basis of Bodywork, Energetic, and Movement Therapies*, Nature's Own Research Association, Dover, NH.

Oster, G. 1983, 'Muscle sounds. Contracting muscle generates distinct sounds that are not heard only because the human ear is insensitive to low frequencies. Such sounds are now studied for their possible usefulness in science and medicine', *Scientific American*, vol. 250, pp. 108-114.

Oster, G. & Jaffe, J.S. 1980, 'Low frequency sounds from sustained contraction of human skeletal muscle', *Biophysical Journal*, vol. 30, pp. 119-127.

Otteskog, P., Ege, T. & Sundqvist, K.G. 1981, 'A possible role of the nucleus in cytochalasin B-induced capping', *Experimental Cell Research*, vol. 136, pp. 203-213.

Pethig, R. 1979, *Dielectric and Electronic Properties of Biological Materials*, John Wiley and Sons, New York.

Pienta, K.J. & Coffey, D.S. 1991, 'Cellular harmonic information transfer through a tissue tensegrity-matrix system', *Medical Hypotheses*, vol. 34, pp. 88-95.

Pollack, G.H. 2001, *Cells, Gels and the Engines of Life.* Ebner & Sons, Seattle, WA.

Pollard, T.D. & Weihing, R.R. 1974, 'Actin and myosin and cell movement', *CRC Critical Reviews of Biochemistry*, vol. 2 pp. 1-65.

Porter, K.R. 1984, 'The cytomatrix: A short history of its study', *Journal of Cell Biology*, vol. 99, no. 1, pt. 2:3s-12s.

Porter, K.R. & Tucker, J.B. 1981, 'The ground substance of the living cell. The high resolution that has been achieved with the high-voltage electron microscope reveals the microtrabecular lattice: A system of gossamer filaments that support and move the cell organelles', *Scientific American*, vol. 244, 56-67.

Rapraeger, A.C. & Bernfield, M. 1982, 'An integral membrane proteoglycan can bind the extracellular matrix directly to the cytoskeleton', *Journal of Cell Biology*, vol. 96, p. 125a.

Rolf, I.P. 1962. 'Structural integration. Gravity: an unexplored factor in a more human use of human beings', *Journal of the Institute for the Comparative Study of History, Philosophy and the Sciences*, vol. 1, pp. 3-20.

Rosenberg, F. & Postow, E. 1969, 'Semiconduction in proteins and lipids—Its possible biological import', *Annals of the New York Academy of Sciences*, vol. 158, pp. 161-190.

Scott, J.A. 1984, 'The role of cytoskeletal integrity in cellular transformation', *Journal of Theoretical Biology*, vol. 106, pp. 183-188.

Seeman, P., Niznik, H.B. 1990, 'Dopamine receptors and transporters in Parkinson's disease and schizophrenia', *FASEB Journal*, vol. 4, pp. 2737-2744.

Soifer, D. (ed). 1975, 'The biology of cytoplasmic microtubules', *Annals of the New York Academy of Sciences*, vol. 253, pp. 1-848.

Staufenbiel, M. & Deppert, W. 1982, 'Intermediate filament systems are collapsed onto the nuclear surface after isolation of nuclei from tissue culture cells', *Experimental Cell Research*, vol. 138, pp. 207-214.

Steinberg, M.S. 1958, 'On the chemical bonds between animal cells: A mechanism for type-specific association', *American Naturalist*, vol. 92, pp. 65-81.

Szent-Györgyi, A. 1941a, 'Toward a new biochemistry?' *Science,* vol. 93, pp. 609-611.

Szent-Györgyi, A. 1941b, 'The study of energy levels in biochemistry', *Nature,* vol. 148, pp. 157-159.

Tomita, S. & Riggs, A. 1970, 'Effects of partial deuteration on the properties of human hemoglobin', *Journal of Biological Chemistry,* vol. 245, pp. 3104-3109.

Walsh, T.P., Clarke, F.M. & Masters, C.J. 1977, 'Modification of the kinetic parameters of aldolase on binding to the actin-containing filaments of muscle', *Biochemical Journal,* vol. 165, pp. 165-167.

Webster's Seventh New Collegiate Dictionary, 1965, G&C Merriam Co., Springfield, MA.

Wessels, N.K., Spooner, B.S., Ash, J.F., Bradley, M.O., Ludvena, M.A., Taylor, E.L., Wrenn, J.T. & Yanoda, K.M. 1971, 'Microfilaments in cellular and developmental processes', *Science,* vol. 171, p. 135.

Properties of the living matrix

Nature is neither kernel nor shell, she is everything all at once.   Goethe

When we look at any one thing in the world, we find it is hitched to everything else.   John Muir

Our story has reached a turning point. Albert Szent-Györgyi, one of the leading scientists of his day, saw something in nature that opened up a new line of inquiry for him. He started to look for a <u>high-speed system </u>that can communicate and energize the rapid and subtle reactions that characterize life. This research was absolutely unprecedented. Biologists at the time were comfortable with the idea that the nervous, hormonal, and chemical systems control everything in the body. There was no need to look further. Prof's research created few waves in the academic community. He was not bothered by this.

Prof had realized that the biological communications he was looking for must be carried out by tiny and highly mobile entities. Molecules and ions were simply too large and bulky to carry information rapidly. The nervous system was simply too slow. What he was looking for had to be very small and highly mobile. Subatomic particles such as electrons and protons were ideal candidates. Not only do these entities have small mass and therefore little inertia, they have wave aspects that enable them to move with a velocity that approximates that of light. To study these entities, he had to take his research team to the quantum world.

AHEAD OF HIS TIME

A thesis of this book is that pioneers are people who see the world as everyone else sees it but think new thoughts. Prof's 1941 suggestion of electronic mobility in organic polymers has been confirmed and vastly extended to the man-made polymers, plastics. The research on plastics led to a Nobel Prize in Chemistry in 2000. Alan Heeger, Alan MacDiarmid, and Hideki Shirakawa were acknowledged for this key discovery that they published, with their colleagues, in 1977 (Shirakawa et al. 1977). Phenomenal applications are on the horizon:

- Video displays, electronic circuits, and efficient solar collectors of any size and shape that are so lightweight and flexible they can be rolled up
- The grocery store with a disposable plastic memory chip printed on every item; on your way out the entire contents of your shopping cart are read out and you are given your bill, without having to unload your cart
- Memory chips printed on mail and packages that can call out their locations at all phases of their journey
- Inkjet or laser printers that can print the circuits on any surface

Nature invented the semiconducting polymer, and Albert Szent-Györgyi discovered that it existed in living systems. Only recently have engineers and polymer scientists begun to catch up.

At the time of Prof's insight, most biologists were bewildered by the implications of quantum mechanics, or they were just plain not interested. I asked many colleagues about Prof's work, looking for a logical and scholarly critique. There were none, but most agreed that things would be a lot easier for all concerned if Prof would simply retire. We will see, however, that this was the beginning of a period of research that historians one day may consider Prof's most important work. We can be confident of this because of some breakthroughs that occurred in other areas.

LIVING CIRCUITS

Scientists researching the mystery of regeneration began to appreciate that their problem boiled down to locating the electronic circuitry that controls biological form. Some considered this to be a solved problem, but we soon shall see that those who have looked deeply into the mystery of development have found more questions than answers.

A conference entitled "Mechanisms of Growth Control, Clinical Applications," was held on September 26 to 28, 1979, at the State University of New York Upstate Medical Center. The conference brochure had the cover logo shown in Figure 9-1. An electronic circuit was superimposed over the body of a salamander, a popular animal for research on regeneration because of its remarkable capacity to replace amputated body parts.

At the time, biologists studying regeneration were intrigued by the electronic properties of the tissues of salamanders that enabled them to do something that humans are unable to do: regenerate their limbs. One of my neighbors at Woods Hole (Massachusetts), Phil Person, and his colleagues, discovered that electrical stimulation sometimes could induce regeneration of a leg in a mammal, although the results were not consistent (Libbin et al. 1979; Person et al. 1979). Remarkably, Person et al. also were able to demonstrate that at least one third of the heart of the common newt

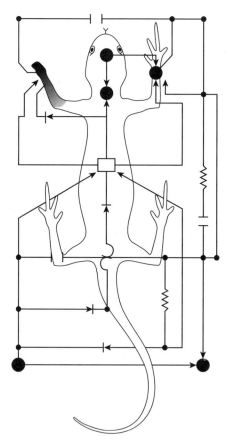

Figure 9-1 Logo for the conference entitled, *Mechanisms of Growth Control, Clinical Applications,* held on September 26-28, 1979, at the State University of New York Upstate Medical Center. An electronic circuit is superimposed over the body of a salamander, a popular organism for research on regeneration. (Reproduced in *Energy Medicine: the scientific basis,* Oschman, Figure 3.6, page 56, Churchill Livingstone, Edinburgh, 2000, with permission from Elsevier Science.)

Notoopthalmus viridescens can be completely regenerated within 24 hours of excision (Philpott et al. 1975). This observation was confirmed by McDonnell and Oberpriller (1984).

Robert O. Becker was a leader in research on regeneration and was the most prominent biologist to take up the study of the electronic properties of living matter. He did a number of experiments that confirmed Szent-Györgyi's ideas about the electronic or semiconductor nature of living tissues. For a summary of this work, see *The Body Electric* (Becker & Sheldon 1985).

Solid-state biochemistry is the biochemistry of liquid crystals

The living matrix is best described as a *liquid crystal*. These are materials that are intermediate between solids and liquids and display properties of both. It is not generally appreciated that virtually all of the body structure is composed of liquid crystals capable of sustaining quantum coherence (Ho 1993). We will discuss quantum coherence later in the book, after exploring some of the relevant background information.

Some of the key crystalline materials in the living matrix are shown in Figure 9-2. They include the membranes of all cells; the contractile apparatus in muscle; the connective tissue found in bones, tendons, ligaments, and cartilage; the arrays of microtubules in sensory cells; and deoxyribonucleic acid (DNA).

Liquid crystals are highly ordered materials that have properties of both liquids and solids. Their study leads in the same direction that Szent-Györgyi was heading.

> *Liquid crystallinity gives organisms their characteristic flexibility, exquisite sensitivity and responsiveness, thus optimizing the rapid, noiseless intercommunication that enables the organism to function as a coherent, coordinated whole.* Ho (1999)

Brilliant work on the liquid crystalline properties of connective tissue was performed by Mae Wan Ho and her colleagues in Britain. They developed a noninvasive optical method for observing the dynamic changes in coherence of the liquid crystalline material of small organisms (Ho & Lawrence 1993; Ho & Saunders 1994; Newton, Hafegee & Ho 1995; Ross et al. 1997). A study by Ho and Knight (1998) emphasized the importance of the water associated with the collagen molecules comprising the acupuncture meridian system. We will discuss more about this later in the book.

Modern developments have followed directly from Prof's insights, which were far ahead of their time and, as often occurs, virtually ignored by the scientific community.

Solid-state biochemistry is also wet biochemistry

Although the discipline in physics that investigates the living matrix is termed *solid-state biochemistry,* the material being studied is far from solid. It is quite wet. The living matrix is composed of molecules that are completely dependent upon water for their stability and dynamic properties.

Prof was deeply appreciative of the properties of water that give rise to the remarkable phenomenon we refer to as *the living state.* His perspective can be reiterated again and again, because it is the basis of a new science of life that we are just beginning to appreciate:

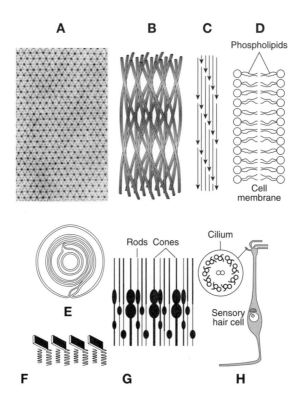

Figure 9-2 It is not generally appreciated that most of the tissues in the living body are composed of highly regular arrays of molecules, otherwise known as *crystals*. **A,** Crystalline packing of actin and myosin molecules as shown in a cross section of muscle. **B,** Crystalline packing enables DNA molecules to fit in the small space within the cell nucleus. **C,** Connective tissue, which forms the bulk of the tissues in the body, is a highly ordered array of collagen molecules. **D,** Cell membranes are also liquid crystals composed of phospholipids. **E,** Myelin sheath of nerve. **F,** Light-sensitive molecules in the chloroplast of a leaf. **G,** Photosensitive cells forming rods and cones in the eye. **H,** Microtubules in sensory and motile cilia found throughout the animal kingdom. (A and E are originally from Bloom & Fawcett's *Textbook of Histology*, 12th ed, 1994, Fig. 10-21, page 280, and Fig. 11-21 from page 335, respectively, with permission from Hodder Arnold, London. B is from Kamien RD 1996 Liquids with Chiral Bond Order. *Journal of Physics II* France 6:461, and is reproduced by permission from RD Kamien and *J. Phys II*. F is an illustration by Allen Beechel. Courtesy Johanna M. Beechel and the Estate of Allen Beechel, from *Scientific American*, December 1974. C, D, E, G, H are from *Energy Medicine: the scientific basis*, Oschman, Churchill Livingstone, Edinburgh, 2000, Figs. 3.4D, 3.4A, 3.4D, 3.4F, and 3.4G, respectively, with permission from Elsevier Science.)

Molecules do not have to touch each other to interact. Energy can flow through . . . the electromagnetic field. . . . The electromagnetic field along with water forms the matrix of life. SZENT-GYÖRGYI (1957)

Semiconductor crystals in living systems

The reason that Szent-Györgyi's suggestion applies to the subject of this book is as follows. The "great number of atoms . . . arranged with regularity in close proximity, as for example in a crystal lattice" is none other than the living matrix with all of its interconnected components extending throughout the body. The crystalline nature of the living matrix arises from the way the component connective tissue and cytoskeletal and nuclear molecules organize themselves into very regular parallel arrays (see Figure 9-2).

Connective tissues and components of the cytoskeleton and nuclear matrix sometimes are referred to as *quasi-crystalline,* meaning they behave *as if* they were crystals. The arrangement provides great tensile strength, flexibility, and interconnectedness, and it allows for the "emergent properties" we seek to appreciate.

Collagen is a highly versatile construction element because it can be mixed with an elastic protein, elastin, and with various ground substances to make composite materials having a wide range of physical and electronic properties. These materials include the superficial fascia, myofascia, the fascia covering the bones, nerves, blood vessels, the bones themselves, tendons, ligaments, the capsules surrounding organs, joints, cartilages, and all of the intervening sheets of fibrous tissues.

There is a substantial field in physics that concerns itself with the study of highly ordered systems. This is *solid-state physics,* which examines the physical properties of solids and the special properties that arise when atoms or molecules become associated in a regular periodic arrangement as in crystals.

Solid-state physics has been successful in developing powerful theoretical concepts and measuring tools that can be used to study a wide range of problems, including those that arise in the study of living systems.

The importance of crystallinity

In Prof's statement made in 1941, mention is made of *a great number of atoms arranged with regularity in close proximity, as for example in a crystal lattice.* A crystal cannot be described in terms of its constituent atoms alone. A crystal can contain within it a variety of entities (such as electrons, protons, phonons, plasmons, holes, excitons, solitons, polarons, and conformons) that arise as *emergent* or *collective* properties of the crystal system. When the crystal is broken into its constituents, these peculiar particles and entities disappear, or at least their properties are drastically altered. Moreover, the important, even vital, collective properties cannot be predicted from study of the system's components, taken one by one.

This is a point worthy of emphasis. We are discussing vital phenomena that distinguish life from nonlife. We are discussing phenomena that, when optimized, lead to the condition we refer to as *health* and that, when they exist in less than an optimal state, can lead to discomfort, disease, or disorder. Certainly optimum athletic or artistic performance cannot "emerge" in the absence of these elusive "emergent" phenomena. Optimum performance, whether it is in the therapeutic encounter or in the Olympics, can only "emerge" from an organism that transcends "the sum of its parts."

If we turn to physics, we find that these extraordinary collective properties or modes, which make all the difference, are best described in terms of quantum field theory (QFT). We will take up this phenomenon in Chapter 25.

COOPERATIVE AND COLLECTIVE PHENOMENA

To the solid-state physicist, the crystallike nature of connective tissue provides a property of great interest, a high degree of order. This regularity greatly enhances the ability of the material to conduct, process, and store energy and information of various kinds. One can look for, and find, in connective tissue many of the solid-state properties that usually are associated with mineral crystals and with the crystals that are used to make modern electronic and computer systems. The solid-state properties of these materials account for *cooperative* or *collective* phenomena. These are whole-system properties that arise because each individual component of the system is modified in its behavior as a consequence of being a part of a collective group.

A simple expression of this is the statement that, within the living matrix, the total force acting on any individual atom at any time, and therefore its behavior, will depend on the positions and activities of all of the other atoms in the organism. This is a collective or unifying concept that applies to any organism or any part of it, regardless of whether nervous or hormonal systems are present.

Synergetics

Study of cooperative or collective phenomena has led to a number of important advances in physics and to the development of a multidisciplinary study known as *Synergetics* (Haken 1975, 1983a, 1983b). This field is concerned with systems composed of many subsystems, each of which may have different properties. From the synergetic point of view, an organism is created from the cooperation of subsystems in space and time.

Cooperative effects often arise spontaneously—they self-organize. Order emerges from disorder, often in a single event known as a *phase transition.* New macroscopic structural or behavioral patterns emerge abruptly in response to a small change in a certain parameter, such as temperature, light, a spoken word, an electromagnetic field, or a touch. This happens often in therapeutic encounters in all medical traditions, but it is not always predictable.

An ordered system may become chaotic or disordered in response to a small disturbance (called a *perturbation* by physicists).

Synergetics seeks the general principles and equations governing phase transitions as they occur in fields ranging from quantum mechanics to computer science to ecology to climatology to sociology or economics. Similar laws may govern the action of an enzyme, wound healing, metastasis, the stock market, or the formation of public opinion.

When general synergetic principles are known, predications can be made about the overall behavior of a complex system composed of many subsystems, even without knowing everything about all of the subsystems.

The significance of emergent properties such as consciousness

Again we have reached an important part of the discussion that is worthy of emphasis. Some of the most remarkable but least understood aspects of human functioning are emergent properties that may be difficult to analyze logically. Consciousness is one of these emergent properties. Neurophysiologists look for consciousness in the operation of massively interconnected nerve nets. Cell biologists look for consciousness in the patterns of proteins deposited in massive arrays of microtubules or other cytoskeletal elements within neurons or other cells. Neurochemists look for consciousness in the patterns of neurotransmitters and neurohormones. Quantum physicists and quantum psychologists look for consciousness as an emergent property of space, matter, and waves.

Each of these approaches can shed light on the nature and mechanism of consciousness, but none can provide a complete picture. It can be more productive to view consciousness as a synergetic, cooperative, or collective property of a system that *emerges* from a multilayered set of subsystems.

As an example, Fröhlich (1969) pointed out that a complete solution of the wave equations describing the dynamic behavior of all of the atoms in a molecule would provide "such an immensity of irrelevant information that selecting from it the features of interest would be prohibitive." The reason for this is that the number of states or conditions of a system increases exponentially with the number of parts it contains. Going from one atom to two atoms increases the number of possible states of a molecule about 100-fold. Going to a molecule composed of three or more atoms creates a situation that is prohibitively complex for quantal analysis, hence the generalizations of Synergetics help us understand the behavior of complex systems.

A logical analysis of consciousness or other emergent properties is entirely dependent on neurological consciousness, the place where logic resides; however, we are developing a concept of another kind of consciousness that is far more sophisticated than neurological consciousness. It is far more sophisticated because it is able to take in, store, and process far more information than we can consciously know about. This is because our knowing in the usual sense is neurophysiological knowing. Our knowledge and discussions of this other kind of consciousness take place in the realm of neurophysiological consciousness and therefore are limited. Perhaps there are ways of connecting the wisdom of these two domains. Perhaps this connection exists all of the time, but only now are we consciously aware of it.

It is interesting to consider disease, disorder, and healing as phase transitions in an intricate synergetic system that is none other than the living matrix. Specific treatments for a disorder work some of the time, but it is difficult to find an approach that works for every condition, and some situations are extremely difficult to treat. Consciousness, a poorly understood emergent property, has definite influences in the processes we refer to as *disease* and *disorder*. In other words, emergent properties can combine to yield additional emergent properties.

Singular points

Shang (1989) compared acupuncture points with developmental organizing centers and singular points. A *singular point* is a place where a very small change in one parameter will cause a huge change in another. In the past, such phase transitions in living systems were examined by a combination of thermodynamic, general systems, and information theories, but these approaches proved inadequate. One reason is that biological systems frequently show phase transitions that achieve something that many scientists have been reluctant to accept. Living systems simply do not obey the second law of thermodynamics. They are syntropic (Szent-Györgyi 1974) rather than entropic, and they have a tendency to perfect themselves. They might even be called *purposeful* (Haken 1973).

An important discovery arising from the study of cooperative phenomena is that giant coherent oscillations can be set up in individual macromolecules. We will discuss this further when we consider energy flow in living systems.

Piezoelectricity and regulation of form

Piezoelectricity is an example of a collective property arising from the way atoms aggregate in a crystal. Connective tissue is piezoelectric (Greek for *pressure electricity*); therefore, it generates electricity when it is compressed. Deformations of bones (Bassett & Becker 1962; Black & Korostoff 1974; Fukada & Yasuda 1957; McElhaney 1967), teeth (Braden et al. 1966), tendons (Fukada 1974), blood vessel walls (Fukada & Hara 1969), muscles (Fukada & Ueda 1970), and skin (Levine, Lustrin & Shamos 1969) all give rise to electric fields as a result of the piezoelectric effect. The piezoelectric constant for a dry tendon, for example, is nearly the same as that for a quartz crystal (Pickup 1978).

A number of scientists concluded that signals generated by the piezoelectric effect are not trivial by-products or artifacts; they are essential biological communications that "inform" neighboring cells and tissues of the movements, loads, compressions, and tensions arising in different parts of the body. These signals join with those generated by other physiological processes, such as nervous signals, muscle potentials and sounds, and glandular secretory signals, to create a veritable symphony of oscillating electric fields that travel a certain distance through the living matrix. The cells and tissues then use this information to adjust their activities concerned with maintenance and nourishment (Oschman 1990). For example, this mechanism accounts for the fact that movement and exercise maintain the skeleton, whereas long periods of bed rest or space travel in zero-gravity conditions lead to loss of bone mass (Vaughn 1970). Lipinski (1977) discussed the significance of piezoelectricity in relation to acupuncture and other approaches. The fully "integrated" body may be a body that is entirely free of restrictions to the flow of signals.

Electrons, holes, and excitons

Szent-Györgyi's 1941 statement refers to electrons as belonging not to single atoms but to the whole system. In other words, certain electrons within a crystalline substance are free to migrate from atom to atom or molecule to molecule, thereby conveying energy and information from place to place throughout the continuum.

The migration of electrons is the province of electronics, and Szent-Györgyi was convinced that the living state is an electronic phenomenon. Indeed, he wrote a book entitled *Bioelectronics* (Szent-Györgyi 1968). Specifically, he suggested that the proteins of the body are semiconductors. These are substances whose electrical conductivity lies between that of a conductor, such as a metal, and an insulator, such as rubber or plastic. Semiconductors are the materials upon which our modern electronics industry has been built, and they have made possible the wonders of the computer with which this book is being written.

The phenomena of the living state are, Szent-Györgyi stated, too rapid and subtle to be explained by the slow fluxes of ions across nerve membranes or through intercellular spaces. Some other form of rapid communication must be taking place within the tissues. He proposed that this was none other than the migration of electrons, protons, and "holes" through the semiconducting protein fabric of the body. (When a material gives up an electron, the vacancy left behind behaves like a mobile particle with a positive change and a mass slightly larger than an electron. The cavity is called a *hole*, and it moves through a crystal much like a positive charge moves.) Figure 9-3 shows the difference between electron and hole conduction as they occur in N- and P-type semiconductors, respectively.

A fascinating aspect of this idea is that semiconductors have important properties that enable them to accomplish far more than conductors. Conductors convey information (as in your phone line) or energy (as in the cord to your toaster). Semiconductors can do both, In addition, they have the ability to *process* energy and information in sophisticated ways, that is, to switch, store, delay, modulate, amplify, filter, detect, or rectify (allow to pass in one direction but not in the other). In addition, semiconductor networks can have read-only and programmable memory and logic circuits that evaluate the information flowing through them. In a living organism, such circuits can make and process decisions that lead to actions. The actions selected are determined by a combination of genetic programming that gives rise to the proteins that are built into the circuitry, the memory of previous activities, and current information from local and distant sources.

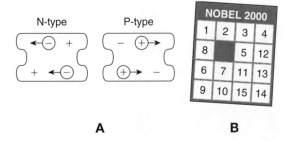

A   **B**

Figure 9-3 A, In N-type semiconductors, energy and information are carried by moving electrons. In P-type semiconductors, energy and information are carried as "holes," which are spaces or gaps in the electronic structure. The holes behave as though they were positive charges because they are the absence of negative charges. **B,** A well-known game is a simple model for the movement of a "hole" because none of the pieces can move unless there is one space. In a semiconducting polymer such as a protein, each movable piece represents an electron that can jump into a space or "hole" vacated by another electron. (**A** is used courtesy of *The Radio Amateur's Handbook,* 55th ed, 1978. **B** is from the Nobel Foundation's description of the 2000 Nobel Prize in Chemistry, awarded to Alan J. Heeger, Alan G. MacDiarmid, and Hideki Shirakawa for the discovery of conductive plastics.)

When an excited electron and a hole remain coupled, they can move about within a crystal almost as though they were a single uncharged particle, an electron-hole pair called an *exciton*. Although it resembles a particle, the exciton spreads through the whole crystal like a wave (Craig & Walmsley 1968; Davydov, Kasha & Oppenheimer 1962; Jeffries 1975; Wolfe & Mysyrowicz 1983).

In other words, the living matrix gives rise to the physical body we can see and touch. In addition, it has invisible properties of an electronic signal processing system. The units that carry information can be electrons, holes, excitons, or other particles.

Semiconductor devices are capable of handling large amounts of *power*. Here we might find the bioenergetic equivalents of switches, relays, circuit breakers or fuses, actuators, and transformers. Semiconductor systems may initiate activities by directing energy flows. Informational functions and power distribution roles overlap, because the amounts and locations of power consumption provide information on actions that are taking place. We shall take up this subject again in later chapters, after we have examined some more pieces of the puzzle we are assembling.

A WATER NETWORK

Now we examine the entire living matrix system at the submicroscopic level. Surrounding every molecule in the body, without exception, is an invisible physical framework composed of water and ions. If by some trick we could make all of the cells and molecules comprising the living matrix disappear, without disturbing the water molecules and ions associated with the surfaces of all of those cells and molecules, we would be left with a complete framework of water with the same configuration as all of the parts of the physical body. Of course, it would be impossible to do this, because as soon as we removed the molecules the water structure would dissipate. Likewise, if we removed the water, all of the molecules would fall apart. Indeed, they would actually *fly* apart, because the electrical charges on the atoms within the molecules would strongly repel each other. Water neutralizes these opposing forces.

> [W]ater . . . is half of the living machinery, and not merely a medium or space-filler. . . it is a mistake to talk abut proteins, nucleic acids or nucleoproteins and water, as if they were two different systems. They form one single system which cannot be separated into its constituents without destroying their essences. SZENT-GYÖRGYI (1960)

One way solid-state physicists describe this situation is by stating that the protein macromolecules *imprint* their structure on the spatial configuration of the water molecules. Collins (1991, 2000) described the vital role water has played in the evolution of living molecular structure.

Layers of water

We now know that every molecule in the body has associated with it a highly organized film of water and ions. This chainlike hydration layer is organized by electrical fields surrounding the charged groups on proteins. The film is several layers thick, and each layer has different properties. Some water molecules become trapped within a molecule and are so tightly bound that they cannot be removed, even by heating the substance to a high temperature in a vacuum. The next water layer, close to the surface of a molecule, is highly ordered, and subsequent layers become less structured.

Pollack (2001) summarizes convincing research documenting the "clinginess" of water to surfaces. For example, place two glass microscope slides face to face. It is easy to separate them. Add a drop of water to the slides, and it becomes virtually impossible to separate them.

To understand the forces involved, Israelachvili and colleagues measured the actual force required to pull wet surfaces apart (Horn & Israelachvili 1981; Israelachvili & McGuiggan 1988; Israelachvili & Wennerström 1996). The closer two wet surfaces get, the harder it is to pull them apart; however, the force-separation relationship is not linear. They discovered regularly spaced peaks and valleys of force. The spacing between the force peaks was equal to the diameter of the sandwiched water molecules. The first layer sticks to the surface, the second layer sticks to the first, and so on. In the experiments by Israelachvili, about eight to ten peaks were discerned, corresponding to as many layers of water. Szent-Györgyi (1972) performed some experiments showing that this effect extends outward about 500 nm, implying hundreds of adherent layers of water.

For a long time, biologists thought of water as a filler or suspending medium in which the important and active biomolecules floated about until they happened to have meaningful encounters. As a ubiquitous solvent, water was, for the most part, taken for granted. We now know that the thin aqueous films covering every molecule of the organism form an intricate, dynamic, and highly structured subsystem with profoundly important properties. Water and ions associated with molecules have special properties because of their relationships. They form continuous interacting chains that influence macromolecular structure and function and allow for transfer of energy and information.

A closer look at water

The study of water in and around cells has a long and controversial history. One of the problems is that most of our information about cells and molecules is based on studies of tissues that were broken open and separated into parts. Information obtained from this approach cannot be used to reconstruct a model of the whole cell and can be misleading when analyzing the state of water in living systems.

The reason water molecules form an ordered film over the surfaces of molecules and cells is because each water molecule is dipolar, that is, its electrical field is unbalanced. Water molecules tend to rotate and align with the lines of force of an electric or magnetic field. Each dipolar water molecule tends to rotate less, to be held in a certain position or orientation, when it is close to a charged region of a molecule. The word *tends* is important here because the effect is a statistical one. This means that the water is not held absolutely rigidly next to a charged region but instead has a *propensity* to be oriented along the electrical gradient. The degree of this tendency depends upon distance and the strength of the orienting field, which depends on the electronic structure of the molecule and that molecule's cooperative relations with its neighbors.

The essential role of water in holding molecules together and regulating their activities is not widely appreciated, but it is extremely important. For example, a standard biochemical technique for disaggregation of a protein involves replacing the water with another molecule such as urea. When hemoglobin is placed in a solution of urea, it splits into two half-molecules, each of molecular weight half that of the intact hemoglobin molecule. If the subunits are returned to water, the active hemoglobin molecule reforms.

Hemoglobin is one of many important aggregations that form "spontaneously" in aqueous solution. We will give more examples of this *self-assembly* process later and will suggest that acupuncture meridians and points also form in this manner.

DNA as an example

An example of the importance of water in determining the structure and properties of a molecule is provided by DNA (Figure 9-4). The usual representation of DNA structure, reproduced in countless texts, is physically impossible. A structure such as this would fly apart due to the electrostatic repulsion between the highly charged phosphate groups. What holds the molecule together in nature are the ions and water molecules that associate with and neutralize the charged regions.

The essential role of water and ions in DNA structure and function has been worked out in great detail by Enrico Clementi and his colleagues at International Business Machines (IBM). They have built up beautifully complete quantum mechanical models of the matrix of relationships among DNA, water, and ions (which are called *counterions* when they associate with a charge on a macromolecule).

Using the largest and fastest IBM computers available at the time, Clementi created virtual models of one turn of the DNA helix within the computer. In these models, the atoms comprising a molecule move about under the influence of their own kinetic energy and the forces exerted on them by surrounding ions and water. With this *molecular dynamic approach,* the model never converges on a single conformation. Instead, the molecular model in the computer continuously changes and moves about in space. The various atoms within the molecule are visualized as clouds of electrons that are readily polarized by the electric fields of nearby water and ions. The DNA structure changes 3,000,000 times per second.

When a counterion or water molecule approaches a protein or a nucleic acid, the shapes of the electron clouds within the molecule are altered. This, in turn, affects the electrical properties of the approaching counterion or water molecule. In practice, the energy structure of the system is recalculated again and again, millions of times per second, to see what position the water and ions will take up in relation to the DNA molecule. Each time another water or ion is added to the system, the entire structure must be recalculated to accommodate the influence of its new neighbor. The calculation is repeated again and again as the water molecule is drawn closer and closer to the DNA.

Figure 9-4, *B,* is a view looking down on the DNA helix and showing the probability distribution of Na^+ counterions around one part of the helix. This is the ionic distribution in the presence of 447 water molecules, which are not shown. Note that the

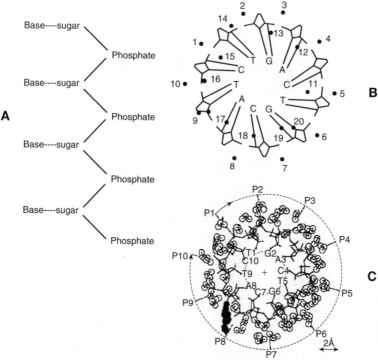

Figure 9-4 A, The nucleic acid backbone of the DNA molecule as it often is represented in textbooks. The different bases are side chains on the repeating sugar-phosphate polymer chain.
This image is convenient to describe the remarkable story of how genetic information is stored in the molecule; however, the structure as shown is unstable. In the absence of counterions and water, the mutual repulsion of the highly charged phosphate groups would make the molecule fly apart.
B, A view looking down on the DNA helix and showing the probability distribution of Na^+ counterions around one part of the helix. This is the ionic distribution in the presence of 447 water molecules, which are not shown. **C,** Water molecules associated with the first 10 phosphate groups of a DNA molecule, looking down the axis. This is the configuration in the presence of 447 water molecules (counterions not shown). The water clusters have two remarkable features. They enclose the phosphate groups, and they form hydrogen-bonded filaments such as the one shown in black at P8. (A is Figure 9, p. 206, from Oschman, J.L., 1984. Structure and properties of ground substances. *American Zoologist* 24:199-215, and is reproduced by permission of *American Zoologist.* B is from Clementi, E. 1981. Computer simulations of complex chemical systems: Solvation of DNA and solvent effects in conformational transitions. *IBM Journal of Research and Development* 25:324, and is republished by permission of IBM Technical Journals. C is Fig. 3B from Clemente E. and Corongiu G. (undated). Solvation of DNA at 300 K: Counter-ion structure, base-pair sequence recognition and conformational transitions. A computer experiment. Proceedings of the Second sSUNYA Conversation in the Discipline Biomolecular Stereodynamics, Volume I. ISSBN 0-940030-00-4, edited and used by permission of Ramaswamy H. Sarma and Adenine Press, New York.)

counterions are not points but are irregularly shaped volumes that represent the most probable locations of the counterions (Clementi 1981).

Figure 9-4, *C*, shows the water molecules associated with the first 10 phosphate groups of a DNA molecule, again looking down the axis. This is the configuration in the presence of 447 water molecules. The clusters of water molecules have two remarkable features. They enclose the phosphate groups, and they form hydrogen-bonded filaments. One of these filaments is emphasized in the drawing by blackening-in the atoms in the water molecules. Careful analysis of these filaments shows that they extend all along the DNA helix. Water molecules inside the core of the helix are associated with the sugar units and base pairs, but these are not shown in the diagrams (Corongiu & Clementi 1981).

The elaborate network of water filaments associated with DNA, revealed in Clementi's studies, is profoundly important in any consideration of the structure and function of DNA or any other macromolecule. The filaments are statistically stable and meaningful structures. Although individual hydrogen bonds are weak compared to covalent or ionic bonds, there are so many hydrogen-bonded filaments around a molecule such as DNA that the water stabilizes the overall structure.

Communication in water

What sorts of communication can take place in such a system? Pure water dissociates into hydrogen and hydroxyl ions:

$$H_2O \rightleftharpoons H^+ + OH^-$$

From determinations of electrical conductivity and other physical measures, it was discovered that at 22° C the concentration of hydrogen ions in pure water is one ten-millionth (10^{-7}) of a gram per liter. This amounts to one dissociated water molecule in every 500 million. The number of hydroxyl ions is equal to that of the hydrogen ions. This concentration of protons (hydrogen ions) and hydroxyl ions seems deceptively low.

To evaluate the potential of any particle to function as a carrier of information, one must know how well it is conducted. Conductivity depends on two quantities: the number of mobile units and their effective mobility. What happens in water is that hydrogen ions, H^+, are rare, because they are rapidly hydrated to the hydronium ion H_3O^+. However, the apparent rate of movement of the relatively large H_3O^+ ion in an electrical field is many times faster than that of the smaller Na^+ and K^+ ions. Kohlrausch (cited in Piccardi 1962) gives the following relative mobilities at 18° C: H^+ 315, OH^- 174, Na^+ 33, K^+ 65, NH_4^+ 64, Cl^- 65, Br^- 67, 1/2 SO_4^- 68. The high mobility of H^+ arises because protons are rapidly transferred along chains of hydrogen-bonded water molecules. The positive charge can move a considerable distance with little or no actual movement of the water molecules themselves. Of this situation, Szent-Györgyi said that "water was the only molecule he knew that could turn around without turning around" (Pethig 1979).

Studies of the flow of protons through the hydration layer around proteins have led to some interesting discoveries. Aiello et al. (1973) described how the network of water molecules and the protons migrating within the network create a communicating subsystem with its own properties and characteristics. The water-proton subsystem is capable of determining the macroscopic behavior of the system as a whole. The jumping of a proton from one water molecule to the next, across the gap or space of the hydrogen bond, involves a phenomenon called *tunneling,* a term we encountered in

Chapter 1 when we were discussing the SQUID (superconducting quantum interference device) magnetometer.

Tunneling is used to account for the statistically improbable movement of a particle across an energy barrier. It is thought that this protonic tunneling is responsible for stabilizing the hydrogen bond. A thorough description of the role of tunneling in stabilization of structure is beyond the scope of this chapter, because it requires visualization of the various transitions in phase space or the space of momenta as opposed to geometric space (Aiello et al. 1973). The mechanism involves a positive feedback between the geometry of the molecule and the motions of the protons. The motion of an individual proton is dependent upon both local conditions and the state of other protons in the system, that is, it is a cooperative phenomenon.

When a macromolecule has charged groups with a uniform repeat distance that is a multiple of a significant distance in liquid water, a *feedback stabilization* will take place. The macromolecule *imprints* its geometry on the water structure, and the water, in turn, influences the stability of the macromolecular array. In other words, the stabilization involves a cooperative geometrical-motional feedback that is imposed by and upon the imprinting macromolecule.

This phenomenon is particularly significant when a macromolecule is part of a relatively rigid three-dimensional pattern (i.e., a quasi-crystalline arrangement) so that random motions cannot disrupt the build up of an ordered water subsystem. Interestingly this is precisely the situation with collagen. Berendsen (1962) and other investigators (Chapman, Danyluk & McLauchlan 1971; Chapman & McLauchlan 1969) showed that the spacing of amino acids along the collagen molecule is ideal for creating icelike chains of water molecules. This arrangement is shown in Figure 9-5.

The living matrix is a communication network, and the organized water molecules surrounding the physical fabric are an intimate part of that network, serving as a proton-conducting subsystem.

PROTICITY

The flow of energy in the form of protons has been termed *proticity* by Mitchell (1976). *Electricity* refers to the flow of electrons, and *proticity* refers to the flow of protons from the various sites where they are generated in cells. According to Mitchell:

> A generator of proticity plugged through the membrane at any point can act as a source of protonic power that can be drawn upon by a suitable consumer unit plugged through the membrane at any other point, even when the producer and consumer units are quite far apart on a molecular scale . . . the use of proticity for power transmission in biology is unique in its ubiquitous occurrence and wide range of application. (Mitchell 1976)

A "generator of proticity" that has been identified by biochemists is the *mitochondrion,* the *powerhouse of the cell.* When isolated from a cell, mitochondria profusely extrude protons. In an intact system, the cytoskeleton attached to the mitochondria, and especially the water system associated with the cytoskeleton, could conduct the energetic protons to other places within and outside of the cell.

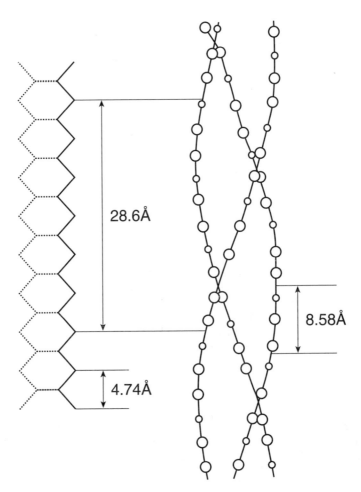

Figure 9-5 Regular array of water molecules *(left)* associated with collagen molecule *(right)*. The dotted lines at the left represent the icelike bonds between adjacent water molecules in the chain. (From Berendsen, H.J.C. 1962, 'Nuclear magnetic resonance study of collagen hydration', *Journal of Chemical Physics,* vol. 36, pp. 3297-3305.)

Proticity is conducted along the DNA water filaments described by Clementi et al. Communication between the hydration and counterion layers and the DNA molecule is continuous. No change can take place in one without a corresponding change in the other. This interchange results in millions of different structures forming and reforming each second. The water molecules profoundly influence the properties of the DNA, and vice versa. One of the influences water has is to greatly increase the electronic conductivity within DNA or other macromolecules (Pethig 1979, p. 293).

Cupane, Palma, and Vitrano (1974) studied the role of the water-proton subsystem in the function of a protein, hemoglobin. The rationale of their experiments involved partial replacement of the network of water molecules surrounding the protein with heavy water, deuterium oxide (^2H_2O). If the water layer has an influence on the functional activity of the protein, replacement of the water with a different molecule should alter the activity. Other studies had predicted that replacement of the hydrogen atoms with deuterium should confer additional stability to the protein because deuterium bonds are stronger and longer than hydrogen bonds (Tomita & Riggs 1970). The effects of deuterium on the protein were determined by studying the reaction of the hemoglobin with oxygen. The property of hemoglobin that was measured was the cooperative interaction between heme units that alters the affinity for oxygen. Oxygen uptake and release are influenced by the protein configuration, which can be "closed" or "open." Cupane et al. found the relationship they expected between a measure of cooperativity in oxygen uptake (Hill's constant) and the percentage deuteration; hence, the water system plays a key role in the functioning of the molecule.

COUPLED OSCILLATIONS, RESONANT TRANSFER, AND ELECTRODYNAMIC COUPLING

Szent-Györgyi had other ideas about energy transfer in living systems that have become widely accepted. One of these he termed *coupled oscillations*. The phenomenon can be demonstrated with two pendulums hanging from the same surface and connected by a rod, string, or spring (Figure 9-6, *A*). If one of the pendulums is put into motion, it will swing for a while and then slowly come to a stop, while the other begins to oscillate (Lai 1984; Walker 1985). The motion will be passed back and forth between the coupled harmonic oscillators. If there were no friction in the system, the transfer would occur without loss and the motions would continue forever. The energy transfer is most efficient if the pendulums have the same structure or natural frequency so that they can resonate. *Resonance* is the term used in physics to describe the strong coupling of systems with the same natural frequency. Two tuning forks with the same frequency (e.g., A-sharp) will resonate even if separated by some distance. Hit one tuning fork to make it vibrate, and the other will vibrate as well.

If a protein has two or more identical side chains, these groups can act as coupled oscillators, and energy will be passed back and forth between them. For example, amino acid side chains can rotate about the bonds that link them to the polypeptide backbone. An example is given in Figure 9-6, *B*, which shows two lysines linked by a single peptide bond. The peptide linkage, composed of the six atoms in the shaded region, lie in a plane. The central C-N bond has some double bond attributes and therefore is relatively rigid. This is known as *resonance stabilization* and is important in determining the three-dimensional conformation of polypeptides. In contrast, the links with amino acids side chains, in most (but not all) cases, allow free rotation, as shown by the arrows in Figure 9-6, *B*.

Vibrational energy can be absorbed by a rotatable amino acid, which will spin about its bond with the polypeptide chain. Like the situation with the pendulum shown in Figure 9-6, *A*, the energy can undergo resonant transfer to an identical rotatable amino acid at some distant point on the protein.

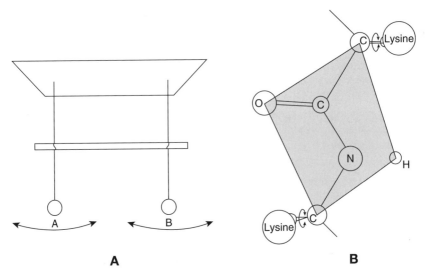

Figure 9-6 Coupled oscillations. **A,** Two pendulums connected by a rod are hung from the same surface. If one of the pendulums is put into motion, it will swing for a while and then slowly come to a stop, while the other will begin to oscillate. The motion will be passed back and forth between the coupled harmonic oscillators. **B,** Part of a polypeptide chain with two lysines linked by a peptide bond. The bond is composed of the six atoms in the shaded region, which lie in one plane and are relatively inflexible. The lysines can rotate about the bonds joining them to the polypeptide. The lysines act as coupled oscillators. Rotational or vibrational energy can pass back and forth between identical rotatable amino acids, much like energy can be passed back and forth between the two pendulums shown in **A.** (A is from Walker J 1985 Strange things happen when two pendulums interact through a variety of interconnections. *Scientific American* 253(4): 176-180.)

Coupling can occur between separate molecules if they are not far apart. Szent-Györgyi termed this *electrodynamic coupling.* In this situation, rotations or vibrations of a charged amino acid create an electromagnetic field that propagates in all directions. This field can couple the motions of a particular amino acid with that of others, much like the vibration of an electron in an antenna of a radio transmitter can resonate with electrons in receiving antennas many miles away. The phenomenon is illustrated in Figure 9-6 of Oschman (2000).

In the article by Pienta and Coffee (1991) proposing a tissue tensegrity-matrix system, it is suggested that the cytoskeleton acts as a coupled harmonic oscillator that conveys signals from the cell periphery to the nucleus and thence to the DNA.

LONGER-RANGE INTERACTIONS: FRÖHLICH OSCILLATIONS

A more detailed model of coupling between separate molecules was developed by Fröhlich and colleagues (Fröhlich 1968a, 1968b, 1969, 1970, 1972, 1973, 1975a, 1975b, 1977, 1978, 1980, 1981, 1986; Fröhlich & Kremer 1983; Genzel et al. 1976; Grundler, Keilmann & Fröhlich 1977; Webb, Stoneham & Fröhlich 1977). If two large molecules are capable of giant dipole vibrations at certain frequencies and if the medium separating them has appropriate dielectric properties, resonance-like interactions may take place, even if the molecules are far apart. The mechanism involves longitudinal vibratory modes, which can be stable in large molecules. Once a molecule becomes strongly excited it may tend to continue to vibrate because it cannot lose energy by emitting radiation. In particular, there can be persistent long-range phase-correlated motions (coherent vibrations), especially in molecules that are strongly polar, as many biomolecules are.

Fröhlich presents a specific example. Physicists are fascinated with cell membranes because they are highly ordered arrays of strongly polar phospholipid molecules and can support huge electric fields, on the order of 10^5 volts/cm. The huge electrical fields keep the molecular arrays under a high degree of stress or tension; therefore, the molecules tend to vibrate strongly and the vibrations last a long time, like the ringing of a huge bell. Oscillations of the membrane field could have frequencies on the order of 10^{11} Hz, corresponding to far infrared light. Fröhlich suggested that some of the large molecules within a cell resonate with the membrane electrical oscillations. Hence, the cell as a whole, and a tissue composed of a number of such cells, could have a stable resonant frequency that would be a collective property of the whole assembly. Long-range phase-correlated vibrations between the components of such an assembly could constitute a communication system regulating certain cellular behaviors, such as cell division, up to the point when growth is complete. At that stage, Fröhlich suggests, the cells might become densely packed together in such a way that the membrane oscillations cease. If the resonant excited mode is the stimulus for cell division, the dense cellular packing would cause the disappearance of the stimulus and growth would stop.

Fröhlich presented this scheme as a highly speculative contribution to a conference in Versailles in 1967. Since then a number of studies have confirmed various aspects of Fröhlich's proposals. One study determined that cells absorb signals oscillating in the region predicted by Fröhlich. Using laser Raman spectroscopy, Webb and Stoneham (1977) discovered resonances in the range from 10^{11} to 10^{12} Hz in living bacteria. Subsequent studies by Rowlands et al. (Rowlands, Sewchand & Enns 1982a, 1982b; Rowlands, Sewchand & Skibo 1983; Rowlands et al. 1981) revealed that the quantum mechanical interactions Fröhlich had predicted were taking place in erythrocytes. Pohl and colleagues (Pohl 1979, 1981; Rivera, Pollock & Pohl 1985) used microdielectrophoresis to study the fields around cells and found that cell division is associated with high-frequency oscillations of the sort predicted by Fröhlich. Finally, Popp et al. (1981) and Mamedov, Popov, and Konev (1969) reviewed and confirmed earlier work, going back to Gurwitsch and colleagues (Gurwitsch, Grabje & Salkind 1923; Gurwitsch & Gurwitsch 1959) in the late 1920s, showing ultraviolet luminescence by living systems. Popp et al. concluded the following:

- Ultraweak photon emission occurs in virtually all animal and plant species examined, with the exception of some algae, bacteria, and protozoa.
- Radiation intensity ranges from single to a thousand photons per square centimeter per second.
- Radiation is spread over the region from infrared to ultraviolet.
- Radiation is more intense from proliferating cell cultures than in those in which growth has ceased. The emissions occur mainly before mitosis.

- Emission is particularly intense in cells that are dying, regardless of the cause.
- All agents tested influence the photon output.
- Emission spectrum depends on the species studied.

Much of this information was brought together by the finding reported by Popp et al. that chromatin and DNA are important photon emitters. All of the results can be explained by suggesting that DNA is a photon store and that the photons are released as the DNA unwinds during replication (Rattemeyer, Popp & Nagl 1981).

Much challenging work remains to be done to describe how the solid-state properties of living matter account for the various phenomena we associate with life. In some areas of biology, though, energy transfer between molecules is well documented. This is particularly so in chloroplasts and mitochondria.

ELECTRONIC CONDUCTION IN CHLOROPLASTS

If a photon, ejected by the sun, interacts with an electron of a molecule on our globe, then the electron is raised to a higher energy level, to drop back, as a rule within 0.00000001 to 0.000000001 sec, to its ground state. Life has shoved itself between the two processes, catches the electron in its high- energy state, and lets it drop back to the ground level within its machinery, using the energy thus released for its maintenance. (Szent-Györgyi 1959)

Each cell in the leaf of a plant has many chloroplasts within it. Inside of the chloroplast are highly folded membranes that form a series of stacks called *thylakoids*. Each disk-like element in the thylakoid is called a *granum*.

On the surfaces of the grana membranes are arrayed countless chlorophyll molecules, which act as antennas that trap photons that have journeyed the 92,000,000 miles from the sun. The membranes within the chloroplast are so highly folded and layered that it is improbable that a photon will traverse the thickness of a leaf without encountering a chlorophyll molecule.

The chlorophyll molecule is the first site where energy is captured. A photon from the sun will interact with an electron within a chlorophyll molecule and excite that electron to a higher energy band, called the *conduction band* (see Figure 8-5). The energized electron becomes highly mobile in the conduction band. It can jump from one chlorophyll molecule to the next to the next, moving rapidly over the surface of the grana membrane.

It has been discovered that the energized electron can do one of two things: if it reaches a specialized chlorophyll molecule known as the *reaction center*, its energy will enter the metabolic machinery of the plant. If it does not reach a reaction center within a very small fraction of a second, the electron will fall back down to its ground state, and the photon will be given off again. These transitions are shown in Figure 9-7, *A*. The two photons coming in on the left side of the drawing are captured by chlorophylls and quickly reach reaction centers. The photon arriving at the top jumps from chlorophyll to chlorophyll without reaching a reaction center and is re-emitted to the lower right. The reaction center chlorophyll molecules are denoted by white dots in their centers. All of these events take place in less than one millionth of a second.

Energy transfer in the chloroplast illustrates an important principle that may be significant in communication systems. A biological system can absorb light energy, move it about in the form of mobile excited electrons, and then re-emit the energy in the form of light.

Once an excited electron reaches a reaction center, it is handed on to other molecules that draw off its energy and store it in a chemical bond of, for example, a sugar molecule. The sugar molecule is very stable—it can be stored for centuries without giving off its energy. When an animal eats the plant (or sugar extracted from a plant) the chemical bonds are systematically broken, releasing the energy. It is this energy, which once came from the sun, that we animals use to power all of the things we do.

Energy from light absorbed by phycocyanin in green algae is efficiently transmitted by electrodynamic coupling to chlorophyll (Arnold & Oppenheimer 1950). Blue, red, and green algae use the same mechanism to transfer energy from one pigment molecule to another (Duysens 1952).

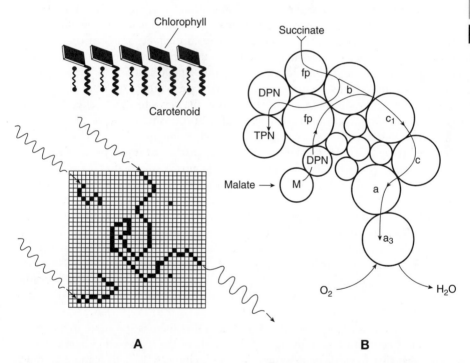

A　　　　　　　　　　　　**B**

Figure 9-7 A, Chlorophyll molecules are like antennas, arrayed on the surface of a granum in a leaf. A photon from the sun interacts with an electron in a chlorophyll molecule, exciting an electron to a higher energy level. The electron becomes mobile and is transferred from one chlorophyll molecule to another until it reaches a reaction center, at which point the energy of the electron is passed on to the metabolic machinery of the plant. If the excited electron does not reach a reaction center within a short time, its energy is re-emitted as another photon. **B,** The flow of energy through mitochondrial electron transport chain within the "oxysome" according Chance, Estabrook and Lee (1963). (A is illustration by Allen Beechel. Courtesy Johanna M. Beechel and the Estate of Allen Beechel, from Scientific American, December 1974.)

MITOCHONDRIA

The site within the animal cell where the energetic exchanges take place is the mitochondrion. It is within this organelle that energy from breaking down sugar molecules is converted into a mobile and readily available form, the adenosine triphosphate (ATP) molecule. Tissues such as muscle, which are involved in conversion of large amounts of energy, are packed with mitochondria. Within each mitochondrion, excited electrons are rapidly handed from molecule to molecule, as shown in Figure 9-4, *B* (Chance, Estabrook & Lee 1963).

These examples are presented because they document a phenomenon of energy transfer that probably is not confined to chloroplasts and mitochondria but that can take place throughout the structural fabric of the organism.

ENERGY FLOWS

The reader who is steeped in the Western biomedical tradition may question the suggestion of energy flow through the living matrix system, including flow through the water network associated with this continuum or through acupuncture channels. The reason to doubt such a suggestion is that the metabolic pathways providing energy for muscle contraction and other activities are firmly established. The high-energy phosphate bonds in ATP and creatine phosphate (CP) are hydrolyzed within cells, providing the energy that causes the actin and myosin filaments to slide past each other to generate all movements of parts of cells or of the organism as a whole. Elucidation of these biochemical pathways is one of the great achievements of modern biochemistry; however, there is a fundamental but not widely publicized unsolved problem in the field of muscle energetics. Biochemists who study muscle contraction have endeavored to prove that the hydrolysis of ATP and CP is adequate to account for all the energy production (heat plus work) of muscle contraction. After years of effort it has been possible to account for only part of the energy output of muscle contraction in this manner. Two reviews on this subject concur that some extra energy is produced during muscle contractions under a variety of conditions and by muscles from a variety of different animals (Curtin & Woledge 1978; Holmscher & Keen 1978). This unexplained energy is not an artifact of inaccurate calculations or measurements but rather indicates the existence of a source of biological energy that is not yet understood by Western science.

We present this example because it indicates that there is still more to be learned about energetic pathways in living systems. Study of the solid-state properties of the connective tissue/cytoskeleton system and proticity may lead to a more complete understanding of biological energetics than we now have. It would not be surprising if some of the energy used for muscle contraction arises as a cooperative phenomenon (Haken 1973) within the connective tissue/cytoskeleton system as a whole. Proticity is also likely to play an active role.

Western physicians and scientists who have observed practitioners of the martial arts, which were developed in parallel with acupuncture, frequently are astonished by the energetic phenomena that can be produced (Eisenberg 1985). Ancient knowledge of human energetics has long been used for self-defense, offensive combat, and healing. Indeed, the mechanisms involved in martial arts may involve a "primitive" property of all protoplasm, used by simple organisms such as protozoa, to avoid predators.

A study conducted in Japan showed that practitioners of traditional health and martial arts exercises, including Qi Gong, yoga, meditation, and Zen, are able to emit strong pulsating magnetic fields from the palms of their hands (Seto et al. 1992). The fields were measured with a magnetometer consisting of two 80,000-turn coils and a high-sensitivity amplifier. The fields detected were about 10^{-3} gauss, which is about 1,000 times stronger than normal human biomagnetic fields, which are about 10^{-6} gauss.

Seto et al. were unable to determine the source of the strong biomagnetic emissions they detected. They performed one study in which the electric field on the surface of the hand was recorded at the same time as the strong biomagnetic emissions. No corresponding bioelectric field could be detected. Seto et al. concluded that the biomagnetic field was not due to current flow within the tissues.

This study had important implications in terms of correlating an ancient healing technology with modern science. The fact that the so-called *Qi emission* was detected with a simple coil magnetometer suggests that at least one aspect of the elusive Qi energy may be none other than biomagnetism. The fact that a relatively insensitive magnetometer can be used to detect this ancient form of *healing energy* indicates that it is a strong effect and therefore is available for further study without the expense of obtaining and operating a SQUID magnetometer.

A few years ago, such a phenomenon probably would have been completely ignored by scientists because there was no theory available to account for it. However, from what we have seen so far in this book, there are a number of biophysical systems that could be involved in developing strong biomagnetic emissions. The various practices used by the subjects in the study by Seto et al. (Qi Gong, yoga, meditation, Zen) gradually may lead to some sort of coherence in the arrangement of the connective tissue/cytoskeletal elements in the forearm, the whole body, and/or the associated water molecules. Perhaps the "Qi emission" is none other than a cooperative phenomenon produced by electronic or protonic conduction through the crystalline-like parts of the body fabric. Relaxation of tensions in the connective tissue, and reorganization of the fibers into a more structurally coherent arrangement or some other simple consequence of various practices, may be the key to bringing about this cooperative state.

Fröhlich described how giant coherent oscillations can be set up in individual macromolecules. The high degree of order in the connective tissue may allow such cooperative effects to be manifested on a macroscopic scale to produce a very strong biomagnetic field. The fact that Seto et al. could not detect current flow on the surface of the skin of the hand during strong "Qi emission" indicates that the field may not have been produced by flow of electricity. However, the electrical flow could be internal and never reach the skin surface. Another source could be a flow of proticity through the water network in the quasi-crystalline array of connective tissue macromolecules in the tendons, ligaments, and bones of the arm or, indeed, of the whole body. Certainly we have seen that living systems have the capacity to generate "emergent" phenomena, and the strong Qi emissions may be an example.

Coherence and healing

If these speculations are correct, the next question is what function coherent biomagnetic emissions would serve in healing. The *healing power* of projected fields may arise from their ability to entrain similar coherent modes in the tissues of a patient. Perhaps such entrainment enhances the evolutionarily ancient communication and regulation systems involved in wound healing and defense. The martial arts techniques that weaken an opponent may involve projecting disruptive fields into points in the body's energy system that are sensitive nodes in a solid-state power distribution system.

Tensegrity

Hints about the biomedical importance of the Tensegrity perspective were introduced in *Energy Medicine: The Scientific Basis* (Oschman 2000, particularly Chapter 4). Tensegrity is a concept developed by R. Buckminster Fuller, who noticed that natural structures use a balance of tension-resisting elements, often called *tendons,* and compression-resisting elements, often referred to as *struts.* In contrast, man-made structures usually are composed primarily of compression members, with tensional elements used only where they cannot be avoided. The human body is a Tensegrity system because it is a continuous network of tensional elements (ligaments, tendons, fascia, muscles) with compressional elements (bones) being discontinuous. Bones usually are considered compressional elements, but this is incorrect. Bones contain both tensional and compressional units. A more general definition of Tensegrity is given in Pugh (1976).

The Tensegrity structure developed by Buckminster Fuller and colleagues (Pugh 1976) has emerged as a useful model for the structure of the nucleus, the cell, and the organism as a whole (Figure 9-8). This model also can be applied to the microscopic structure of space (Smolin 2001; Solit n.d.).

Key work on biological Tensegrity has been done by Donald Ingber et al., who discerned the tensegrous geodesic architecture of cells and tissues (Ingber 1993). This is a design for a living molecular scaffold that is simultaneously porous, flexible, strong, and adaptable, and has a huge surface area. These cytoskeletal scaffolds mediate the internal structural transformations that enable cells to move, change shape, grow, and do all of the other things they do. This includes the vital activities of renewal, repair, and defense against disease.

> . . . [T]he generic molecular machinery and master switches that are found in all cells have been placed under the control of a higher authority: cell and nuclear architecture.
>
> INGBER (1993)

Direct applications of the science of Tensegrity are leading to exciting new materials and devices that incorporate innovations inspired by nature.

> Nature is impressive in its ability to create living materials with an internal microarchitecture that is optimally structured for its function. These materials are then manufactured using environmentally sound fabrication strategies, including molecular and cellular self-assembly. The Biomimicry movement is more than "nanotechnology" since living systems integrate design and

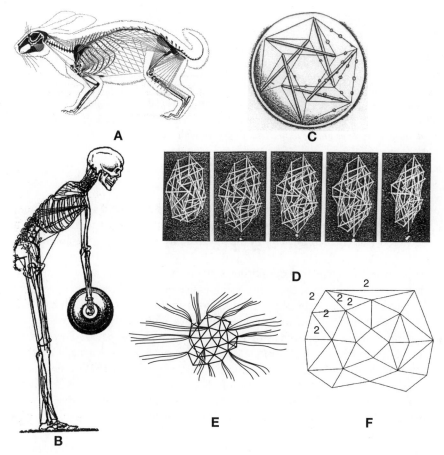

Figure 9-8 **A,** The whole body visualized as a Tensegrity structure. This drawing of the Tensegrity system in the rabbit was created by replacing each muscle-tendon unit with a single straight line. (From a figure drawn from life or redrawn from Young, J.Z. 1957, *The Life of Mammals,* Oxford University Press, New York, with permission from Oxford University Press.) **B,** The tensegrity system in the whole human body. **C,** The cell's matrix, held together by tension, can be represented as a Tensegrity structure. (From *Energy Medicine: the scientific basis,* Oschman, Figure 4.2, page 63, Churchill Livingstone, Edinburgh, 2000, with permission from Elsevier Science. Originally from Levine 1985. Copyright 1979 by Scientific American. All rights reserved.) **D,** The cell nucleus can be depicted as a Tensegrity structure that changes its properties when its shape is changed. (Reprinted with permission from Wang N, Butler JP, Ingber DE 1993. Mechanotransduction across the cell surface and through the cytoskeleton. *Science* 260:1124-1127, Copyrighted 1993, American Association for the Advancement of Science.) **E,** Cellular geodome structure composed of actin, tropomyosin, and alpha-actinin. This structure forms around the nucleus during the part of the cell cycle when the cell stops moving and begins to flatten prior to mitosis. (From Lazarides & Revel, 1979. *Scientific American,* vol. 240, pp. 100-113.) **F,** Tensegrous geometry of space, composed of a spin network. This geometry is thought to exist at an extremely small scale known as the Planck scale, about 20 orders of magnitude (i.e., a factor of 1,020) smaller than an atomic nucleus. The units of length are called Planck units, which are approximately 10^{-33} cm. Events take place at around a Planck time, about 10^{-33} second. (Originally from Rovelli and Smolin, 1995. 'Spin networks in quantum gravity', *Physical Review* vol 52, pp. 5742-5759.)

construction on many different size scales. It is through this architecture of life—how nanome-ter-sized molecular building blocks are connected and arranged on the micron to centimeter size scales—that living tissues exhibit their exquisite material properties and biochemical functions as well as their incredible abilities to sense, respond, learn, and repair. If we could harness but a mere portion of these natural powers, we would revolutionize the way we heal, the machines and devices we construct, and the processes we use to manufacture.

The field of biomimetics is in its infancy. Up to now, most biomimetics approaches set out to reproduce specific designs and structures found in living creatures. The future of biomimetics lies in development of our ability to understand and emulate the very nature of the design process itself. (Ingber 2002)

A HYDRATED VIBRATORY CONTINUUM

It now is realized that the entire dynamic living matrix is capable of creating, sustaining, propagating, and processing a variety of vibratory entities as messages and as energy. Here we use the word *energy* in the sense of physics: energy is the ability to do work.

A superficial analogy is that the fibrous systems in the body are a set of interconnected stringed instruments, each with a resonance or a set of resonances and harmonics that are constantly changing as the tensions on the strings alter. Cellular events of all kinds, changes in shape, changes in membranes, motility, signal processing, and pathological changes, set up signals that propagate through the matrix. These signals are altered by growth factors, hormones, carcinogenesis, trauma—virtually anything that happens within and around the organism (Pienta and Coffey 1991). As with a set of stringed instruments, the sum total of their functioning resembles a musical arrangement performed by an orchestra. The tones and harmonies played by our internal orchestra are the tones of our health and our life as a whole—they can be melodious or discordant.

Memories and traumatic memories are recorded in the living matrix. Although trauma affects the nervous system, the nervous system is not the one and only part of the body that is affected. In the model of consciousness that is emerging from the puzzle we are working on here, we will see that trauma and traumatic memories arise as joint or cooperative aspects of both brain and body. Trauma resolution must take into consideration the properties of the living matrix.

Life is not possible without the framework of water molecules that holds the entire assembly together. That there is a special relationship between water and all components of the living matrix is demonstrated in Figure 9-5, which shows that the spacing of charges along the collagen molecule is ideal for structuring a surrounding water matrix. All parts of the living matrix are in an intimate and precise relationship with water, as is shown in Figure 9-9.

For future research into the way water and life exist interdependently, Collins (1991, 2000) has developed a graphic imaging technique that enables one to visualize how water performs its vital function of regulating the shape and forms of natural molecules. Research into homeopathy is focusing on the clustering of water molecules around dissolved compounds, and the graphic methods Collins has developed are invaluable for studying these shells (Figure 9-10).

Hydrogen bonds

If we are to comprehend nature, work with nature, live with nature, and explore our inner potentials, we must learn more about how to investigate and engage the vibrant living fabric. We must appreciate that the high degree of crystallinity, or liquid crystallinity, mentioned earlier, is everywhere associated with an equally ordered and vibrant array of water molecules. These water molecules hold the system together by extremely weak forces known as *hydrogen bonds*. Individually these bonds are extremely weak, but there are so many of them that they combine to essentially hold the organism together. In later chapters we will look closely at what these water molecules are doing. They appear to form their own coequal communication system that is intimately involved with all of the other communications in the body.

AN EXCITABLE MEDIUM

The living matrix is an excitable medium. By this we mean that its components are everywhere poised to absorb incoming energy and propagate signals. In media of this kind, a disturbance of the local conditions can propagate as a solitary pulse. Waves moving in excitable media are different from other kinds of waves, such as waves on the ocean. The main difference is that every element of the excitable medium, when triggered by an encroaching wave or other stimulus, becomes a source of energy to maintain the propagation; hence, spreading waves are not attenuated as they propagate. They do not dissipate and disperse.

When we think of excitable media, attention usually is focused on nerve cell membranes; however, many if not all biological membranes and tissues, including layers of cells called epithelia, and some chemical reactions share the property of *excitability*. Figure 9-11, *A* shows waves of spiraling chemical change in a reagent developed by the Russian researchers Zaikin and Zhabotinsky (1970).

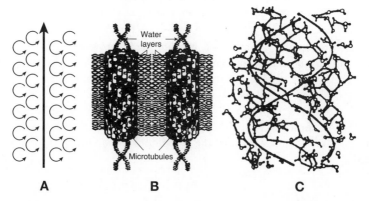

Figure 9-9 Hydration of the major components of the living matrix. A, Collagen molecule; B, microtubules in the cytoskeleton; C, DNA. (B is modified from Figure 5 on page 104 of Hameroff SR 1994 Journal of Consciousness Studies 1(1):91-118. C is from Figure 7, page 57, in Clemente E and Corongiu G 1981 B-DNA's structure determination of Na+ counter-ions at different humidities, ionic concentrations and temperatures. IBM IS&TG Research Report POK-04, dated September 23, 1981, Adenine Press, Schenectady, NY.)

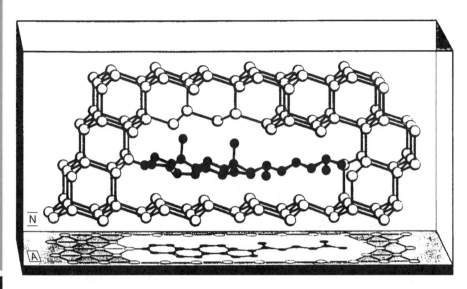

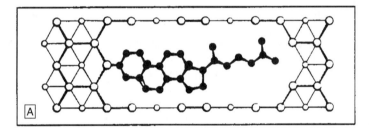

Figure 9-10 Hydration shell of cholesterol shown by ingenious graphical method developed by Collins (1991). (Reproduced with generous permission of JC Collins.)

Another spectacular example of excitability is provided by the slime mold *Dictyostelium discoideum,* whose amoeboid cells can spread over a wide area on the ground. The cells synthesize and hoard cyclic adenosine monophosphate (c-AMP) molecules, but they can release them in an abrupt "sneeze." The flood of molecules causes the neighbors to sneeze as well, and the disturbance propagates in a pattern such as the one shown in Figure 9-11, *A.* Meanwhile, each cell that has discharged its AMP begins to accumulate more of the compound and eventually is capable of being triggered again.

TEMPORAL ANATOMY

The study of excitable media and the kinds of waves they can sustain is extremely important and has received much attention. A.T. Winfree has made an extensive series of investigations in this area (Winfree 1984, 1987, 2001). Various kinds of natural and artificial media are excitable, and their study has given rise to a new science of temporal anatomy. This science studies periodic behavior and its pathologies, as can

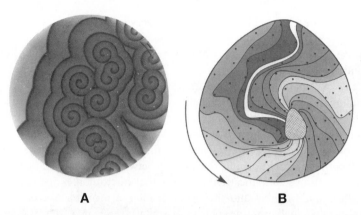

A   **B**

Figure 9-11 Excitable media. **A,** Waves of spiraling chemical change in a reagent developed by the Russian researchers Zaikin and Zhabotinsky (1970). **B,** Wave recorded with a plaque electrode array placed on the surface of a ventricle. The pattern is divided into isochrons, which are curves connecting points representing the same timing of a rhythm. This heart is beating more than four times per second (260 beats/min) because the activation wave is "snagged on a hole" of depressed excitability, a lethal condition called *ventricular flutter.* The "hole" probably was caused by a myocardial infarction. The "hole" is shown as a stippled region at the apex (bottom) of the heart. This inexcitable diseased tissue was surgically removed minutes after the recording was made. (A is from Winfree, A.T. 1987: *When Time Breaks Down: The Three-Dimensional Dynamics of Electrochemical Waves and Cardiac Arrhythmias,* Princeton University Press, Princeton, NJ. B is from Downar, E., Parson, I.D., Mickleborough, L.L., Yao, L.C., Cameron, D.A., Waxman, M.B. 1984. On-line epicardial mapping of intraoperative ventricular arrhythmias: initial clinical experience. *J Am Coll Cardiol* 4(4):703-714, with permission from Elsevier Science.)

occur in the heart that has become arrhythmic (Figure 9-11, *B*). The heart is a three-dimensional excitable medium, and the study of its behavior is challenging and extremely important from the clinical perspective.

Temporal anatomy is important for the study of circadian rhythms, the choreography of muscular contractions in running, breathing, the heartbeat, and so on. Pacemakers occur in the heart and in other organs, including the stomach, intestine, kidney, uterus, and ovaries. The way in which all of these rhythms are coordinated in an athletic or artistic performance of any kind has to be one of the challenging frontiers of applied regulatory biology.

Winfree views the heart as "the most elaborately engineered clock on the planet. Its supreme sophistication arose during evolution in part because of the fact that there is no backup system to take over in the event of a failure, yet its normal coordination can come unraveled in a great variety of ways."

A FIBEROPTIC SYSTEM?

Some researchers have envisioned the protein fabric in the body as a fiberoptic system. Fascinating evidence has been published by Pankratov (1991) in Moscow. Pankratov projected light on acupuncture points and found that it reemerged from other points along the same meridian. This fascinating discovery suggests that the meridians (see Figure 8-7) are the main channels through which light is preferentially conducted in the body. Pankratov cites work on plants that showed that they, too, have light channels (Mandoley & Briggs 1984).

A testable explanation for Pankratov's results, as well as the effects of phototherapies including color puncture, is that light is converted into solitons in sensory cells. Because the light-absorbing reactions are reversible, it is likely that solitons arriving at the ends of microtubules, such as in retinal cells, can be converted back into light. These hypotheses are highly speculative but definitely worthy of further study.

From the information available, the analogy between proteins and fiberoptic wave guides may be an oversimplification. It is true that microtubules and collagen fibers are long, thin, hollow tubes and therefore capable, in principle, of serving as light pipes; however, we shall examine solitons or solitary waves in a later chapter and see that there is evidence that they are likely entities for conduction of energy through tissues. Soliton propagation follows nonlinear rules as opposed to ordinary linear optics. Soliton conduction allows far more sophisticated signal processing to take place within the matrix. Soliton conduction in the living matrix, if verified, does not disprove any of the ideas we have already discussed, such as the electronic and protonic conduction suggested by Szent-Györgyi. In actuality, the soliton/quantum coherence concept we will discuss later incorporates all of these phenomena.

Ultimately, if light is to activate healing processes, the light must in some way speak to these cells directly, to the pathways that enable the cells to "whisper" to each other, to the pathways by which cells migrate or crawl about in the organism, or to all of these (Adey 1993; Oschman 2001).

From the biology of the situation, it is suggested that the immediate effect of light is on the communication pathways. Specifically, it is proposed that light stimulates the flow of solitons, which are waves of energy and information that travel rapidly through the protein fabric of the body. The flow of solitons opens gates and switches and organizes the dynamic living matrix pathways. Cells can then "whisper" to each other using their own "languages." These whisperings orchestrate the repair of traumas of all kinds. Light, electromagnetic fields, sounds, solitons, bosons, and chemicals, including nitric oxide, all are part of vital communications, but undoubtedly there are others.

The manner by which light-stimulated soliton transmission can open up communications throughout the body is a topic for further research. Perhaps it is an effect on the integrins spanning cell membranes (see Figures 8-3 and 8-4). Integrins are essential components of the living matrix communication pathway and have been implicated in a wide range of disorders. Another hypothesis has arisen from the work of Albrecht-Buehler (1992), who has developed concepts of "vision" at the cellular level, involving the microtubules acting as the "nerves" of cells. Albrecht- Buehler found that light alters the stability of the radial array of microtubules surrounding the centrosomes, which therefore appear to be the light "detectors" within cells. The centrosomes, in turn, are important in regulating cell division and other cytoplasmic processes. In any case, the diversity of clinical problems that are approachable through phototherapy seem to indicate system-wide biological effects on the communications vital to the healing process.

REFERENCES

Adey, W.R. 1993, 'Whispering between cells: Electromagnetic fields and regulatory mechanisms in tissue', *Frontier Perspectives,* vol. 3, pp. 21-25.

Aiello, G., Micciancio-Giammarinaro, M.S., Palma-Vittorelli, M.B. & Palma, M.U. 1973, 'Behavior of interacting protons: The average-mass approach to its study and its possible biological relevance', in eds H. Haken & M. Wagner, *Cooperative Phenomena,* Springer-Verlag, New York, pp. 395-403.

Albrecht-Buehler, G. 1992, 'Rudimentary form of cellular "vision"', *Proceedings of the National Academy of Sciences USA,* vol. 89, pp. 8288-8292.

Arnold, W. & Oppenheimer, J.R. 1950, 'Internal conversion in the photosynthetic mechanism of blue-green algae', *Journal of General Physiology,* vol. 33, p. 423.

Bassett, C.A.L. & Becker, R.O. 1962, 'Generation of electric potentials by bone in response to mechanical stress', *Science,* vol. 137, pp. 1063-1064.

Becker, R.O. & Sheldon, G. 1985, *The Body Electric: Electromagnetism and the Foundation of Life,* William Morrow & Company, New York.

Berendsen, H.J.C. 1962, 'Nuclear magnetic resonance study of collagen hydration', *Journal of Chemical Physics,* vol. 36, pp. 3297-3305.

Black, J. & Korostoff, E. 1974, 'Strain-related potentials in living bone', *Annals of the New York Academy of Sciences,* vol. 238, pp. 95-120.

Braden, M., Bairstow, A.G., Beider, I. & Ritter, B.G. 1966, 'Electrical and piezoelectrical properties of dental hard tissues', *Nature,* vol. 212, pp. 1565-1566.

Chance, B., Estabrook, R.W. & Lee, C.P. 1963, 'Electron transport in the oxysome', *Science,* vol. 140, pp. 379-380.

Chapman, G.E., Danyluk, S.S. & McLauchlan, K.A. 1971, 'Evidence for a chainlike hydration structure for collagen', *Biophysical Journal,* vol. 11, p. 186a.

Chapman, G.E. & McLauchlan, K.A. 1969, 'The hydration structure of collagen', *Proceedings of the Royal Society Series B,* vol. 173, pp. 223-234.

Clementi, E. 1981, 'Computer simulations of complex chemical systems: Solvation of DNA and solvent effects in conformational transitions', *IBM Journal of Research and Development,* vol. 25, pp. 315-326.

Collins, J.C. 1991, *The Matrix of Life,* Molecular Presentations, East Greenbush, NY.

Collins, J.C. 2000, *Water: The Vital Force of Life,* Molecular Presentations, Kinderhook, NY.

Corongiu, G. & Clementi, E. 1981, 'Simulations of the solvent structure for macromolecules. I. Solvation of B-DNA double helix at T=300 K', *Biopolymers,* vol. 20, pp. 551-571.

Craig, D.P. & Walmsley, S.H. 1968, *Excitons in Molecular Crystals,* W.A. Benjamin, New York, p. 2.

Cupane, A., Palma, M.U. & Vitrano, E. 1974, 'Dependence of co-operativity in the reaction of haemoglobin with oxygen on deuterium oxide concentration and temperature', *Journal of Molecular Biology,* vol. 82, pp. 185-192.

Curtin, N.A. & Woledge, R.C. 1978, 'Energy changes in muscular contraction', *Physiological Reviews,* vol. 58, pp. 690-761.

Davydov, A.S., Kasha, M. & Oppenheimer, M. 1962, *Theory of Molecular Excitons,* McGraw-Hill, New York, pp. 165-166.

Duysens, L.N.M. 1952, Transfer of excitation energy in photosynthesis, Thesis, University of Utrecht.

Eisenberg, D. 1985, *Encounters with Qi: An American Doctor's Firsthand Observations of Qi (Chee), "Vital Energy," and the Qi Gong Masters Who Use It to Heal,* Penguin, New York.

Fröhlich, H. 1968a, 'Long-range coherence and energy storage in biological systems', *International Journal of Quantum Chemistry*, vol. II, pp. 641-649.

Fröhlich, H. 1968b, 'Bose condensation of strongly excited longitudinal electric modes', *Physics Letters*, vol. 26A, pp. 402-403.

Fröhlich, H. 1969, 'Quantum mechanical concepts in biology', in ed. M Marios, *Theoretical Physics and Biology*, North Holland, Amsterdam, pp. 13-22.

Fröhlich, H. 1970, 'Long range coherence and the action of enzymes', *Nature*, vol. 228, p. 1093.

Fröhlich, H. 1972, 'Selective long range dispersion forces between large systems', *Physics Letters*, vol. 39A, pp. 153-154.

Fröhlich, H. 1973, 'Collective behavior of non-linearly coupled oscillating fields: With applications to biological systems', *Collective Phenomena*, vol. 1, pp. 101-109.

Fröhlich, H. 1975a, 'The extraordinary dielectric properties of biological materials and the action of enzymes', *Proceedings of the National Academy of Sciences USA*, vol. 72, pp. 4211-4215.

Fröhlich, H. 1975b, 'Evidence for Bose condensation-like excitation of coherent modes in biological systems', *Physics Letters*, vol. 51A, pp. 21-22.

Fröhlich, H. 1978, 'Coherent electric vibrations in biological systems and the cancer problem', *IEEE Transactions Microwave Theory and Techniques*, vol. MTT-26, pp. 613-617.

Fröhlich, H. 1980, 'The biological effects of microwaves and related questions', *Advances in Electronics and Electron Physics*, vol. 53, pp. 85-152.

Fröhlich, H. 1981, 'Coherence in biological systems', *Collective Phenomena*, vol. 3, pp. 139-146.

Fröhlich, H. 1986, 'Coherent excitations in active biological systems', in eds F. Gutmann & H. Keyzer, *Modern Bioelectrochemistry*, Plenum Press, New York, pp. 241-261.

Fröhlich, H. & Kremer, F. 1983, *Coherent Excitations in Biological Systems*, Springer-Verlag, New York.

Fukada, E. 1974, 'Piezoelectric properties of organic polymers', *Annals of the New York Academy of Sciences*, vol. 238, pp. 7-25.

Fukada, E. & Hara, K. 1969, 'Piezoelectric effect in blood vessel walls', *Nippon Seirigaku Zasshi Journal of the Physiological Society of Japan*, vol. 26, pp. 777-780.

Fukada, E. & Ueda, H. 1970, 'Piezoelectric effect in muscle', *Japanese Journal of Applied Physiology*, vol. 9, p. 844.

Fukada, E. & Yasuda, I. 1957, 'On the piezoelectric effect of bone', *Nippon Seirigaku Zasshi Journal of the Physiological Society of Japan*, vol. 12, pp. 149-154.

Genzel, L., Keilmann, F., Martin, T.P., Winterling, G., Yacoby, Y., Fröhlich, H. & Makinen, M.W. 1976, 'Low-frequency Raman spectra of lysozyme', *Biopolymers*, vol. 15, pp. 219-225.

Grundler, W., Keilmann, F. & Fröhlich, H. 1977, 'Resonant growth rate response of yeast cells irradiated by weak microwaves', *Physics Letters*, vol. 62A, pp. 463-466.

Gurwitsch, A., Grabje, S. & Salkind, S. 1923, 'Die Natur des spezifischen Erregers der Zellteilung', *Wilhelm Roux' Archiv für Entwicklungsmechanik der Organismen*, vol. 100, p. 11.

Gurwitsch, A.G. & Gurwitsch, L.D. 1959, *Die Mitogenetische Strahlung*, VEB G. Fischer, Jena.

Haken, H. 1973, 'Synergetics: Towards a new discipline', in eds H. Haken & M Wagner, *Cooperative Phenomena*, Springer-Verlag, New York, pp. 363-380.

Haken, H. 1975, 'Cooperative effects in systems far from thermal equilibrium and in nonphysical system', *Review of Modern Physics*, vol. 47, p. 67.

Haken, H. 1983a, *Synergetics: An Introduction*, 3rd ed, Springer, Berlin.

Haken, H. 1983b, *Advanced Synergetics. Instability Hierarchies of Self-Organizing Systems and Devices*, Springer-Verlag, Berlin; see also the Springer Series in Synergetics.

Ho, M.-W. 1993, *The Rainbow and the Worm: The Physics of Organisms*, World Scientific, Singapore.

Ho, M.-W. 1999, Coherent energy, liquid crystallinity and acupuncture, talk presented to British Acupuncture Society, 2 October 1999. Available: http://www.i-sis.org/acupunc.shtml.

Ho, M.-W. & Knight, D. 1998, 'The acupuncture system and the liquid crystalline collagen fibers of the connective tissues', *American Journal of Chinese Medicine*, vol. 26, pp. 251-263.

Ho, M.W. & Lawrence, M. 1993, 'Interference colour vital imaging: A novel noninvasive microscopic technique', *Microscopy and Analysis*, Sept., p 26.

Ho, M.W. & Saunders, P.T. 1994, 'Liquid crystalline mesophase in living organisms', in eds M.W. Ho, F.A. Popp & U. Warnke, *Bioelectrodynamics and Biocommunication*, World Scientific, Singapore, pp. 213-227.

Holmscher, E. & Keen, C.J. 1978, 'Skeletal muscle energetics and metabolism', *Annual Review of Physiology*, vol. 40, pp. 93-131.

Horn, R.G. & Israelachvili, J. 1981, 'Direct measurement of structural forces between two surfaces in a nonpolar liquid', *Journal of Chemical Physics*, vol. 75, pp. 1400-1411.

Ingber, D.E. 1993, 'The riddle of morphogenesis: A question of solution chemistry or molecular cell engineering?' *Cell*, vol. 75, pp. 1249-1252.

Ingber, D.E. 2002, Available: http://www.zohshow.com/News/Newsbytes/tidbits030800.htm.

Israelachvili, J.N. & McGuiggan, P.M. 1988, 'Forces between surfaces in liquids', *Science*, vol. 241, pp. 795-800.

Israelachvili, J.N. & Wennerström, H. 1996, 'Role of hydration and water structure in biological and colloidal interactions', *Nature*, vol. 379, pp. 219-225.

Jeffries, C.D. 1975, 'Electron-hole condensation in semiconductors: Electrons and holes condense into freely moving liquid metallic droplets, a plasma phase with novel properties', *Science*, vol. 189, pp. 955-964.

Kohlrausch, F. (1840-1910), cited from Piccardi, G. 1962, *The Chemical Basis of Medical Climatology*, C.C. Thomas, Springfield, Illinois, p. 23.

Lai, H.M. 1984, 'The recurrence phenomenon of a resonant spring pendulum', *American Journal of Physics*, vol. 52, pp. 219-223.

Levine, L.S., Lustrin, I. & Shamos, M.H. 1969, 'Experimental model for studying the effect of electric current on bone in vivo', *Nature*, vol. 224, pp. 1112-1113.

Libbin, R.M., Person, P., Papierman, S., Shah, D., Nevid, D. & Grob, H. 1979, 'Partial regeneration of the above-elbow amputated rat forelimb. II. Electrical and mechanical facilitation', *Journal of Morphology*, vol. 159, pp. 439-452.

Lipinski, B. 1977, 'Biological significance of piezoelectricity in relation to acupuncture, Hatha yoga, osteopathic medicine and action of air ions', *Medical Hypotheses*, vol. 3, pp. 9-12.

Mamedov, T.G., Popov, G.A. & Konev, V.V. 1969, 'Ultraweak luminescence of various organisms', *Biofizika*, vol. 14, pp. 1047-1051.

Mandoley, D.F. & Briggs, W.R. 1984, 'On the light path through plants', *Scientific American*, August, pp. 90-98.

McDonnell, T.J. & Oberpriller, J.O. 1984, 'The response of the atrium to direct mechanical wounding in the adult heart of the newt, *Notophthalmus viridescens*. An electronmicroscopic and autoradiographic study', *Cell Tissue Research*, vol. 235, pp. 583-592.

McElhaney, J.H. 1967, 'The charge distribution on the human femur due to load', *Journal of Bone and Joint Surgery*, vol. 49A, pp. 1561-1571.

Mitchell, P. 1976, 'Vectorial chemistry and the molecular mechanics of chemiosmotic coupling: Power transmission by proticity', *Biochemical Society Transactions*, vol. 4, pp. 399-430.

Newton, R.H., Hafegee, J. & Ho, M.W. 1995, 'Colour-contrast in polarized light microscopy of weakly birefringent biological specimens', *Journal of Microscopy*, vol. 180, pp. 127-130.

Oschman, J.L. 1990, 'Bioelectromagnetic communication', *BEMI Currents*, vol. 2, pp. 11-14.

Oschman, J.L. 2000, *Energy Medicine: The Scientific Basis*, Churchill Livingstone, Edinburgh.

Pankratov, S. 1991, 'Meridians conduct light', *Raum und Zeit,* vol. 35, pp. 16-18.

Person, P., Libbin, R.M., Shah, D. & Papierman, S. 1979, 'Partial regeneration of the above-elbow amputated rat forelimb. I. Innate responses', *Journal of Morphology,* vol. 159, pp. 427-438.

Pethig, R. 1979, *Dielectric and Electronic Properties of Biological Materials,* John Wiley & Sons, New York.

Philpott, D., Corbett, R., Person, P. & Becker, R. 1975, 'Aspects of myocardial regeneration in *Notoopthalmus viridescens,*' in ed. G.W. Bailey, *33rd Annual Proceedings of the Electron Microscopy Society of America,* Las Vegas, NV.

Pickup, A.J. 1978, 'Collagen and behaviour: A model for progressive debilitation', *International Journal of Research on Communication Systems, Medical Science,* vol. 6, pp. 499-502.

Pienta, K.J. & Coffey, D.S. 1991, 'Cellular harmonic information transfer through a tissue tensegrity-matrix system', *Medical Hypotheses,* vol. 34, pp. 88-95.

Pohl, H.A. 1979, Micro-dielectrophoresis of dividing cells II, Research Note 90 from the Quantum Theoretical Research Group, Oklahoma State University, Stillwater.

Pohl, H.A. 1981, 'Do cells in the reproductive state exhibit a fermi-Pasta-Ulam-Fröhlich resonance and emit electromagnetic radiation?' *Collective Phenomena,* vol. 3, pp. 221-244.

Pollack, G.H. 2001, *Cells, gels and the engines of life,* Ebner & Sons, Seattle, WA.

Popp, F.A., Ruth, B., Bahr, W., Böhm, J., Grass, P., Grolig, G., Rattemeyer, M., Schmidt, H.G. & Wulle, P. 1981, 'Emission of visible and ultraviolet radiation by active biological systems', *Collective Phenomena,* vol. 3, pp. 187-214.

Pugh, A. 1976, *An Introduction to Tensegrity,* University of California Press, Berkeley.

Rattemeyer, M., Popp, F.A. & Nagl, W. 1981, Evidence of photon emission from DNA in living systems', *Naturwissenschaften,* vol. 68, pp. 572-573.

Rivera, H., Pollock, J.K. & Pohl, H.A. 1985, 'The ac field patterns about living cells', *Cell Biophysics,* vol. 7, p. 43.

Ross, S., Newton, R.H., Zhou, Y.M., Hafegee, J., Ho, M.W., Bolton, J. & Knight, D. 1997, 'Quantitative image analysis of birefringent biological materials', *Journal of Microscopy,* vol. 187, pp. 62-67.

Rowlands, S., Sewchand, L.S., Lovlin, R.E., Beck, J.S. & Enns, E.G. 1981, 'A Fröhlich interaction of human erythrocytes', *Physics Letters,* vol. 82A, pp. 436-438.

Rowlands, S., Sewchand, L.S. & Enns, E.G. 1982a, 'A quantum mechanical interaction of human erythrocytes', *Canadian Journal of Physiology and Pharmacology,* vol. 60, pp. 52-59.

Rowlands, S., Sewchand, L.S. & Enns, E.G. 1982b, 'Further evidence for a Fröhlich interaction of erythrocytes', *Physics Letters,* vol. 87A, pp. 256-260.

Rowlands, S., Sewchand, L.S. & Skibo, L. 1983, 'Conversion of albumin into a transmitter of the ultra long-range interaction of human erythrocytes', *Cell Biophysics,* vol. 5, pp. 197-203.

Seto, A., Kusaka, C., Nakazato, S., Huang, W., Sato, T., Hisamitsu, T. & Takeshige, C. 1992, 'Detection of extraordinary large bio-magnetic field strength from human hand', *Acupuncture and Electro-Therapeutics Research International Journal,* vol. 17, pp. 75-94.

Shang, C. 1989, 'Singular point, organizing center and acupuncture point', *American Journal of Chinese Medicine,* vol. 17, pp. 119-127.

Smolin, L. 2001, *Three Roads to Quantum Gravity,* Basic Books, New York.

Shirakawa, H., Louis, E.J., MacDiarmid, A.G., Chiang, C.K., & Heeger, A.J. 1977, 'Synthesis of electrically conducting organic polymers: Halogen derivatives of polyacetilene (CH)n', *Chemical Communications,* vol. 33, pp. 579-603.

Solit, M. n.d., Available: http://www.fnd.org.

Szent-Györgyi, A. 1957, Bioenergetics, Academic Press, New York, p. 139.

Szent-Györgyi, A. 1959, 'Introductory paper', *Faraday Society Discussions,* vol. 27, pp. 111-114.

Szent-Györgyi, A. 1960, *Introduction to a Submolecular Biology,* Academic Press, New York, p. 93.

Szent-Györgyi, A. 1968, *Bioelectronics,* Academic Press, New York.

Szent-Györgyi, A. 1972, *The Living State: With Observations on Cancer,* Academic Press, New York.

Szent-Györgyi, A. 1974, 'Drive in living matter to perfect itself', *Synthesis,* vol. 1, pp. 14-26.

Tomita, S. & Riggs, A. 1970, 'Effects of partial deuteration on the properties of human hemoglobin', *Journal of Biological Chemistry,* vol. 245, pp. 3104-3109.

Vaughn, J.M. 1970, *The Physiology of Bone,* Clarendon Press, Oxford, England.

Walker, J. 1985, 'Strange things happen when two pendulums interact through a variety of interconnections', *Scientific American,* vol. 253, pp. 176-180.

Webb, S.J., Stoneham, M.E. & Fröhlich, H. 1977, 'Evidence for non-thermal excitation of energy levels in active biological systems', *Physics Letters,* vol. 63A, pp. 407- 408.

Webb, S.J. & Stoneham, M.E. 1977, 'Resonances between 10^{11} and 10^{12} Hz in active bacterial cells as seen by laser Raman spectroscopy', *Physics Letters,* vol. 60A, pp. 267-268.

Winfree, A.T. 1984, 'Wavefront geometry in excitable media', *Physica,* vol. 12D, pp. 321-332.

Winfree, A.T. 1987, *When Time Breaks Down: The Three-Dimensional Dynamics of Electrochemical Waves and Cardiac Arrhythmias,* Princeton University Press, Princeton, NJ.

Winfree, A.T. 2001, *The Geometry of Biological Time,* 2nd ed, Springer, New York.

Wolfe, J.P. & Mysyrowicz, A. 1983, 'Excitonic matter. A conduction electron can combine with a positively charged "hole" in a semiconductor to create an exciton: Excitons in turn can form molecules and liquids, and a new phase of matter may be attainable', *Scientific American,* vol. 250, pp. 98-107.

Zaikin, A.N. & Zhabotinsky, A.M. 1970, 'Concentration wave propagation in two-dimensional liquid-phase self-oscillating systems', *Nature,* vol. 225, pp. 535-537.

10 The living matrix and acupuncture

Whilst a typical chart of the acupuncture channels . . . illustrates only the
superficial pathways of the twelve primary channels, we should remember that the
channel network is considerably more complex than this, and there is no part of
the body, no kind of tissue, no single cell, that is not supplied by the channels.
Like a tree, the trunk and main branches define the main structure, whilst ever finer
branches, twigs, and leaves spread out to every part.

Deadman, Al-Khafaji, and Baker (1998)

In the past, academic biomedicine has placed acupuncture with other unexplained or
anomalous phenomena that seem to lie outside of the current medical paradigm. This
situation is changing rapidly simply because many people like acupuncture, and
acupuncture has given relief from chronic problems when other approaches have
failed. Moreover, the most conservative scientist admits that our present understanding of physiology and medicine is tentative and incomplete. The history of science
shows that today's anomalies often become tomorrow's orthodoxies.

All of the material presented so far may relate to acupuncture theory. Those who
work with the human body as an energetic system need no convincing about the fact
of the meridian system. As a paradigm, meridian theory makes possible many remarkable therapeutic interactions that lead to the resolution of a vast array of physical and

emotional ailments; however, to bring acupuncture into mainstream biomedicine it would be beneficial to have logical connections between standard physiology and biophysics and the meridian system. I believe it will be possible to accomplish this to a degree. Certainly the effort will reveal new insights into living structure and function; however, it probably would be a mistake to believe that we are close to a thorough scientific explanation of the nature of Ch'i in all its aspects.

Acupuncture theory offers a set of potentially important hypotheses that are open to test using scientific methods. These hypotheses relate to phenomena of great medical importance, particularly regulatory biology, the defense and repair mechanisms in the body, and the mechanisms involved in recognition of self and nonself. The importance of these phenomena is that they form the basis of both prevention and recovery from disease or tissue damage. They also may be part of the key to understanding the so-called *chronic and degenerative diseases,* which are so costly both financially and in terms of human suffering.

The overall motivation for our inquiry is to determine what the theories for the various alternative and complementary modalities have to offer in our quest for a deeper understanding of life's mechanisms and to see how this knowledge might be applied to our health care system and to human potential in general.

The effects of acupuncture on the neuropeptide system and application of this knowledge to the treatment of addiction and detoxification provide examples of a successful integration of traditional and modern medicine. Other such connections are emerging.

In particular, solid-state physicists have discovered a number of cooperative or collective phenomena that could provide a basis for rapid movements of information and energy within living systems. Some of these phenomena were discussed in the last chapter. Acupuncture is based on the existence of energetic phenomena that seem to have similar properties. In contrast, the Western physiologist usually is satisfied that the nervous, hormonal, and immune systems are *the* communication systems in the living body. It is the task of the biophysicist to resolve this dilemma, to see if living systems possess other signaling systems that contribute to whole-body integration and coordination and to see how such systems may relate to approaches such as acupuncture.

THE NEED FOR THEORY

The meteoric rise in the popularity of acupuncture has led to a need for basic and clinical research. The National Center for Complementary and Alternative Medicine has undertaken large-scale trials to determine the safety and effectiveness of several acupuncture techniques; however, history has shown repeatedly that a good outcome with a clinical trial has little impact on clinical practice if there is no viable theory to account for the results.

At the same time, a poor outcome for a clinical trial may have little or no impact on clinical practice. For example, a recent study by scientists at Duke University compared St. John's wort, a popular herbal remedy for depression, with a placebo pill having no medicinal properties and with the drug sertraline, which has been marketed as Zoloft.

The multicenter study was conducted according to the most rigorous double-blinded, placebo-controlled standards available to the scientific community. The study cost $6,000,000 and was begun in 1997. The results, reported by Davidson (2002), show that the placebo worked better than the drug and that the drug and St. John's wort were less but about equally effective. The results of this study will be discussed for a long time. It is not clear whether the herbal treatment or the study failed. Certainly the results seem to prove that participating in a clinical trial can be good for you!

Before research is done to determine the clinical efficacy of a method, it is helpful to describe the mechanisms involved as much as possible in terms of existing scientific paradigms. Experience has shown that acceptance of any medical technique requires both proof of clinical benefit and explanatory mechanisms that mesh with orthodox thinking. Believable and well-documented theories can greatly facilitate clinical testing and add to its impact if the outcome is positive. Before invoking out-of-the-mainstream concepts to account for phenomena, we need to check what ordinary science has to say.

IMPLICATIONS OF ACUPUNCTURE FOR REGULATORY BIOLOGY

The state we refer to as *health* is maintained by a constellation of regulatory systems. Any deviation from normality immediately brings into operation a set of repair and defense mechanisms that endeavor to restore form and function to the organism. Regulation is a fundamental concept in both Western and Eastern approaches to the human body.

The acupuncture model of regulation involves a set of channels or meridians that function in communication. The meridians comprise "an invisible network that links together all the fundamental substances and organs" of the body (Kaptchuk 1983). The channels are unseen, but they convey nourishment, strength, and communications that unify all the parts of the organism. Acupuncture theory aims to discern the interconnections among substances, organs, and meridians.

The Western biomedical description of the same regulatory phenomena involves the nervous system, hormones, metabolic pathways, and circulation. We have a good understanding of the nervous system and the mechanisms by which signals arise, are propagated, and have their effects. We also know that the vascular system carries nourishment and regulatory messages, hormones, into every nook and cranny of the body. The messages regulate well-understood metabolic systems and cellular activities. The relations among the neuronal, hormonal, and metabolic systems have been intensively studied. This information goes a long way toward explaining how the parts of the organism are unified and how they accomplish what they do.

WHY LOOK FOR OTHER COMMUNICATION SYSTEMS?

Why, then, should we look for another energetic and communication system as proposed in the theory of acupuncture? If such a system does, in fact, exist, why has it not been discovered by Western physiologists? If it exists, what are its functions in relation to orthodox regulatory systems, how does it carry out these functions, and how is it formed? The following sections will begin to examine these questions.

We shall present four reasons, from the point of view of Western biomedicine, why it might be useful to search for additional communication systems in the body. The first reason is that there are still some major unsolved problems in regulatory biology. The second reason has to do with the conservative nature of the evolution of regulations. The third reason concerns cells and subcellular components that are not under direct influence of the nervous system but that must nevertheless have some way of being in communication with the rest of the body. The fourth reason has already been mentioned: there is a solid-state physical basis for additional communication systems in organisms.

So far, these communication systems have not been fitted into Western biomedicine. When this happens, it will advance our understanding of both Western and Eastern approaches and even help us solve problems that are beyond the present scope of either method alone.

Unsolved regulatory problems

Our knowledge of how unity is achieved and maintained in organisms, in healthy and in diseased states, is incomplete. We have a detailed understanding of some of the communicating systems in the body; others remain to be discovered. Although we can take some pride in the progress we have made, we do not really know if our present understanding is nearly complete or if we have just scratched the surface.

In the following sections we will see some indications of communicating systems that have a sound physical basis but that have not been explored in Western medicine to see how they function, what they accomplish, and how they relate to the systems we already know about. One challenge is to see if any of these communication systems correspond to phenomena that have been described from the subjective point of view by practitioners of various alternative approaches such as acupuncture. Following are some specific examples of vital regulations that still are difficult to understand.

Mysteries of growth, development, and maintenance of form. The integrations among metabolic growth processes in the embryo exist before nerves and hormones are formed (Adolph 1957). How is this accomplished? We know much about the structure of the genetic material (deoxyribonucleic acid [DNA]) and how the components of the living body (proteins) are assembled (polymerized) from their basic building blocks (amino acids). We remain largely in the dark as to the blueprint that directs the assembly of the genetically determined parts into a whole organism. We are unsure of what maintains the overall form of the organism in the midst of a constant flux of chemical constituents, with tissue components being continually replaced.

The mysteries of wound healing. The healing of an injury anywhere in the body can involve the coordinated participation of tissues that are some distance away from the site of injury. We do not fully understand how this is accomplished.

The mystery of regeneration. Related to all of the problems just listed is the issue of regeneration of limbs and organs. Salamanders can regenerate amputated limbs and damaged organs; frogs and higher forms cannot. Why humans are unable to accomplish what salamanders can do is perhaps the regulatory question with the most profound clinical significance of all, because its solution would take us from prosthesis to biological replacement of limbs and organs. This would be the greatest of medical revolutions and is a major inspiration and challenge for medical research.

Whole-system mysteries. The common thread running through the unsolved problems just listed, and others that may concern us, is how unity is achieved and maintained in biological systems. We are limited, in part, because of our reductionist approach, which focuses on one component at a time. Biochemistry, for example, is largely based on the study of fractions of tissues that have undergone homogenization and centrifugation. In the process of isolating and purifying an interesting component for ease of study, the cellular and tissue environment of that component is destroyed.

Today many of our major health care problems are whole-system disorders and involve regulatory mechanisms that are incompletely understood. Acupuncture theory relates to whole-system regulations and therefore could provide Western biomedicine with extremely useful insights and perspectives.

Evolution of integrations

A second reason for looking for communications beyond the nervous and circulatory systems concerns the evolution of regulations. Defense, repair, and recognition of self and nonself are regulations that take place in the simplest organisms, including protozoa and sponges.

An ability to respond to environmental stimuli developed long before there were nerves, hormones, and circulatory systems, or even tissues and organs. Evolution of more complex organisms required the development of more intricate regulatory and communication mechanisms. There is good evidence, though, that the original control processes were never abandoned and replaced. Mechanisms that worked extremely well for bacteria and protozoa are retained by modern mammalian cells. These mechanisms may have continued to evolve and become more sophisticated, or they may have already been honed to virtual perfection. What is clear, though, is that they now coexist with the more modern systems that arose from the older systems. These modern developments are the nervous and endocrine systems that are a major focus of modern biomedicine.

Rasmussen surveyed cellular regulations in living systems and reached some important conclusions. He asked whether the primitive control devices developed by simple organisms were replaced as more sophisticated ones evolved, or whether the primitive solutions to regulatory problems were "of such simple elegance and evolutionary adaptability that they survived through the millennia by being modified, embossed, and adapted to new functions to meet new evolutionary demands but retaining throughout their basic properties" (Rasmussen 1981).

Rasmussen's careful survey of regulations supports the latter. The slime mold and the liver cell use similar control elements. Receptor proteins in the amoeba and the mammalian brain are comparable in structure. White blood cells seek out and destroy invading bacteria with the same efficiency that an amoeba uses to obtain nourishment. Evolution has been conservative.

Neurobiology, endocrinology, and biochemistry have been so successful that we have been lulled into believing that we do not need to search for other possible regulatory mechanisms. We shall see that this situation is changing. There are good reasons to suspect that acupuncture and other unconventional approaches may be mediated by vital and evolutionarily ancient regulatory systems. Some of these undoubtedly are based on the cooperative solid-state properties of molecules and molecular arrays.

Local and global communications

A third reason for looking beyond conventional mechanisms is that the nervous system, although quite pervasive, does not reach into every nook and cranny of the organism. A substantial number of our living cells and tissues lie outside the sphere of direct influence of the nervous system. Hormones carried by the blood eventually can find their way to, and influence, every cell in the body, but this process is far from instantaneous. It relies on diffusion, an inexorable but extremely slow and random process by which a molecule involved in delivering an important message eventually may, by chance, bump into the appropriate receptor on a cell located far away from the signal source.

Total physiological integration implies rapid communication with every cell, organelle, and molecule, rapid responses, and rapid feedbacks.

Emerging solid-state concepts of regulation

A physical basis for high-speed submolecular communications is emerging from studies of cell and tissue structure and from studies of the cooperative synergistic solid-state properties of biomolecules. Research in this direction may provide a scientific basis for some of the phenomena that take place in the practice of acupuncture, the martial arts, complementary and conventional medicine, and cutting-edge athletic and artistic performance. This is really the theme of this book.

Designing and constructing a communication system. Total physiological integration implies total interconnection, which means, specifically, that virtually every part knows precisely, or at least sufficiently, what every other part is doing. The body must possess systems capable of rapid and subtle regulations, both at the local level and at the level of the organism as a whole. The ability of an impaired or deficient organism to adapt, within reason, to missing or damaged parts shows that there are redundant pathways and adjusting mechanisms that can maintain wholeness even when primary information pathways are compromised.

Let us suppose that we have been given the task of designing the ideal communication system for integrating the parts and processes of the human body. Let us also suppose that this system must extend into the regions within the body that are *not* in direct synaptic contact with the nervous system and that are not directly adjacent to blood vessels. Preferably the system would not end at the cell surface but would penetrate into the cytoplasm of every cell in the body and connect to the various organelles, including the cell nucleus and its remarkable contents. To aid our hypothetical task, we have at our disposal a wide range of natural materials with useful solid-state properties and millions of years of evolutionary selection to test different combinations for their effectiveness, keeping only the assemblies that perform extremely well.

What properties would we design into such a system? What electronic, photonic, and quantum mechanical tricks might we use to miniaturize this system and make it simple, automatic, and virtually flawless in operation? How could we assemble the components of this system, create many copies of the key circuit elements, and position them at the correct places within the organism? What programs or algorithms would we incorporate into the system to direct its operations?

It would be ideal to use the living matrix, with all of its extensions into the cells and organelles, as a bioelectronic, bioinformational, cybernetic, bioenergetic, synergetic system. To the major extracellular protein component of this system, collagen, we might add suitable molecules that are readily manufactured by cells. Following a basic design and construction principle that is used throughout nature, we could incorporate molecules that can self-assemble into intricate microstructures. These molecules, along with collagen, then could form the biological equivalents of integrated circuits and microprocessor arrays extending virtually throughout the domain of the organism, extending to all tissues, cells, and even the interiors of organelles. The crystalline nature of the living system is ideal for extremely rapid flow of energy and information, signal processing, and localized memory functions.

A distributed network. In the terminology of information theory, such a system is called a *distributed network*. The living organism has certain attributes that are similar to those found in a network of computers working on a common problem or activity. In both situations we are dealing with a number of physically separate microprocessors that are linked by communication channels so they can work together. Each node in the system has certain processing and memory capabilities and is programmed to carry out its part of the overall process while it receives and sends control messages over the communication network.

Node algorithms are the programs that govern local activities, and the assembly of all of the algorithms in the system is called a *distributed network protocol* (Segall 1983). Some of the problems such protocols must deal with, in a computer network and in a living system, include the routing of information, network synchronization, shortest path determinations, testing of connectivity, routing-path updating, and common channel coordination. As in an ensemble of computers, living networks must be able to accommodate the failure of individual or groups of nodes. In the living system this may be caused by damage, disease, or temporary or permanent loss of a part of the system. "The protocols must be able to work under arbitrarily changing network topology" (Segall 1983).

The general approach used to develop and validate distributed network protocols for computer systems may apply to networks in living systems:

- Each link is bidirectional.
- All messages are control messages.

- Link protocols ensure that each message transmitted from a node arrives correctly at another node and that messages either arrive in the same order as they were sent or that the message sequences are kept track of.

- Additional protocols exist for dealing with channel errors. These are detection/retransmission or correction algorithms.

- Messages received at a node are identified and tagged as to source.

- Messages are placed in a line or queue to await further processing.

- The processor takes the message at the head of the queue, processes it, and discards the message when processing is complete.

- Each node has an identification and knows the identity of all of the other nodes actually or potentially in the network.

- Each node knows about adjacent links, but it does not necessarily know the topology of the network as a whole or which nodes are currently functioning.

- A protocol can be started by any node or by several nodes when a "start" message is received.

- An assumption sometimes used in computer networks is that once a node has begun to operate its algorithm, it cannot receive a "start" message. Modern systems are able to store a variety of "start" messages and act upon them at a later time.

The terminology used to describe protocols for computer networks can give us some realistic insights into how living systems may be able to communicate and function in a coordinated fashion with little error. For example, when a node receives a piece of information that must be transmitted to all other nodes, a simple approach is to flood the network, that is, the node transmits a message to all of its neighbors, and these neighbors send the same message to their neighbors. Protocols for such transfers must consider message priorities; if a message has to be propagated throughout the system as soon as possible, other messages received at the same time must be disregarded or stored for later processing and transmission.

Feedback also is required—the node initiating a transmission to all other nodes must be informed that the signal has been received at all of the other nodes. According to Adolph (1982), "Feedback is an essential part of integration. . . . Information from action is always present." Messages flow through a system in two waves, one from the initiating node into the network and the second from the network back to the node to acknowledge the signal.

Connectivity tests are protocols used to enable a node to determine which nodes are connected to it. *Shortest path determinations* enable communications to proceed through the smallest number of nodes to reach another node. *Routing-path updating protocols* allow communication pathways to change and improve with each cycle of operation, without the formation of unwanted loops that would create redundancies.

Of particular interest to diseased or disordered conditions in living systems are *topological change protocols* that sense and accommodate failures of links or nodes. These protocols inform the network of the new topological situation and reset the system variables accordingly. For example, shortest path determinations must be revised at regular intervals. When a link or node is repaired or recovers, the topological change program must be run again.

There are obvious similarities and differences between a network of computers and a biological network. We noted earlier that link protocols in a computer network keep track of the origins and timing of messages. This requirement arises because information may flow between two points via a variety of different pathways operating at different transmission rates. In living systems, a variety of types of signaling processes, transmission velocities, and routings are present, and operating systems determine the additional information that is represented by the order of message arrival.

In an organism of any appreciable size, the local networks probably will be distinct from long-distance channels. The latter would include elements that maintain signal strength and clarity so that information can be transferred over relatively long distances.

At certain points or nodes along the network, we would require interfaces to handle the essential interactions between long-distance communications and local traffic. In the interest of simplicity of design, construction, and control, it would be desirable to locate these interfaces at the same places where signal strength and clarity are maintained, and where signals are processed to make decisions. Information, by itself, is useless unless it leads to choices between different possible actions, and decisions are of no consequence without some form of energy to convert them into actions. Hence we integrate power distribution functions into our nodal network. Finally, to enable our system to respond quickly to changes in the internal and external environment, we place our system near the surface of the body and install sensory elements.

Figure 10-1, *A*, shows the four major functions we just discussed assembling themselves at a node. We envision this self-assembly to be comparable to other sorts of self-assembly that have been described, such as the assembly of ribosomes (Figure 10-1, *B*), bacteriophages (Figure 10-1, *C*), and collagen fibrils (Figure 10-1, *D*). The meridian channel then assembles itself (Figure 10-1, *E*) in relationship to the node, periodically inserting additional notes at appropriate places. Assembly of meridian components within the living matrix may be comparable to the attachment of enzymes to the cytoskeleton, as discussed earlier in Chapter 8.

Studies of collagen, ribosomes, phage particles, and other structures have revealed some basic principles involved in biological self-assembly (Goel & Thompson 1988):

1. *Subassembly.* Functional units are assembled in stages rather than all at once. Major subcomponents are assembled from a relatively small number of parts, and these subcomponents then organize into larger structures.

2. *Energetics.* Once the pieces are made, self-assembly does not require input of metabolic energy, as it is driven by thermal motions and charge interactions.

3. *Sequential assembly.* Components will assemble in a certain order. One component must be installed before the next can attach. This enables certain proteins that have several possible conformations to be fitted into a structure in such a way that the most useful conformation is selected.

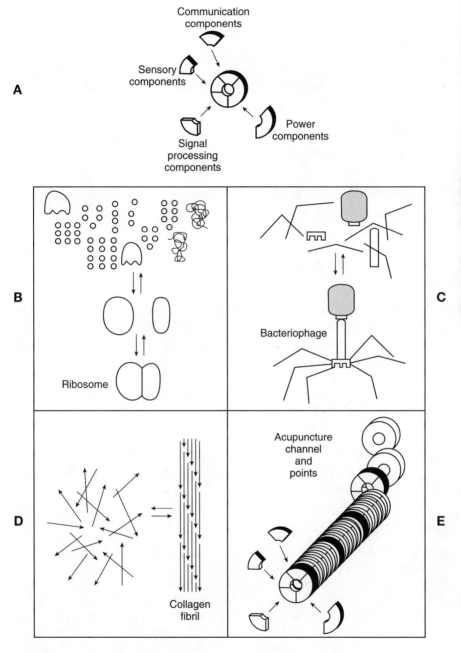

Figure 10-1 A, Self-assembly of major functions at a node in a hypothetical communication system. B, Reversible self-assembly of ribosomes. C, Reversible self-assembly of bacteriophage. D, Self-assembly of collagen molecules into a fibril of connective tissue. E, Self-assembly of an acupuncture channel and points.

All of these principles of self-assembly arise as a natural consequence of cells being able to synthesize certain components. Once these components have been synthesized and delivered to the assembly site, no further input of energy or information from the cell is required. Thermal agitation causes the components to move about; far-field and near-field interactions bring them into alignment and connect them in specific ways according to the charge configuration on the surfaces of the components. Once proximity is achieved, covalent, ionic, and hydrogen bonding will stabilize the structure. The relation of the components to the water matrix will be essential in bringing them together and stabilizing them. Water plays a vital role in the entire process. This point cannot be overemphasized.

Self-assembly

Self-assembly is an extremely important concept in biology that has not received the attention it deserves. Behind each of the self-assembly principles listed above is a set of unanswered questions. For example, we mention thermal motions and charge interactions as providing the energy for assembly. What is the nature of the long-range interactions that attract and align molecules so that they can form more specific bonds with each other?

Fröhlich has developed a sophisticated model to explain the attraction between an enzyme and its substrate. This model also could account for the long-range attraction that orients molecules and brings them closer together during the self-assembly process. The long-range attraction is a consequence of coherent modes of electrical vibrations. A metastable state is achieved in which polar groups are stretched to the point that the molecule as a whole exhibits a very strong electric dipole moment that will interact with similar molecules over a very long range. A consequence of the interaction of an excited molecule with a distant neighbor will be a similar excitation of the second molecule. When the two molecules are excited at the same frequency, their dynamic behavior will be dominated by mutual attraction. This is a synergetic, cooperative, collective, resonant, long-range ordering of molecular behaviors of the sort mentioned in Chapter 9.

Electronic analogies

Having envisioned the assembly of four major subsystems at a node in a communication system, we now can design each subsystem from available components. For example, the long-distance *communication functions* might use operational and other types of amplifiers, attenuators, feedback and relaxation oscillators, pads of various kinds for matching input and output impedances, signal choppers and restorers, rectifiers, and delay lines. For definitions of the various electronic components mentioned here as examples, consult a dictionary or encyclopedia of electronics (Graf 1970; Mandl 1966).

These are presented only as examples of the kinds of electronic functions that might be self-assembled and integrated at a node. *Signal processing* **would involve a different set of components, perhaps including various sorts of filters, voltage or**

frequency doublers and dividers, bridges and gating circuits, signal limiters and differentiators, read-only and random-access types of memory, logic gates, and pads. The various *sensory functions* at a node might involve elements designed as transducers, detectors, thermistors, converters, photoconductive cells, discriminators, variable frequency oscillators, photoelectric and photovoltaic cells, antennas, and clippers. Finally, *power distribution functions* also might be integrated at a node. These might include the biological equivalents of circuit breakers, drivers, tanks, doublers and dividers, rectifiers, power amplifiers, magnetic and photonic amplifiers, and switches.

The hypothetical communication and integration system we have been asked to design is beginning to take on a specific form. The assembly of some of these components is shown in Figure 10-2.

We have used electronic components as examples in this discussion, although nature probably does not restrict itself to conventional electronics. Computer designers realize that the speed of signal propagation and processing is limited by the time required for electrons to flow through wires from one component to another. Modern technology transcends such problems by replacing electricity with light and sound,

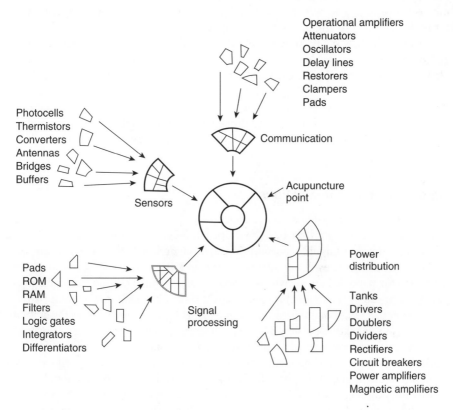

Figure 10-2 Hypothesis for self-assembly of electronic components into different functional elements at a node in a communication system.

which often can be more rapidly manipulated in smaller spaces. In spite of these advances, our technology is still far from creating the level of microminiaturization and sophisticated control present in living tissues.

Life does not discriminate in capitalizing on physical laws. No physical, chemical, or quantum technologies are "off limits" for evolutionary testing and incorporation into the living matrix. For example, if integrated electrooptic, magnetooptic, or acoustooptic circuitry is useful in biological communications, sophisticated versions will be found in nature. In years to come, solid-state biophysics and quantum biology will continue to emerge as productive tools for the study of regulatory biology. We probably will find the biophysical correlates of quantum wires, quantum dots, optical resonators, semiconductor diodes, quantum wells, and free-electron lasers (Saleh & Teich 1991; Weisbuch & Vinter 1991; Yariv 1989). Such systems could have important technological applications because of their high speed and capacity, low energy and low noise operation, coherence, ease of coding and processing, immunity from electromagnetic interference, and low toxicity. Moreover, the ability of the components to assemble themselves is a distinct manufacturing advantage.

Some scientists have raised the possibility that living systems may use superconducting and plasma phenomena (Little 1964; Sedlak 1979). To our knowledge, this has not been thoroughly demonstrated, but it is a possibility worthy of further exploration for both its biological and technological implications.

It would be challenging to determine the nature of the communication and integration systems in organisms and the details of how they are manufactured, assembled, and interdigitated with each other to establish both global and local unity of function. Meeting this challenge would be rewarding, because it would tell us many things about how the body operates and how an integrated circuit capable of handling many different functions can be miniaturized and self-assembled from basic biological building blocks. Knowledge of such a system would have many biomedical and technological implications.

Ontogeny of communications

The speculations and analogies of the previous section give some idea of how a communication system might have been developed and perfected during a long period of evolution and how it might be assembled from biomaterials. How might a system such as this arise during the evolution, development, and maturation of a simple organism before the formation of the nervous and hormonal systems? Do sponges, for example, have the equivalent of acupuncture meridians? Presumably acupuncture meridians do not form in the adult but are present in some form at early stages, even in the embryo.

It has been stated that the flow of energy through a system tends to organize the system, and this may apply equally to the formation of solid-state communication networks in organisms. If an energetic and informational network forms by self-assembly of quantum electronic devices, it is likely that the assembly process will be influenced by the flows of energy during ontogeny. Jaffe (1982) studied electrical current flows through developing systems, beginning with very early stages. The developing oocyte has patternless stages that give way to prepatterns and then to visible patterns. The prepatterns arise from some self-organizing mechanisms and "become ever more difficult but never impossible to rearrange."

There are a number of ideas about the "self-organizing mechanisms" that operate during development. Jaffe's research over the years has shown how electrical currents may contribute to the global self-organization of structures during development. A sophisticated tool in this research is a highly sensitive vibrating probe system that can be used to measure minute currents and voltages at the surfaces of cells and organisms. The vibrating probe has a 10- to 30-mm diameter platinum ball at its tip. The tip vibrates at a rate of several hundred Hertz between two points in the medium and detects the tiny voltage differences between those points as a result of current flows. With this system the current density can be determined in microscopic regions near the surfaces of cells and tissues (Jaffe & Nuccitelli 1974).

Because of the polar nature of proteins, other macromolecules, and water, they will segregate and align along current gradients by self-electrophoresis and electroosmosis. Currents of 1 to 100 mA/cm^2 are common throughout development in a variety of species. Fucia eggs, for example, have a substantial current, carried by calcium ions, which plays a key role in establishing the pattern of future development. Jaffe's review of developmental currents and gradients relates these phenomena to other possible mechanisms of pattern formation in a variety of systems.

At the beginning of this chapter we asked, if acupuncture meridians exist, why have they not been discovered by Western scientists. From the information and ideas presented so far, there is an obvious explanation. The meridian system may be composed of ordinary connective tissue and cytoskeletal elements and therefore is not identifiable as an anatomically separate system. The structures comprising the acupuncture communication network may be laid down along lines of current flow of the sort described by Jaffe in the egg and early embryo. The properties of the meridians may be a consequence of invisible submolecular solid-state cooperative phenomena, a consequence of energetic properties rather than observable macroscopic structures.

To summarize, we suggest that neither the Jing nor the Luo, nor all of their countless branches (see to Figure 8-7, are mystical entities but instead are material pathways located within the connective tissue/cytoskeletal fabric. At the microscopic level, this branching network reaches to the surfaces of cells and extends across those surfaces, where it is continuous with the cytoplasmic matrix, nuclear matrix, mitochondrial matrix, and the interiors of other organelles. The long-sought substance of the meridians consists of an intricate set of protein and other molecules, as well as an adhering film of water.

LOW-RESISTANCE PATHWAYS

We have begun to develop a possible correspondence between acupuncture and modern biomedical perspectives on the nature of the meridian system in relation to energetics and information flow in the body. In view of the likely solid-state properties of the meridians, as revealed by their shininess and ability to act as low-resistance pathways (Bergsmann & Woolley-Hart 1973; Reichmanis, Marino & Becker 1975, 1976), it

133

is not surprising that there is a field of electroacupuncture, in which small currents are passed along the acupuncture meridians to balance the system. Studies of Tiller (1982, 1987), Tiller and Cook (1974), Motoyama (1975, 1980), Voll et al. (1993), and others (reviewed by Scott-Mumby 1999) led to the development of sophisticated devices for diagnosis and treatment using electrical and electronic properties of the meridians.

From the information summarized here, it seems likely that the discovery of meridians as low-resistance pathways to the flow of electricity is but a first hint of the vital and dynamic electronic properties of the living meridian network. The flow of electricity is the veritable "tip of the iceberg" in relation to biological electronics. Meridian components may superficially resemble resistors, but more sophisticated functions will emerge when they are studied as transistors and integrated circuits. They may be electrical conduits, but they will not be fully understood until they are studied as electronic, photonic, and acoustic microprocessors.

It is not surprising that some acupuncturists use magnets and lasers to influence the meridians, because these sources of energy have specific and well-documented effects on semiconductor systems. One of the well-known phenomena that could be related to magnet therapy is the Hall effect, which was discovered in 1879. A magnetic field at right angles to a semiconductor will exert a deflecting force that causes moving charges to drift to one side or another of the main axis of current flow. This drift will produce a transverse Hall voltage that can be used to determine the nature of the charge carriers.

Given the acoustooptic properties of certain semiconductors, it also is not surprising to find that some practitioners think sounds can enhance their treatments.

Specifically, we suggest that the living matrix/acupuncture meridian system may be a distributed communication and energetic network. The meridians may be channels for long-distance physiological communications, and the acupuncture points may represent nodes in the system that are responsible for local distribution of signals arriving from other parts of the body, for inserting local news into the global network, for maintaining the strength and clarity of signals, for processing signals into decisions, and for powering those decisions into certain kinds of actions. Hence we might find at the acupoints switches, amplifiers, couplers, filters, gates, and even memory elements. These are properties that could be looked for, so this is an explicit statement of an hypothesis to be tested or refuted.

The recent literature indicates that parts of this hypothesis were thought of by a number of investigators looking at specific components of the living matrix. For example, Hameroff et al. (Hameroff 1987; Hameroff, Rasmussen & Mansson 1988) have cited a dozen groups that published models of microtubular information processing.

There is a growing realization that the cytoskeleton contains the cell's "nervous system" and that information is processed by the cytoskeleton. Microtubules are polymers of globular protein subunits (tubulin) that pack together into a spiral array that forms a tube. Each microtubule subunit can exist in two or more conformations; therefore, the system can store information. Microtubules are electrets, oriented assemblies of dipoles, and are predicted to have piezoelectric properties. We shall discuss cytoskeletal memory in more detail in Chapter 22.

Another place to look for subnodes in an information processing system is at the cell surface. Some years ago we reviewed the literature on the interface between the cell and the extracellular fabric and listed the components that were being studied. At the time, fibronectin had been discovered to have a key role in linking the cell interior and exterior. Since then, additional components of this system have been discovered, including vinculin and talin, which are located inside of the cell membrane (Burridge et al. 1987). Additional information on the various kinds of cell surface molecules can be found in Chapter 8.

To show how acupuncture points might be assembled, refer again to Figure 10-1, *E*. Here each point is envisioned as an integrated microcircuit with a number of functions as suggested in the hypothesis. Components of the channel also are shown as part of the assembly. Not shown are lateral connections (Luo) and their branches that shuttle information and energy back and forth between the primary channels and the surrounding tissues. The acupuncture meridian proper is hollow to accommodate the flow of fluids within it, as has been indicated by a number of studies using radioactive tracers (Kovacs et al. 1991).

SOME CONCLUSIONS

Techniques used in acupuncture, in the other so-called *alternative approaches to the human body* and in Western medicine, including surgery, have one thing in common. All involve interactions with the structural fabric of the body, which we have termed *the living matrix*. Accumulating evidence indicates that this remarkable and dynamic living material has a variety of solid-state cooperative properties that may explain phenomena that previously were regarded as inexplicable or anomalous.

Chapters 8 through 11 summarize some of the concepts that exist in the biophysical literature that can be used to develop testable and refutable hypotheses regarding the nature of the acupuncture meridian system. Some of these concepts, such as the dynamic properties of the cellular and extracellular matrices, self-assembly, and electron transport in chloroplasts and mitochondria, are well accepted or are becoming accepted in the mainstream of biological research. Others, such as the solid-state electronic, photonic, protonic, and quantum mechanical properties of the living fabric, are at the cutting edge of current biophysics and have not become incorporated into what we might term *orthodox* biology. Progress in this direction has been slowed by the experimental and theoretical complexity introduced by the high water content of biological systems. Water is the most common biological material. Water is a small molecule, but nevertheless it is a substance of great complexity and remarkable properties. The difficulties posed by water are now beginning to be surmounted by the development of suitable analytical techniques and by more sophisticated and detailed understandings of the intimate role water plays in all biological structures and functions. The subtle properties of water may hold the key to many of the mysteries of biology and medicine.

What is most significant in this discussion is that there are unsolved problems regarding how unity of structure and function is achieved and maintained within the organism, and how the cells and tissues within the organism communicate with one another. A prediction is that biophysical study of acupuncture and other energetic therapies will lead to a deeper understanding of regulatory biology, particularly with regard to the medically important processes involving defense, recognition of self and nonself, and tissue repair. These are evolutionarily ancient regulatory mechanisms that are "of such simple elegance and evolutionary adaptability that they survived through the millennia by being modified, embossed, and adapted to meet new functions to meet new evolutionary demands but retaining throughout their basic properties" (Rasmussen 1981). We may discover that the deceptively simple elegance of acupuncture results from its ability to stimulate these evolutionarily ancient regulatory processes, which are not a primary focus of modern biomedical research.

Among other beneficial outcomes, biophysical research could lead to answers to profound problems that have not been solved by either Eastern or Western medicine alone. One of the most important of these is the regeneration of damaged or missing tissues.

Biophysical inquiry also has the potential to contribute to physics. Where in the physics laboratory can one see in operation the perfectly coordinated working together of a wide range of physical principles, all directed toward a cooperative action? Living systems are ideally suited to handle and interconvert virtually all of the forms of energy that are known to science or that may be discovered in the future. To accomplish these conversions, living systems undoubtedly use quantum mechanical tricks that are as yet undreamed of by the physicists who study inanimate matter.

The considerations raised here explain why, if acupuncture meridians exist, they have not been firmly established by Western scientists. The meridian system may be composed of ordinary connective tissue and cytoskeletal elements that can be found everywhere in the body and therefore cannot be distinguished as an anatomically separate system. The properties of the meridian system may arise as a consequence of invisible submolecular solid-state cooperative phenomena, or quantum wave phenomena, rather than from observable macroscopic anatomical structures. For the most part, energy flows are invisible, and the main axes of energy flow in soft tissues may not have an anatomically visible counterpart. Perhaps the best way of identifying the meridians is by their energetic properties, as acupuncturists have been doing for several thousand years.

Biophysical study of acupuncture is in its infancy. When one embarks on an exploration of a new scientific territory, there is an initial phase that scientists affectionately refer to as *stamp collecting*. This is the process of gathering together all of the relevant information about the subject, the phenomena that are to be examined, the questions that are to be asked, and the likely places to look for answers. This book is the beginning of a stamp collection. Its goal will have been achieved if a few are stimulated to think about the subject in new and useful ways.

MULTIPLE WORKING HYPOTHESES

Recall Chamberlin's method of multiple working hypotheses outlined in the Prologue. A goal of our inquiry into both acupuncture and Western biomedicine is to solve problems that have been resistant to conventional ways of thinking, to stimulate independent and creative insight, to discover new truths, or to make new and useful combinations of existing information. Chamberlin urged the method of multiple working hypotheses to "distribute the effort and divide the affections." The method involves bringing up every rational explanation of a phenomenon and developing every tenable hypothesis about it, as impartially as possible.

From Chamberlin's perspective, this chapter has done no more than add new members to an interesting family of ideas about the nature of the acupuncture meridians. I am aware that others are convinced that the meridians correspond to nerves (Mann 1973), the lymphatics, or the spaces along the vascular nerve bundles in the limbs (Kovacs et al. 1991). Others firmly believe that none of these explanations come even close to appreciating the reality of the phenomenon of meridians. Perhaps time will show that each of these perspectives holds a clue to the puzzle we are working on.

REFERENCES

Adolph, E.F. 1957. 'Ontogeny of physiological regulations in the rat', *Quarterly Reviews of Biology,* vol. 32, pp. 89-137.

Adolph, E.F. 1982, 'Physiological integrations in action', a supplement to *The Physiologist,* vol. 25, April.

Bergsmann, O. & Woolley-Hart, A. 1973, 'Differences in electrical skin conductivity between acupuncture points and adjacent skin areas', *American Journal of Acupuncture,* vol. 1, pp. 27-32.

Burridge, K., Beckerle, M., Croall, D., & Horwitz, A. 1987, 'A transmembrane link between the extracellular matrix and the cytoskeleton', in ed. C. Waymouth, Molecular Mechanisms in the Regulation of Cell Behavior, *Modern Cell Biology,* vol. 5, pp. 147-149.

Davidson, J.R.T. 2002, 'Hypericum depression trial study group. Effect of *Hypericum perforatum* (St. John's wort) in major depressive disorder: A randomized, controlled trial', *JAMA,* vol. 287, pp. 1807-1814.

Deadman, P., Al-Khafaji, M. & Baker, K. 1998, *A Manual of Acupuncture,* Journal of Chinese Medical Publications, East Sussex, UK, p. 11.

Goel, N.S. & Thompson, R.L. 1988, 'Movable finite automata (MFA): A new tool for computer modeling of living systems', in ed. C. Langton, *Artificial Life, SFI Studies in the Sciences of Complexity,* vol. VI, Addison-Wesley, Redwood City, CA, pp. 317-340.

Graf, R.F. 1970, *Modern Dictionary of Electronics,* Howard W. Sams, Indianapolis, IN.

Hameroff, S.R. 1987, *Ultimate Computing: Biomolecular Consciousness and Nanotechnology,* Elsevier, North Holland.

Hameroff, S., Rasmussen, S. & Mansson, B. 1988, 'Molecular automata in microtubules: Basic computational logic of the living state?', in ed. C. Langton, *Artificial Life, SFI Studies in the Sciences of Complexity,* vol. VI, Addison-Wesley, Redwood City, CA, pp. 521-553.

Jaffe, L.F. 1982, 'Developmental currents, voltages, and gradients', in *Developmental Order: Its Origin and Regulation,* eds S. Subtelny & P.B. Green, Alan R. Liss, New York, pp. 183-215.

Jaffe, L.F. & Nuccitelli, R. 1974, 'An ultrasensitive vibrating probe for measuring steady extracellular currents', *Journal of Cell Biology,* vol. 63, pp. 614-628.

Kaptchuk, T.J. 1983, *The Web That Has No Weaver,* Congdon and Weed, New York, p. 77.

Kovacs, M., Gotzens, A., Garcia, F., Garcia, N., Mufraggi, D., Prandi, D., Setoain, J. & San Roman, F. 1991, 'Experimental study on radioactive pathways of hypodermically injected technetium-99m', *Journal of Nuclear Medicine,* vol. 33, pp. 403-407.

Little, W.A. 1964, 'Possibility of synthesizing an organic superconductor', *Physical Review,* vol. 134, pp. A1416- A1424.

Mandl, M. 1966, *Directory of Electronic Circuits: With a Glossary of Terms,* Prentice-Hall, Inc, Englewood Cliffs, NJ.

Mann, F. 1973, *Acupuncture. The Ancient Chinese Art of Healing and How It Works Scientifically,* Vantage Books, New York, pp. 5-26.

Motoyama, H. 1975, *How to Measure and Diagnose the Functions of Meridians and Corresponding Internal Organs,* The Institute for Religious Psychology, Tokyo, Japan.

Motoyama, H. 1980, 'Electrophysiological and preliminary biochemical studies of skin properties in relation to the acupuncture meridian', *I.A.R.P. Research for Religion and Parapsychology,* vol. 6, pp. 1-36.

Rasmussen, H. 1981, *Calcium and cAMP as Synarchic Messengers,* John Wiley & Sons, New York.

Reichmanis, M., Marino, A.A. & Becker, R.O. 1975, 'Electrical correlates of acupuncture points', *IEEE Transactions on Biomedical Engineering,* Nov., pp. 533-535.

Reichmanis, M., Marino, A.A. & Becker, R.O. 1976, 'D.C. skin conductance variation at acupuncture loci', *American Journal of Chinese Medicine,* vol. 4, pp. 69-72.

Saleh, B.E.A. & Teich, M.C. 1991, *Fundamentals of Photonics,* Wiley, New York.

Scott-Mumby, K. 1999, *Virtual Medicine,* Thorsons/Harper Collins, London, UK.

Sedlak, W. 1979, *Bioelektronika, 1967-1977,* Instytut Wydawniczy Pax, Warszawa.

Segall, A. 1983, 'Distributed network protocols', *IEEE Transactions on Information Theory,* vol. IT-29, pp. 23-35.

Tiller, W.A. 1982, 'On the explanation of electrodermal diagnostic and treatment instruments. Part I. The electrical behavior of human skin', *Journal of Holistic Medicine,* vol. 4, pp. 105-127.

Tiller, W.A. 1987, 'What do electrodermal diagnostic acupuncture instruments really measure?', *American Journal of Acupuncture,* vol. 15, pp. 15-23.

Tiller, W.A. & Cook, W. 1974, Psychoenergetic field studies using a biomechanical transducer. Part I: Basics, in *Proceedings of the A.R.E. Medical Symposium on New Horizons in Healing,* Phoenix Arizona, January, 1974.

Voll, R., Sarkisyanz, H. 1983, *The 850 EAV Measurement Points of the Meridians and Vessels Including Secondary Vessels.* Medizinisch Literarische Verlagsgesellschaft, Uelzen, Germany.

Weisbuch, C. & Vinter, B. 1991, *Quantum Semiconductor Structures: Fundamentals and Applications,* Academic Press, Boston.

Wessels, N.K., Spooner, B.S., Ash, J.F., Bradley, M.O., Ludvena, M.A., Taylor, E.L., Yariv, A. 1989, *Quantum Electronics,* 3rd ed. John Wiley, New York.

11 More clues from acupuncture

Injury has . . . left its imprint in our tissues, even in our cells, in the form of built-in, life-saving reactions, ready to be triggered at an instant's notice. And myriads of wounds have become stepping stones to one of man's greatest creations–the art of healing.   Guido Majno (1975)

One of the key issues facing us is why there seem to be such impenetrable intellectual barriers between Eastern and Western approaches to medicine. We are beginning to discern the nature of these barriers and to see how they can be bridged through the development and testing of logical hypotheses. A logical and testable hypothesis connecting classic acupuncture theory and modern biomedicine obviously would have major implications for world medicine.

For several centuries, the primary focus of Western science has been reductionistic—the study of parts. In contrast, Eastern philosophy for millennia has concerned itself with invisible energetic relations and connections that govern living form and function. Without adequate technology to measure energy fields in and around organisms, termed *Ch'i, Ki,* or *Qi* in East Asian medicine and given a wide range of other names in other traditions, it was easy for biomedicine to dismiss the empirical and intuitive deductions that led to acupuncture theory. There is an aura of mystery and even fear about the invisible forces of nature.

Since biology is the study of living things, it is simultaneously the study of ourselves, making it the most intensely personal of the sciences and one whose philosophy is the most subject to emotionalism and dogma.   BECKER & MARINO (1982)

Means are now available to measure biological energy fields, and an "energy medicine" has been emerging in the West for the past several decades (Williamson 1983; Williamson & Kaufman 1981). The practical methods of manipulating Ch'i, developed by acupuncturists, are providing invaluable clues for the advancement of Western energy medicine. Acupuncturists' manipulations and theory are being formulated in precise scientific language and are being developed into testable hypotheses.

A fundamental connection between superficially different approaches is developing because acupuncture researchers are critically examining parts of the clinical puzzle, and Western scientists are exploring whole-system principles, physiological integration, cybernetics, and information theory. The distinguished physiologist E.F. Adolph has written eloquently about physiological integration (Adolph 1979). In terms of nonlinear mathematical modeling, Western science is recognizing that living matter is composed of multiply-connected systems, as opposed to simply-connected ones (Barnsley 1993). The significance of this development for health care cannot be underestimated, because nonlinear systems are capable of undergoing large-scale changes as a result of minute perturbations.

An exciting development is the description of a common denominator, the living matrix continuum, which accounts for many phenomena in both Eastern and Western approaches to medicine. Every branch of science has contributed to our understanding of the living matrix, but recognition of its systemic continuity has been slow because of the ways biological and biomedical inquiries proceed and the ways we communicate the process and its results.

BIOLOGICAL INQUIRY

When a new molecule or structure or system is discovered, its properties are documented. A functional role is sought. The most obvious function captures the attention of investigators, supporting evidence is gathered, and a suitable descriptive name is given. The new information is incorporated into the literature, it is taught to students, and satisfied investigators move on to new territory. Often this process closes the door to further inquiry. Years or decades or centuries may pass before a system is reinvestigated in terms of entirely different functional roles. When new information or ideas emerge, they usually are resisted because they are perceived as a threat to prevailing subdisciplines, paradigms, or lines of inquiry.

In general, it has been far easier for scientists to relate to mechanical or architectural functions of molecules and tissues. The invisible energetic relations are more subtle and take longer to discern. In spite of this, most of what we have learned about atomic and molecular structure is, in fact, derived from the energetic relations that are studied by spectroscopy. As we have seen, *biomagnetism,* the study of magnetic fields produced by tissues and organs, and *magnetobiology,* the study of the effects of magnetic fields on living things, now are emerging as branches of biomedicine.

In this chapter we give several examples of anatomical systems that are being reexamined for their energetic roles. When connective tissue was first characterized and given its name, it's role as an mechanical framework for the animal body was obvious. Only recently have we begun to consider the connective tissue and fascia as informational and energetic networks that provide a basis for the phenomena of acupuncture and complementary energetic approaches. Fortunately, the original name given to connective tissue is descriptive of its newly found (in Western science) role in conducting energy and information. A historical review has shown that these concepts have been a part of acupuncture theory for millennia (Matsumoto & Birch 1988).

The cytoskeleton was named for its obvious role as a mechanical scaffolding for the cell. Many biologists now refer to the cytoskeleton as "the nervous system of the cell"; however, the name *cytoskeleton* is not a descriptor for such a role. Moreover, the cytoskeleton does not fit with the image most physiologists have of a nervous system.

Microtubules have an established architectural function in determining the shapes of cells. Recently it has been widely recognized that microtubules also are sites of information processing, storage, and recall (memory). The concept that memory is not confined to the nervous system but is spread throughout the various types of cells found everywhere in the body has profound implications for our understanding of the emotional effects of acupuncture and of other energetic approaches, because our individual memories and their emotional contents shape our sense of who we are, as well as what we do, how we do it, and how well we do it. Emotional memories help us quickly eliminate a variety of possible options for any situation, on a moment-to-moment basis (Damasio 1994).

The living matrix is a continuous molecular system that simultaneously conducts energy and information throughout the body, and that regulates growth, form, and wound healing. Every part of the body is a part of this continuous living matrix. It is a system of systems. Memories are stored within this system, and the totality of its operations gives rise to what we refer to as *consciousness* (Oschman & Oschman 1995a, 1995b). This system is accessed by acupuncture and other complementary medical approaches.

WOUND HEALING AS A CONTEXT

Wound healing provides a useful context for our presentation because Western medicine has focused on this process for a long time. For a scholarly historical review, see Majno (1975). For an extensive review of wound healing from the medical perspective, see Marchesi (1985).

Although they do not agree about which methods are most effective, all medical practitioners, from shamans to neurosurgeons, have a common goal—facilitating the repair of injuries. As Majno points out, the wound was the first medical laboratory. Research on treatments to reduce infection and speed healing has gone on for millennia and continues today with the same urgency. For example, healing can be promoted by adding natural growth factors, or genes for those growth factors, directly to a site of injury (Vogt et al. 1994).

Clinicians recognize that it is the body itself that repairs an injury. A goal of research is to determine whether an intervention actually helps—was the patient made better or worse, or was the patient unaffected by the treatment? To answer these questions, we still need to learn more about what flesh is really made of and about how cells and tissues respond to stimulation of all kinds.

Western medical literature indicates that wound healing involves many nonlinear phenomena, with dynamic interactions taking place between local and systemic processes. A wide range of physiological activities are stimulated during wound healing, and all of these must be down-regulated when the healing has been completed. Some activities persist for weeks after injury or stimulation.

Wound healing is interesting from the whole-system perspective because damaged tissue is replaced with tissue of the same kind, rather than with tissue of a different kind. Bones are repaired with bone tissue, skin with skin of the appropriate thickness, mucous membranes with mucous membranes, etc. Hence, wound healing directly involves the elusive "form-giving" forces that biologists have been seeking for centuries.

HYPOTHESES TO EXPLAIN THE EFFECTS OF ACUPUNCTURE NEEDLING

Now we look at a series of hypotheses about the effects of acupuncture needle insertion. First, it is suggested that acupuncture and related methods activate the intricate systems involved in the body's responses to injury and that facilitate tissue repair. The body benefits from this activation, even when no actual injury has taken place. This phenomenon is particularly important for the chronic patient whose disorder may not be detectable by conventional diagnosis. We therefore make the following explicit hypothesis:

The act of puncturing the skin with an acupuncture needle simulates an injury and thereby elicits the local and systemic cascade of regulatory, restorative, repair, and regenerative processes associated with wound healing.

This is not to say that acupuncture produces wounds. It has long been recognized that when a cell's membrane is punctured, torn, or broken, the membrane quickly seals itself so that the cytoplasm does not flow out. In other words, acupuncture is *not* expected to cause permanent damage to cells and tissues.

The resistance of living cells to injury was reported in 1835 by Dujardin (Dujardin 1835). In 1927, Heilbrunn named this response "the surface precipitation reaction" (Heilbrunn 1927). Heilbrunn showed that the response depended upon the presence of calcium ions in the environment. Twenty years later, Heilbrunn and Wiercinski (1947) performed one of the classic experiments in the history of physiology, when they demonstrated that injection of calcium ions into muscle initiates muscle contraction.

There are many parallels between the rupture of a cell and the rupture of a blood vessel. In both cases, the outward flow of fluid is quickly stopped by a process that requires calcium ions. Heilbrunn and others who have carefully studied the behavior of living cells have concluded the following:

There is no sharp line between the response to stimulation and the response to injury. Stimulation and injury produce similar electrical responses in cells and tissues. It is the nature of life to take advantage of injury. The injury reaction is a fundamental reaction of protoplasm. HEILBRUNN 1956

Heilbrunn's observations provide a basis for suggesting that the natural regulatory, restorative, repair, and regenerative mechanisms in the body are activated by the acupuncture needle *as though* the body has been injured and that this initiates a range of beneficial effects. Figure 11-1 shows some of these processes.

Figure 11-1 hardly does justice to the sophistication of the regulatory processes involved in wound healing or to the apparent elegance and specificity of acupuncture in facilitating such processes. Clark (1994) divides wound repair events into inflammation, tissue formation, and tissue remodeling. Needham (1952) describes six morphological events, which can include the massive but controlled proliferation of cells, blood clotting, inflammatory reactions, and wound closure. The latter process involves contraction of modified fibroblasts, called *myofibroblasts,* an event that is physiologically and pharmacologically related to smooth muscle contraction (Gabbiani et al. 1972; Majno et al. 1971).

The splinter hypothesis

We surely are not the first to suggest that inserting an acupuncture needle simulates an injury. Birch (1994) refers to the "splinter hypothesis," that is, an acupuncture needle produces an effect related to that produced when one gets a splinter in the skin. However, the effects of acupuncture are far more specific than those produced by a randomly placed splinter. A second explicit and testable hypothesis:

The skin surface is not uniform. Certain areas are more sensitive to stimuli than others in activating local and systemic activity in the body's regulatory, defense, and repair systems.

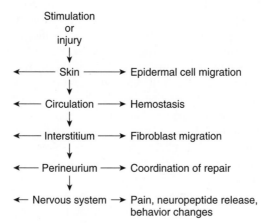

Figure 11-1 Waves of information and activity spread away from the site of a stimulus or injury and initiate local and systemic repair mechanisms.

Acupuncture needling and mechanical stimulation of the living matrix

Recall from the Prologue the method of multiple working hypotheses. The concept presented above, that acupuncture *simulates* an injury and thereby activates repair processes, is one of several hypotheses to join our family of hypotheses. Now we look at another hypothesis related to the experiences of the patient and the practitioner during needling.

Careful research has been done on the physical aspects of needling (Langevin, Churchill & Cipolla 2001). The research is providing important clues about the living material into which the needle is inserted: the continuous connective tissue/cytoskeletal system, collectively called the *living matrix*.

A phenomenon that is reported again and again during acupuncture treatments is called *de qi*. It consists of a characteristic sensation, an ache or heaviness in the surrounding area (as perceived by the patient), and a "needle grasp" perceived by acupuncturist.

The needle grasp is noticed by the acupuncturist as he or she manipulates the needle to obtain maximum effect. The needle is either twisted in different directions or moved up and down. For thousands of years, classic texts have described a sort of tug on the well-placed needle. The effect is particularly noticeable as a tenting that takes place when the needle is withdrawn.

The hypothesis to emerge from research on this effect is that the collagen and elastic fibers in the underlying connective tissue wind and tighten around the needle. The resulting mechanical coupling enables needle manipulation to transmit a mechanical signal through the living matrix to nearby cells such as fibroblasts. This signal activates cellular responses (Langevin, Churchill & Cipolla 2001).

Histological studies show that needle rotation is accompanied by a marked thickening of the subcutaneous connective tissue layer in the area around the needle. Electron microscopic studies of the debris found on needles after insertion, manipulation, and removal reveal elastic and collagen fibers entwined around the needle (Kimura et al. 1992). Quantitative studies using a computer-controlled acupuncture needling instrument revealed that the "pullout force" is significantly greater after needle rotation compared with insertion without rotation (Langevin et al. 2001). A consistent observation is that the torque needed to rotate the needle increased continuously as rotation proceeded. It therefore appears that the needle catches on the connective tissue fibers, which then become wrapped around the needle as it is twisted. Histological studies confirmed that collagen bundles become straighter and more nearly parallel to each other after needle rotation. At the same time, fibroblasts become aligned with the collagen fibers and change shape from a rounded to a spindlelike shape. One minute after rotation, the cells show an increased staining for polymerized filamentous actin, confirming that a cellular response has taken place.

The mechanism by which tension on the living matrix could activate cellular processes has been carefully worked out by Ingber and his colleagues in their studies relating architectural changes to biochemical events (Chicurel, Chen & Ingber 1998). The signaling cascade involved in the effects of acupuncture needling and twisting is as follows:

- Needle insertion
- Needle rotation
- Attachment of collagen and elastin fibers to needle
- Mechanical tugging on the extracellular matrix
- Tugging on focal adhesions on cell surfaces
- Activation of focal adhesion kinase
- Activation of intracellular signaling pathways
- Activation of a variety of cellular activities

The research of others, cited in the papers by Langevin and colleagues, have shown a variety cellular responses to mechanical stimulation, including cell contraction, migration, protein synthesis, and formation of myofibroblasts (Desmouliere & Gabbiani 1994). Responses include synthesis and release of growth factors, cytokines, vasoactive substances, degradative enzymes, and structural matrix molecules.

Hence we now have a plausible scientific explanation for the effects of acupuncture needling on the milieu in which wound healing and tissue repair take place. Beneficial changes in the environment for cellular activities, and stimulation of the cells themselves, are logical outcomes. A single needle can influence large numbers of nearby fibroblasts, and these effects can produce a wave of matrix deformation and cell contraction that spreads a distance through the interstitial connective tissue. Patients often describe a slow spreading of de qi sensation along the meridians during needling.

Meridians as anatomy trains

The ancient acupuncture maps of points and meridians may correspond to tracts, or "trains," within the fascia that are particularly conductive for the waves of deformation and activation just described. The fact that the waves of contraction can be detected by the patient indicates that sensory nerve fibers are activated. This may further enhance the spreading of the effects to tissues distant from the site of needling, including the brain and the various organs.

A picture of the way acupuncture stimulation spreads through the body can be obtained from study of the so-called *anatomy trains* or *myofascial meridians,* many of which follow the classic musculotendon meridians (Myers 2001). As described by Myers and by Deane Juhan's Foreword to Myer's remarkable book, the anatomy trains are linkages of fascia and bone that wind through the body, connecting head to toe and core to periphery. These are the major lines within the living matrix that orchestrate the gravitational and muscular forces involved in all movements.

From years of hands-on work, Myers has distilled an enormous amount of anatomical detail into a simple lattice of tensional bands that are involved in every posture, movement, and emotion and that are also the places where musculoskeletal pain and dysfunction arise.

Taken together, the images Langevin and her colleagues have developed to account for the immediate effects of acupuncture needling, the connection between tissue and cell architecture and biochemistry developed by Ingber's group, and the

global interconnectedness illustrated by Myers give us with a powerful set of conceptual tools for the analysis of human form and function and motion and emotion in health and disease. Of equal importance is the significance of all of this new insight for those who wish to take their bodies into new realms of human activity, whether as art, sport, dance, or occupation. Even vaster possibilities arise when one considers the ways these images can contribute to our inner sense of ourselves as integrated beings capable of miracles of adaptation and self-repair. It is through the experience of this network that we are able to contact the full reach, range, and potential of the kinesthetic intelligence that is our birthright.

The wound that does not heal

The idea that a *simulation* of injury could facilitate tissue repair raises an important question. Why should it be necessary to *simulate* an injury when the original injury should have triggered appropriate repair? We believe the answer to this question is the same as the answer to another question of major importance to health care. Why are there disorders that the Western physician is unable to treat because no pathology can be detected? The patient is sick or in pain, and the doctor cannot detect a problem. Frustrating situations such as this have led many patients and physicians to seek alternatives. An acupuncturist, for example, may examine an individual who has been declared to be perfectly healthy and may detect large imbalances that, if left untreated, will develop into problems later on. An acupuncture treatment may resolve a set of symptoms that are related according to acupuncture theory but are not connected in the Western paradigm.

The dilemma of chronic disorder arises in part because the modern physician has a limited set of parameters for assessing injuries, making treatment choices, and determining when healing is complete. In contrast, the various alternative or complementary practitioners have a variety of unorthodox yet successful ways of looking at the same phenomena.

This distinction has many implications for our health care. The acupuncturist, massage therapist, cranial-sacral therapist, Rolfer, or other complementary practitioner recognizes that tissue that is "normal" from the medical perspective can be compromised in a variety of ways. A tissue may lack a certain quality, tone, or vibrancy, or it may reveal subtle traces of unresolved emotional issues. These features may be revealed in patterns of movement, flexibility, range of motion, texture, color, tone of voice, or other subtle but detectable features.

For example, a broken leg is medically healed when the bone has mended. Many years after the injury, the patient may develop back problems that seem to have no immediate cause. A sensitive bodyworker will notice a slight aberration in the individual's gait, and ask if the patient they ever had a broken leg. The practitioner is observing a remnant of an imbalanced movement pattern that arose because the muscles in the immobilized leg became hypotoned and the muscles in the active leg became hypertoned during the healing process.

When a patient has undergone many traumatic physical and/or emotional experiences, compensations and imbalances can accumulate from injuries that appeared to heal but that nonetheless left their imprints upon the tissues. From the emotional perspective, traumatic injuries leave their marks in the form of repressed memories and the reduced physical or emotional flexibility that results from those memories.

For scientific medicine to explore this arena, it will be necessary to measure biophysical, solid-state, electronic, photonic, and other tissue properties that previously have not been considered relevant. It will be necessary to quantify attributes of tissues that holistic practitioners evaluate qualitatively.

It is unscientific for biomedicine to dismiss as invalid therapies that are based upon phenomena that biomedicine cannot quantify. Instead, the scientific process demands that researchers develop appropriate measuring techniques before making judgments regarding efficacy or mechanism.

Cells move

In terms of physiology, wound healing involves the migrations of cells toward a site of injury. White cells move in to fight infection; epithelial cells and fibroblasts crawl into position to replace damaged tissues. These events are triggered by chemical, electrical, and electronic messages that must be conducted through the tissues. Cells migrating into a wound must be replaced by cell division. In some responses, such as clotting or inflammatory reactions, cells are destroyed at the site of the injury and must be replaced by proliferation of stem cells in distant organs, such as bone marrow or liver. An important factor in initiating, sustaining, and then switching off these processes is the organization of the substrate through which the various chemical messengers, nutrients, toxins, and waste products diffuse and through which cells must crawl to a site of injury and/or infection.

Many years ago biologists studying wound healing recognized two major factors that determined the response of the skin to injury: the "proliferative energy" of the skin cells, and the character of the surface over which the epithelial cells move. If the surface is unfavorable, the epithelial cells advance more slowly because of the resistance encountered (Akaiwa 1919).

Tissue that is medically acceptable, that is, no pathology is detectable with biomedical diagnostic tests, can range from barely alive to fully vibrant and radiant. Some practices, such as yoga, meditation, and martial arts exercises, can bring tissues toward a peak of order and aliveness that has not, until recently, had any quantifiable physiological basis. New evidence is revealing additional biophysical parameters that can be measured to quantify these attributes and to understand the mechanisms involved in phenomena that previously have been considered outside of normal science (Seto et al. 1992).

AN EVOLUTIONARY PERSPECTIVE

In discussing wound healing in relation to acupuncture, it is important to keep in mind that human physiological systems are modifications and elaborations upon evolutionarily ancient mechanisms, developed long ago by primitive animals and plants. Even single-celled organisms are able to repair injuries. A complete amoeba, for example, can regenerate from a fragment consisting of only 1/80th of the original animal, as long as the fragment includes the cell nucleus. A complete planarian can be regenerated from a fragment representing 1/280th of the original animal (Vorontsova & Liosner 1960). Repair and regeneration of damaged parts are readily accomplished by simple animals such as sponges, which lack a recognizable nervous system. The ability of lower animals to regenerate missing appendages or even large portions of their bodies shows that regeneration, the most sophisticated response to major injury, has been lost during the evolution of higher forms.

This perspective is important because in the past too much emphasis has been placed on the role of the nervous system in wound healing. Many of the key physiological mechanisms are far more ancient than nerves. Horridge (1968) presented a theory of the origin of the nervous system in which the transmission of electrical impulses by epithelia, such as the skin, *preceded* the appearance of neurons; likewise, spikes and specialized receptors and effectors preceded axons. Neurosecretion evolved into chemical transmission (Horridge 1968).

Further research confirmed the ability of diverse epithelia and even individual nonneural cells to conduct patterns of electrical activity that closely resemble action potentials (Pallotta & Peres 1989; Rink & Jacob 1989; Rosenberg, Reuss & Glaser 1982). These signals are associated with the actions of hormones, epidermal growth factors, and mitogenic factors, and they play various roles in activating wound healing. They are oscillations of the membrane potential caused by fluctuating calcium flows across the cell surface. Cells in the digestive tract can be depolarized, and a wave of excitation propagates from the site of the stimulus. Frog skin will conduct an action potential. The distinction between so-called *excitable* and *nonexcitable cells* is qualitative rather than quantitative. These points are particularly relevant to R.O. Becker's work on the current of injury developed in perineural tissues, to be discussed later.

Injury has left its imprint in our cells and tissues in the form of a multitude of lifesaving reactions. The fossil record indicates that our earliest ancestors did violence to each other (Ardrey 1967; Dart 1949, 1957). This activity shaped the evolution of a neuromuscular system that was able to manipulate weapons and an autonomic nervous system prepared for fight or flight. The martial arts and the need to heal injuries from battle cooperated to shape our tissues and their reactions to injury.

Now we explore various systems involved in the healing of wounds that are likely to be influenced by the insertion of an acupuncture needle. We begin at the body surface and progress inward.

Wound healing—the skin

We begin to consider wound healing by focusing on events in the epidermis. There is evidence that the skin initiates electrical and electronic processes that mediate local and systemic effects of acupuncture and electroacupuncture.

Du Bois-Reymond discovered that currents are produced by small epidermal wounds in human fingers immersed in saline. This was confirmed by Herlitzka (1910). In 1980, Illingworth and Barker (1980) reported that currents are produced by the stumps of accidentally amputated human fingers immersed in saline.

In 1982, Barker, Jaffe, and Vanable (1982) reported a detailed electrophysiological study of the properties of the skin of the guinea pig and human. The guinea pig was studied because it has regions (the glabrous or hairless epidermis) that are free of hair and glands. Guinea pig skin has a battery that is comparable in power to that of frog skin. In frogs, the skin is an osmoregulatory organ, absorbing salts from the surrounding pond water. In mammals the skin battery is thought to function in epidermal wound healing.

The results of the 1982 study by Barker et al. indicate that the skin battery is located in the deeper living layer of the epidermis (stratum germinativum) rather than in the superficial dead layers (stratum corneum). The resting potential of the skin is about 20 to 200 millivolts, with the inside positive.

Barker et al. suggest that the fields in the skin help guide the movements of cells that close the wound, a process known as *reepithelialization,* which restores the epithelial barrier function. That cellular migrations take place has been known for a long time (Peters 1885), and dynamic aspects of these movements were beautifully documented by Lash (1955). Lash injected individual skin cells with carmine granules. Injected cells remained viable and visible. Using a microscope, Lash was able to follow migrations of injected cells during wound closure. He found that there was a wave of mobilization of epidermal cells, which detach from the underlying basement lamina and form a sheet that migrates toward the center of the wound. This migration ceases when the wound is closed.

Much is now known about the cellular mechanisms involved in the migration of epidermal and other cell types. For a recent summary, see Stossel (1994). Responsible for cell crawling are reversible changes in the gel state of the cytoskeletons and reversible attachments to the underlying substrate.

Lash found that the wave of activation of epidermal cells begins near the wound border and spreads into the surrounding tissue at a rate of about 0.4 mm per hour. Barker et al. suggested that the cells migrate in response to the steady lateral fields set up by the skin battery in the region around the wound. These fields decline in strength by about threefold for each 0.3 mm from the wound edge. Hinkle, McCaig, and Robinson (1981) showed that various kinds of mammalian cells will migrate toward the negative pole at field strengths considerably smaller than those present near wounds.

The polarization of the skin battery, with the inside positive, and the current that flows inward at the site of an injury are shown in Figure 11-2. The resulting field is thought to trigger the migration of epidermal cells, fibroblasts, leukocytes, nerve growth, and extension of new blood vessels (angiogenesis) toward the site of injury. For a detailed description of the cell migrations and other activities taking place during wound healing, see Majno (1975).

Also shown in Figure 11-2 is the map of transcutaneous voltages. Note that the voltage varies from place to place on the skin. Because the skin battery is important for activating wound healing, regional variations in the strength of the battery support our second hypothesis that in some areas it is easier to trigger wound healing responses. Moreover, regional differences in skin potential could drive currents through the "ion pumping cords" used by acupuncturists (Matsumoto & Birch 1988).

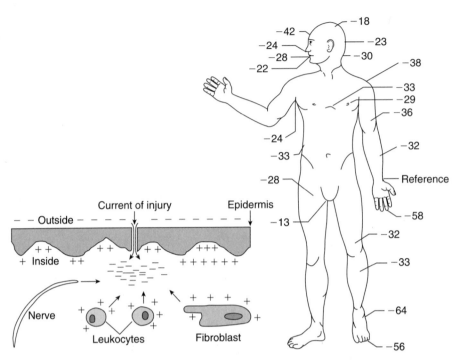

Figure 11-2 The skin battery is thought to activate wound healing. To the left is shown the inward flow of anions produced at the site of a puncture. The resulting field is thought to trigger the migration of epidermal cells, fibroblasts, leukocytes, and nerve growth toward the site of injury. To the right is a map of the transcutaneous voltages in the calm and conscious human. (Right figure is based on the work of Barker, Jaffe, & Venable. 1982, 'The glaborous epidermis of cavies contains a powerful battery.' *American Journal of Physiology* 242:R358-366, used by permission of the American Physiological Society.)

These cords could act to couple separate points and thereby spread the area of stimulation.

In 1971, Lykken (1971) reported studies of the impedance properties of skin in relation to wound healing. The method involves applying square wave pulses to the skin and observing the current waveforms with an oscilloscope. Of interest to our discussion is the discovery that the skin's electronic properties during healing change little during the first 3 to 4 days, and then the skin potential recovers suddenly. The leakage or shunt resistance, thought to be a property of the superficial layers, may not be entirely restored until 1 to 2 weeks after injury.

These findings are significant in relation to acupuncture because they suggest that the cascade of events following puncture of the skin is likely to have both fast and slow components. That acupuncture changes skin electrical properties was established by Pomeranz and others, who found that a current of about 1 microampere flows away from the site where an acupuncture needle has been removed (Pomeranz 1989). The current lasts for 24 to 48 hours and arises from the flow of ions through the acupuncture point.

Taken together, the available evidence indicates that the insertion of an acupuncture needle, a splinter, a small cut, or even mild stimulation by other means all produce long-lasting changes in the electrical and electronic characteristics of the skin and trigger cellular and hormonal activities related to wound repair. Physiologists have devised ingenious methods for quantifying these responses, and those methods could be used to test our second hypothesis, which is that certain areas on the skin have different effects than others.

Skin potentials and emotions

The literature summarized above represents a small part of the research on the electrophysiological properties of the skin. Far more research has focused on psychophysiological phenomena. Of considerable interest are endogenous voltages produced by sweat glands and the ways those voltages are influenced by emotional stimuli (for reviews of this subject, see Edelberg 1977; Fowles 1974; and Venables & Christie 1973). The well-known galvanic skin response, for example, is a major component of the polygraph, or lie-detector test.

Could the well-established connection between emotional state and skin electricity be related to emotional effects of acupuncture? There is a general rule of regulatory and integrative physiology indicating that this is likely. The principle, expressed by Adolph (1979), is that "information from action is always present" in regulatory pathways. Here the action consists of sweating, piloerection, and electrodermal potential change (galvanic skin response). The action is triggered by an emotion. The regulatory pathway is the sympathetic branch of the autonomic nervous system. A well-known example is the flight-or-fight response, in which adrenaline is released into the circulation by the adrenal cortex and triggers a variety of responses, including sweating and piloerection. Applying Adolph's rule, we expect a variety of feedbacks and systemic relational complexity. In the following scheme, the arrows pointing to the right are feed-forward pathways and those pointing to the left are feedbacks.

emotion → autonomic activation → adrenalin release → sweating, piloerection → galvanic skin response

To our knowledge, physiologists have not identified specific feedback pathways linking skin electrical responses back to the emotions that trigger them; however, we can state from experience that acupuncture has relaxing and calming effects. According to Adolph's analysis of physiological regulations, we expect many feed-forward and feedback connections in any regulatory pathway. The noted neurologist A.R. Damasio (1994) defines emotions as links in survival-oriented regulations.

Wound healing—deeper connecting layers

Beneath the epidermis is the dermis or superficial fascial layer of connective tissue. This is the substrate upon and through which the epidermal cell and fibroblast migrations mentioned above take place. It is also a medium through which electrical, electronic, and chemical messages flow from a site of puncture to the cells that are influenced. Activation of the cells, in turn, involves reversible changes in their cytoskeletons, cell movements, and changes in metabolic processes such as collagen fibrogenesis.

We now have a detailed map of the mechanical connections between adjacent epidermal cells and between the epidermis and dermis (Ellison & Garrod 1984). The details were shown in Figure 8-4, *B*. These investigators concluded that "the epidermis and dermis are linked into an integrated structural unit." Tonofilaments extend from cell to cell and from cells into the substrate. We shall see in the next section that the epidermal-dermal connections also form an integrated electronic circuit. The tonofilaments, desmosomes, hemidesmosomes, and anchoring filaments all are labile structures that can retract, dissolve, and reform (Gabbiani et al. 1978; Krawczyk & Wilgram 1973). These processes provide epidermal cells and fibroblasts with the lateral mobility required for reepithelialization. Such cytoskeletal and extracellular connections occur throughout the body.

For a long time we have suspected that the classic acupuncture meridians are none other than the primary channels of the pervasive living matrix continuum. Becker (1990) suggested that the acupuncture points are "booster amplifiers" for communications through the meridian system. We also have suggested that the acupuncture points have other functions, as nodes or microprocessors in a semiconductor network, and that they include communication, sensory, power, and signal processing components (see Chapter 10). The extent to which these properties are concentrated at the nodes or acupoints, as opposed to being distributed throughout the system, are subjects of future research.

Properties of the living matrix in relation to acupuncture

Now we briefly summarize the properties of the living matrix in relation to the acupuncture system. For more details, see Chapters 8 through 10.

All components of the living matrix are *semiconductors*. This means they are able to generate and conduct vibrational information. Where two or more components intersect (semiconductor junctions) there is a possibility of signal processing analogous to that taking place in transistors, integrated circuits, or microprocessors found in computers and other electronic devices. Semiconductor molecules convert energy from one form to another. One way this takes place is through the *piezoelectric* effect. Many of the components are piezoelectric. This means that waves of mechanical vibration moving through the living matrix produce electrical fields, and vice versa; that is, waves of electricity produce mechanical vibrations. Phonons are electromechanical waves in a piezoelectric medium. Piezoelectric properties arise because much of the semiconducting living matrix is highly ordered, or *crystalline*. Specifically, many of the molecules in the body are regularly arrayed in crystal-like lattices. This includes the lipids in cell membranes, the collagen molecules of connective tissue, the actin and myosin molecules in muscles and other cells, and other components of the cytoskeleton, such as microtubules and microfilaments. The high degree of structural order (the matter field) gives rise to a highly ordered or *coherent* electromagnetic field. This electromagnetic field is composed of giant coherent or laserlike oscillations that move rapidly throughout the living matrix and that also are radiated into the environment. These vibrations are called *Fröhlich oscillations*. They occur at particular frequencies in the microwave and visible light portions of the electromagnetic spectrum. A number of scientists have detected these signals.

Water is a dynamic component of the living matrix. On average, each matrix protein has 15,000 water molecules associated with it. Because many of the proteins are highly ordered, as we have just seen, the associated water molecules also are highly ordered. Water molecules are electrically polarized (dipoles) and therefore tend to orient or rotate in a magnetic field. The living matrix organizes the dipolar water molecules in a way that constrains or restricts their ability to vibrate, rotate, or wiggle about in different spatial planes. Water molecules are only free to vibrate or spin in particular directions. The following statement summarizes some recent findings:

> *Water can act as a structural signal transmitter and stabilizer. The water-proton subsystem is imprinted by the appropriate biomolecules. An interplay takes place between geometry and dynamics. . . . [P]rotons behave as a (sub)system having its own properties and characteristics, determining the macroscopic properties of the system as a whole.* AIELLO ET AL. (1973)

Continuity is a characteristic of the system we are describing. The properties just listed are not localized but are spread throughout the organism. Although we may distinguish individual organs, tissues, cells, and molecules, the living matrix is a continuous and unbroken whole. Each part can generate signals, conduct them, and respond to them.

Finally, components of the living matrix undergo reversible polymerization processes, or *gel-sol transitions*. Each time a microtubule or actin filament or other polymeric structure forms, information is encoded into its structure. Before cells divide, their cytoskeletons depolymerize or fall apart, and some of the subunits or fragments assemble into the mitotic apparatus that separates the chromosomes during mitosis. Depolymerization into monomeric units is comparable to erasing a tape recording.

Neurophysiologists have had difficulty understanding how and where memories are stored in the brain (Squire 1987). Hameroff and colleagues have summarized research suggesting that all of the cells in the body can store information in microtubules (Hameroff 1974, 1988; Hameroff, Rasmussen & Mansson 1988). Mathematician Roger Penrose (1994) has elaborated upon this idea. In later chapters we document how the flow of information through the living matrix can deposit memories and how therapeutic movement of energy can release and erase deeply repressed traumatic memories. We also have described how the polymeric or gel nature of ground substances, and their highly charged character, enables the release of toxic materials (insecticides, drugs, solvents) that are stored in tissues (storage excretion). This accounts for the odors that practitioners often detect during bodywork.

Wound healing—the perineural system

Robert O. Becker has pointed out that modern neurophysiology takes into consideration the activity of less than half of the brain. The "neuron doctrine" holds that all functions of the nervous system are the result of activities of the neurons alone. Integration of brain function is regarded as arising because of the massive interconnectivity of the neurons. Most neurophysiological research into consciousness is based upon this premise. This is an incomplete view because a more basic and primitive informational system exists in the neglected perineural connective tissue cells that constitute more than half of the cells in the brain.

To replace the neuron doctrine, Becker proposes a "dual nervous system" consisting of the classic digital (all-or-none) nerve network and an evolutionarily more ancient and primitive but important direct current analog system. The analog system resides in the perineural connective tissues. There is a substantial base of theoretical, electrophysiological, and bioelectromagnetic evidence, developed over many years, to support Becker's conclusions.

The dual nervous system concept is important to acupuncture and other forms of bodywork because of the role the perineural system plays in wound healing. We suspect that the perineural system and the surrounding connective tissue and cytoskeletal matrices are the substrate for the action of virtually all forms of therapy. This system displays the Hall effect (see later), providing a possible basis for the therapeutic action of magnets.

Becker (1990) summarizes the evidence for the dual nervous system and explains how this concept can help us understand phenomena, such as consciousness, that have been so elusive in the past because of the focus on the digital nervous system. The significance of the perineural system can best be described in Becker's own words with reference to Figures 2-4, B, and 8-1.

> If a way were devised to dissolve all of the nerves in the brain and throughout the body, it would appear to the naked eye that nothing was missing. The brain and spinal cord and all of the peripheral nerves would appear intact down to their smallest terminations. This is because the central nervous system is composed of two separate types of cells; the nerve cells, or neurons, and the perineural cells. There are far more perineural cells in the central nervous system than there are neurons. The brain is totally pervaded by glial cells of various types and every peripheral nerve is completely encased in Schwann cells from its exit from the brain or spinal cord down to its finest terminations. Every nerve cell body and its projections of axons or dendrites is covered with perineural cells of one type or another. Despite their ubiquitous presence these cells have been dismissed as supportive or nutritive (the term neuroglia means "nerve glue"). It is my thesis that collectively the perineural cells constitute an information transmission system, more primitive in nature but capable of exerting a controlling influence on the basic functions of the neurons. As such, the two systems, the perineural and the neural, function as a unit—a dual nervous system, with properties and capabilities greater than those of each part. (Becker 1991)

Becker has reviewed 50 years of evidence indicating that the body contains a separate operational system, based on extraneuronal DC electrical currents, that serves to integrate the operation of the digital nervous system.

The perineural direct current (DC) system is responsible for a current of injury, a continuously flowing DC electrical current generated at the site of a wound. Early work on the current of injury was done on plants in 1959 by the Russian Siniukhin (1959). Siniukhin found that the current of injury was involved in the regeneration of parts of plants that had been cut off. Adding a small DC current to the normal injury current increased the rate of regeneration and a current of reversed polarity retarded regeneration. Based on this work, Becker examined the current of injury produced by amputation of limbs in salamanders and frogs. Salamanders are able to regenerate missing limbs, but frogs are not. Becker found the current of injury was negative in polarity in salamanders and positive in frogs. In both cases, the current of injury persisted until all healing was complete, more than 30 days after amputation.

Amputation altered the overall field of the animal body, including the brain. For a current to propagate from the site of an injury to the entire body, conductive pathways must be present. There are three kinds of conduction: metallic conduction, ionic conduction, and semiconduction. Because the body does not contain metal wires, Becker narrowed his search to ionic and semiconducting currents. He used the Hall effect to distinguish between ionic conduction and semiconduction. A magnetic field causes a current flowing in a semiconductor to veer off to one side. The extent to which this happens is determined by measuring a DC transverse Hall voltage at right angles to the current flow. Ions are fairly immobile and produce relatively small Hall voltages, whereas electrons and holes in semiconductors are very mobile and produce larger Hall voltages. In studies on salamander limbs, Becker showed Hall voltages indicative of semiconduction rather than ionic currents. The Hall voltages increased during recovery from anesthesia, indicating that the semiconducting DC current is related to the level of consciousness.

Becker's studies indicated that the system of acupuncture points and meridians described in the ancient Chinese literature and clinical practice consists of real structures that serve as input channels for the total DC system. Research showed that acupuncture points have measurable electrical properties that are different from surrounding skin (Figure 11-3). The meridians have electrical properties indicating they are electrical transmission lines (Reichmanis, Marino & Becker 1975).

Wound healing—other systems

A consideration of the nervous system must take into account its duality—the neurons plus the perineural system. A similar case can be made for the circulatory and lymphatic systems, which are everywhere associated with perivascular and perilymphatic connective tissue cells. Stimulation also is likely to be conducted in various ways into and through the deep fascia, consisting of the myofascia around muscles and the periosteum around bones.

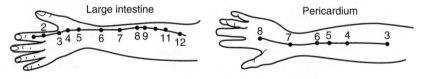

Figure 11-3 Acupuncture meridians and points have measurably different electrical conductances compared with surrounding skin. The meridians have electrical properties indicative that they are electrical transmission lines. (From Reichmanis, M., Marino, A.A. & Becker, R.O. 1975, 'Electrical correlates of acupuncture points', *IEEE Transactions on Biomedical Engineering*, Nov., pp. 533-535. © 1975 IEEE.)

There are indications in the physiological literature that severe trauma can cause large-scale changes in the circulatory system. When a limb is severed by crushing trauma, vessels as large as arteries can spasm so that there is no serious blood loss (Guyton 1971). One explanation for this is that waves of depolarization are conducted along vessels or perivascular tissue a considerable distance from a site of injury.

Systemic effects of acupuncture on the circulation have been reported by Itaya et al. (1987). Effects on the rabbit's cutaneous microcirculation were pronounced after removal of the acupuncture needle. The needling was on the back of the rabbit (corresponding to B17 in the human) and the microcirculatory effects were examined in the ear, using a transparent viewing chamber. Acupuncture had a vasodilating effect and increased the rate of rhythmic vasomotion.

Stimulation of the digital nervous system can trigger the release of neuropeptides, and this is an important aspect of wound healing and of the simulation of injury that we suggest is triggered by needle insertion. This is a subject that is thoroughly discussed elsewhere in the literature. The nervous system has other obvious roles in wound healing, such as behavioral changes. The nervous system's plasticity is exemplified by the limp that serves to take the weight off an injured limb until it heals.

Wound healing—integration of activities

It has long been recognized that a current of injury arises at the site of a wound and spreads through the surrounding tissues. Historically, the injury potential was discovered *before* resting and action potentials, when it was noted that an electrode placed on a damaged portion of a muscle or nerve was negative with respect to the intact portion (Davson 1970).

There is confusion in the literature about which tissues are responsible for conducting the current of injury. Some researchers refer to the current as a property of the skin; others relate it to the perineural system. We suspect that currents of injury are a fundamental characteristic of epithelia and will arise in any tissue (epidermal, vascular, or nervous) that is injured. Moreover, the nature of the currents depends upon the depth and severity of the injury. We suspect that little or no epidermal current of injury arises in blunt trauma in which the skin surface is not damaged. The primary currents of injury in blunt trauma are carried by deeper tissues such as the perineural, myofascial, and even periosteal and perivascular systems. Some of these layers are illustrated in Figure 11-4.

The electrical effects of inserting an acupuncture needle depend upon the depth to which the needle is inserted. Small shallow punctures would be expected to activate only epidermal currents, whereas deeper needling could affect deeper systems. On the other hand, electronic effects are expected to be transmitted from a shallow needle to all layers, through the semiconducting living matrix.

Most acupuncture needles are electrically conductive and therefore will provide electrical continuity between systems at different depths. Even if the needle is not a conductor (as would be true for a wood splinter), a thin film of highly conductive interstitial fluid forms over the surface of any object that has penetrated into the body.

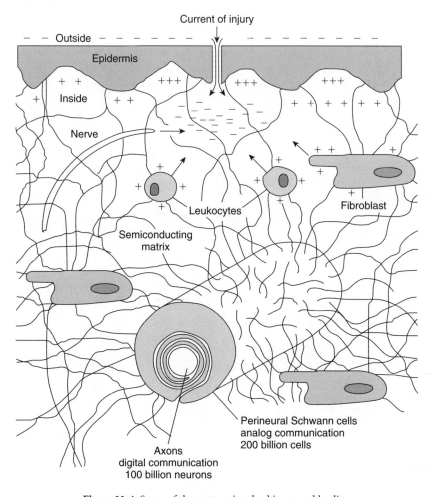

Figure 11-4 Some of the systems involved in wound healing.

This film of fluid also would electrically connect systems at various depths. This concept is illustrated in Figure 11-5, *A*.

We suspect that the various techniques acupuncturists use to "get the Ch'i," such as twisting the needle, moving it up and down, attaching ion-pumping cords, heating the needle, and electroacupuncture, have in common an accentuation and spreading of the activation of wound healing systems documented earlier.

Finally, needle insertion is expected to influence chemical regulatory systems involved in wound healing. The concept of a wound hormone dates back to the work of Wiesner (1892). In 1938, Bonner and English (1938) isolated and purified a wound hormone, "traumatin," from plants. Recent mammalian research has led to the discovery of a family of fibroblast growth factors, as summarized at a 1991 symposium of the New York Academy of Sciences (Baird & Klagsbrun 1991). Elegant research has

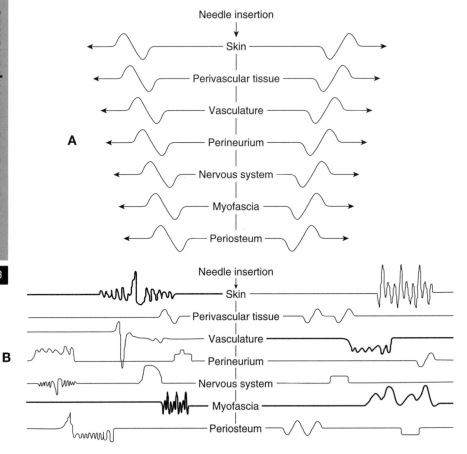

Figure 11-5 A, Waves of energy and information traveling away from a site of acupuncture needle insertion or injury. The waves are traveling in a crystalline semiconducting piezoelectric medium. These waves may be electronic, protonic, and other solid-state phenomena, as well as waves of various chemical activators and inhibitors. There also are waves of adhesion-disadhesion, cell movement, mitosis, and contraction. Each feed-forward pathway has a set of feedback loops. **B,** Hypothetical scheme for various kinds of waveforms encoding activation of different layers as they are perturbed.

documented that a wide variety of stimulating and inhibiting factors activate and integrate wound repair, and then wind down the processes when healing is complete.

Figure 11-5 is schematic and does not do justice to the wealth of activities sent forth from a site of injury or stimulation. It also does not do justice to the wealth of specific techniques acupuncturists have available to trigger particular healing and repair processes.

In Figure 11-5, *B,* waves are depicted traveling away from the site of puncture. These waves correspond in part to pulses of electricity, solitons, proticity (Mitchell 1976), and so on. There is far more to the story, however, because the waves are traveling in a crystalline semiconducting piezoelectric medium and will liberate various forms of energy and involve influences on water molecules and ions associated with protein structures. We conceptualize these waves as electronic, protonic, and other solid-state phenomena, as well as waves of various chemical activators and inhibitors. There also are waves of adhesion-disadhesion, cell movements, mitosis, and contraction. Finally, each feed-forward pathway will have a set of feedback loops.

Information is expected to be transferred through each layer at different velocities and by one or more different mechanisms. If each of these communications represents a different form of Ch'i, Figure 11-5 allows for a minimum of seven different forms. Porkert (1980) lists 32 types.

Many of the events in wound repair are classed as self-assembly processes. Fröhlich (1988) has pointed out that vague concepts of self-assembly, self-organization, and other spontaneous processes "mask many difficulties." To the biologist, the various automatic events that take place in living matter represent remarkable and intricate accomplishments.

CONCLUSION

Wound healing is a remarkable and intricate process involving the integrated and cooperative activities of a variety of systems. Each wound is different, and the body's response must be precisely appropriate if the body is to be fully restored to the way it was before injury.

The common denominator to both Eastern and Western approaches to the human body is a set of systems that are well known to biomedicine but that are not usually thought of as being connected to each other. Study of the interconnections is opening up the scientific investigation of so-called *alternative approaches to medicine,* including acupuncture. Links between systems are emerging from studies of physiological integration—the mechanisms involved when many systems cooperate or synergize to accomplish an activity such as the healing of a wound or injury. Solid-state biophysics is a rich sources of new information about communication and regulatory processes. The phenomena involved in wound healing illustrate how systems work together to restore form and function. The reason physical and emotional well-being seem to intimately related to each other is that they share many common regulatory pathways.

We now recognize that the same living matrix that connects every part of the organism with every other part is the origin, the conductor, and the interpreter of energetic vibrations that are known in the ancient literature as *life force* or *Ch'i.* This is the same signaling system that gives rise to and maintains the form of the organism.

Taken together, the various regulatory messages traveling about within the living matrix comprise what we refer to as *consciousness.* Emotional states arise in part from the ability of the various regulatory messages to interact with traumatic and other kinds of "memories" stored in the connective tissues, in microtubules, and in other polymeric components of the cytoskeletons of all cells in the body, including but not limited to the neurons. A consequence of acupuncture and of many other alternative approaches is the release and resolution of repressed memories (*somatic recall,* see Chapter 22), which have beneficial influences upon the organism.

Two hypotheses were presented at the beginning of this chapter: acupuncture *simulates* injury; and specific areas on the skin surface (acupoints) are more effective than others in initiating local and systemic responses comparable to those taking place during wound healing. A consequence is the restoration of tissues and vital processes that have been compromised by the accumulation of injuries and traumas, both physical and emotional. The hypotheses provide a basis for discussion and research on the effects of acupuncture in terms that are understandable and significant for all approaches to health care. We have seen from the work of Heilbrunn and others that biologists have long recognized that there is a similarity between the ways cells and tissues respond to injury and the ways they respond to stimuli in general. Hence there is a physiological basis for the concept that stimulation of acupuncture points by objects including needles, magnets, laser beams, moxa, and sounds, produce therapeutic consequences for the organism. These consequences ultimately will be quantifiable by study of the solid-state properties of the substrate through which chemical and electronic messages travel from the site of puncture and through which migrate epidermal cells, fibroblasts, and cells of the immune system. Moreover, a variety of elegant experimental techniques and biological models developed for the study of wound healing can be used to quantify the effects of acupuncture. Such research is worthwhile because of its implications to all forms of clinical practice.

REFERENCES

Adolph, E.F. 1979, 'Look at physiological integration', *American Journal of Physiology,* vol. 237, pp. R255-R259.

Aiello, G., Micciancio-Giammarinaro, M.S., Palma-Vittorelli, M.M. & Palma, M.U. 1973, 'Behaviour of interacting protons: The average-mass approach to its study and its possible biological relevance', in *Cooperative Phenomena,* eds H. Haken & M. Wagner, Springer-Verlag, New York, pp. 395-403.

Akaiwa, H. 1919, 'A quantitative study of wound healing in the rat. I. Cell movements and cell layers during wound healing', *Journal of Medical Research,* vol. XL, pp. 311-351, see also earlier papers by Loeb, Addison, and Spain cited by Akaiwa.

Ardrey, R. 1967, *African Genesis,* Dell, New York.

Baird, A. & Klagsbrun, M. (eds.) 1991, The fibroblast growth factor family, in *Annals of the New York Academy of Sciences,* vol. 638.

Barker, A.T., Jaffe, L.F. & Vanable, J.W. 1982, 'The glabrous epidermis of cavies contains a powerful battery', *American Journal of Physiology,* vol. 242, pp. R358-R366.

Barnsley, M.F. 1993, *Fractals Everywhere,* 2nd ed, Academic Press Professional, Boston, p. 26.

Becker, R.O. 1990, *Cross Currents. The Promise of Electromedicine. The Perils of Electropollution,* Jeremy P. Tarcher, Inc., Los Angeles, CA, p. 48.

Becker, R.O. 1991, 'Evidence for a primitive DC electrical analog system controlling brain function', *Subtle Energies,* vol. 2, pp. 71-88.

Becker, R.O. & Marino, A.A. 1982, *Electromagnetism and Life,* State University of New York Press, Albany, NY p. 3.

Birch, S. 1994, *Needles and Fire. Understanding Acupuncture and its Journey West,* Paradigm Publications, Brookline, MA.

Bonner, J. & English, J. 1938, 'A chemical and physiological study of traumatin, a plant wound hormone', *Plant Physiology*, vol. 13, p. 331.

Chicurel, M.E., Chen, C.S. & Ingber, D.E. 1998, 'Cellular control lies in the balance of forces', *Current Opinion in Cell Biology*, vol. 10, pp. 232-239.

Clark, R.A.F. 1994, 'Mechanisms of cutaneous wound repair', in *Dermatology in General Medicine*, vol. I, eds T. B. Fitzpatrick, A.Z. Eisen, K. Wolff, I.M. Freedberg & K.F. Austen, McGraw-Hill, New York, pp. 473-486.

Damasio, A.R. 1994, *Descartes' Error: Emotion, Reason and the Human Brain*, G.P. Putnam's Sons, New York.

Dart, R.A. 1949, 'The predatory implemental technique of Australopithecus', *American Journal of Physical Anthropology*, vol. 7, pp. 1-38.

Dart, R.A. 1957, 'The osteodontokeratic culture of Australopithecus Prometheus', *Transvall Museum Memoir*, no. 10.

Davson, H. 1970, *A Textbook of General Physiology*, 4th edn, Williams & Wilkins, Baltimore, p. 559.

Desmouliere, A. & Gabbiani, G. 1994, 'Modulation of fibroblastic cytoskeletal features during pathological situations: The role of extracellular matrix and cytokines', *Cell Motility and the Cytoskeleton*, vol. 29, pp. 195-203.

Dujardin, F. 1835, 'Recherches sur les organimes inférieurs. III. Sur les prétendus estomacs des animalcules Infusoires et sur une substance appelée Sarcode', *Annales des sciences naturelles. Zoologie et biologie animale*, vol. 2, pp. 364-377.

Edelberg, R. 1977, 'Relation of electrical properties of skin to structure and physiologic state', *Journal of Investigative Dermatology*, vol. 69, pp. 324-327.

Ellison, J. & Garrod, D.R. 1984, 'Anchoring filaments of the amphibian epidermal-dermal junction traverse the basal lamina entirely from the plasma membrane of hemidesmosomes to the dermis', *Journal of Cell Science*, vol. 72, pp. 163-172.

Fowles, D.C. 1974, 'Mechanisms of electrodermal activity', in *Bioelectric Recording Techniques*, part C, eds R.F. Thompson & M.M. Patterson, Academic Press, New York, pp. 232-272.

Fröhlich, H. 1988, 'The genetic code as language', in *Biological Coherence and Response to External Stimuli*, ed. H. Fröhlich, Springer-Verlag, Berlin, pp. 192-204.

Gabbiani, G., Hirschel, B.J., Ryan, G.B., Statkov, P.R. & Majno, G. 1972, 'Granulation tissue as a contractile organ', *Journal of Experimental Medicine*, vol. 135, pp. 719-734.

Gabbiani, G., Chaponnier, C. & Huttner, I. 1978, 'Cytoplasmic filaments and gap junctions in epithelial cells and myofibroblasts during wound healing', *Journal of Cell Biology*, vol. 76, pp. 561-568.

Guyton, A.C. 1971, *Textbook of Medical Physiology*, 4th ed, WB Saunders, Philadelphia, p. 136.

Hameroff, S., Rasmussen, S. & Mansson, B. 1988, 'Molecular automata in microtubules computational logic of the living state?', in *Artificial Life, SFI Studies in the Science of Complexity*, vol. VI., ed. C. Langton, Addison-Wesley, Redwood City, CA, 521-553.

Hameroff, S.R. 1974, 'Ch'i: A neural hologram? Microtubules, bioholography, and acupuncture', *American Journal of Chinese Medicine*, vol. 2, pp. 163-170.

Hameroff, S.R. 1988, 'Coherence in the cytoskeleton: Implications for biological information processing', in *Biological Coherence and Response to External Stimuli*, ed. H. Fröhlich, Springer-Verlag, Berlin, pp. 242-263.

Heilbrunn, L.V. 1927, 'The colloid chemistry of protoplasm. V. A preliminary study of the surface precipitation reaction of living cells', *Archiv Für Experimentelle Zellforschung Besonders Gewebezüchtung*, vol. 4, pp. 246-263.

Heilbrunn, L.V. 1956, *The Dynamics of Living Protoplasm*, Academic Press, New York, p. 64.

Heilbrunn, L. & Wiercinski, J. 1947, 'The action of various cations on muscle protoplasm', *Journal of Cellular and Comparative Physiology*, vol. 29, pp. 15-32.

Herlitzka, A. 1910, 'Ein Beitrag zur Physiologie der Regeneration', *Wilhelm Roux Arch Entwicklungsmech Org*, vol. 10, pp. 126-159.

Hinkle, L., McCaig, C.D. & Robinson, K.R. 1981, 'The direction of growth of differentiating neurons and myoblasts from frog embryos in an applied electric field', *Journal of Physiology (London)*, vol. 314, pp. 121-135.

Horridge, G.A. 1968, 'The origins of the nervous system', in *The Nervous Tissue, vol. I, Structure I*, ed. G.H. Bourne, Academic Press, New York, pp. 28-29.

Illingworth, C.M. & Barker, A.T. 1980, 'Measurement of electrical currents emerging during the regeneration of amputated finger tips in children', *Clinical Physics and Physiological Measurement*, vol. 1, pp. 87-89.

Itaya, K., Manaka, Y., Ohkubo, C. & Asano, M. 1987, 'Effects of acupuncture needle application upon cutaneous microcirculation of rabbit ear lobe', *Acupuncture & Electro-Therapeutics Research International Journal*, vol. 12, pp. 45-51.

Kimura, M., Tohya, K., Kuroiwa, K., Oda, H., Gorawski, E.C., Hua, Z.X., Toda, S., Ohnishi, M. & Noguchi, E. 1992, 'Electron microscopical and immunohistochemical studies on the induction of "qi" employing needling manipulation', *American Journal of Chinese Medicine*, vol. 20, pp. 25-35.

Krawczyk, W.S. & Wilgram, G.F. 1973, 'Hemidesmosome and desmosome morphogenesis during epidermal wound healing', *Journal of Ultrastructural Research*, vol. 45, pp. 93.

Langevin, H.M., Churchill, D.L. & Cipolla, M.J. 2001, 'Mechanical signaling through connective tissue: A mechanism for the therapeutic effect of acupuncture', *FASEB Journal*, vol. 15, pp. 2275-2282.

Langevin, H.M., Churchill, D.L., Fox, J.R., Badger, G.J., Garra, B.S. & Krag, M.H. 2001, 'Biomechanical response to acupuncture needling in humans', *Journal of Applied Physiology*, vol. 91, pp. 2471-2478.

Lash, J.W. 1955, 'Studies on wound closure in Urodeles', *Journal of Experimental Zoology*, vol. 128, pp. 13-28.

Lykken, D.T. 1971, 'Square-wave analysis of skin impedance', *Psychophysiology*, vol. 7, pp. 262-275.

Majno, G. 1975, *The Healing Hand. Man and Wound in the Ancient World*, Harvard University Press, Cambridge, MA.

Majno, G., Babbiani, G., Hirschel, B.J., Ryan, G.B. & Statkov, P.R. 1971, 'Contraction of granulation tissue in vitro: Similarity to smooth muscle', *Science*, vol. 173, pp. 548-550.

Marchesi, V.T. 1985, 'Inflammation and healing', in Anderson's Pathology, 8th ed, eds J.M. Kissane & W.A.D. Anderson, Mosby, St. Louis, pp. 22-60.

Matsumoto, K.S. & Birch, S. 1988, *Hara Diagnosis: Reflections on the Sea*, Paradigm Publications, Brookline, MA.

Mitchell, P. 1976, 'Vectorial chemistry and the molecular mechanics of chemiosmotic coupling: Power transmission by proticity', *Biochemical Society Transactions*, vol. 4, pp. 399-430.

Myers, T. 2001, *Anatomy Trains. Myofascial Meridians for Manual and Movement Therapists*, Churchill Livingstone, Edinburgh.

Needham, A.E. 1952, *Regeneration and Wound-Healing*. Methuen, London, UK.

Oschman, J.L. 1994, 'A biophysical basis for acupuncture', in *Proceedings of the First Symposium of the Society for Acupuncture Research*, presented January 23-24, 1993. Rockville, MD. Published by the Society for Acupuncture Research, Bethesda, MD.

Oschman, J.L. & Oschman, N.H. 1995a, 'Somatic recall. Part I. Soft tissue memory', *Massage Therapy Journal, American Massage Therapy Association, Lake Worth, Florida*, vol. 34, pp. 36-45, 111-116.

Oschman, J.L. & Oschman N.H. 1995b, 'Somatic recall. Part II. Soft tissue holography', *Massage Therapy Journal, American Massage Therapy Association, Lake Worth, Florida*, vol. 34, pp. 66-67, 106-116.

Pallotta, T. & Peres, A. 1989, 'Membrane conductance oscillations induced by serum in quiescent human skin fibroblasts', *Journal of Physiology*, vol. 416, pp. 589-599.

Penrose, R. 1994, *Shadows of the Mind: A Search for the Missing Science of Consciousness*, Oxford University Press, Oxford, UK.

Peters, C. 1885, Ueber die Regeneration des Epithels der Cornea. Inaug. Diss. Bonn. Cited in Barfurth, D. 1991. 'Zur Regeneration der Gewebe', *Archiv für Mikroskopische Anatomie*, vol. 37, pp. 406-491.

Pomeranz, B. 1989, 'Research into acupuncture and homeopathy', in *Energy Fields in Medicine. Study of Device Technology Based on Acupuncture Meridians and Chi Energy*, J.E. Fetzer Foundation, Kalamazoo, MI, pp. 66-77.

Porkert, M. 1980, *The Theoretical Foundations of Chinese Medicine*, The MIT Press, Cambridge, MA, pp. 168-173.

Reichmanis, M., Marino, A.A. & Becker, R.O. 1975, 'Electrical correlates of acupuncture points', *IEEE Transactions on Biomedical Engineering*, Nov., pp. 533-535.

Rink, J. & Jacob, R. 1989, 'Calcium oscillations in non-excitable cells', *Trends in Neuroscience*, vol. 12, pp. 43-46.

Rosenberg, P., Reuss, L. & Glaser, L. 1982, 'Serum and epidermal growth factor transiently depolarize quiescent BSC-1 epithelial cells', *Proceedings of the National Academy of Sciences USA*, vol. 79, pp. 7783-7787.

Seto, A., Kusaka, C., Nakazato, S., Huang, W., Sato, T., Hisamitsu, T. & Takeshige, C. 1992, 'Detection of extraordinary large bio-magnetic field strength from human hand', *Acupuncture and Electro-Therapeutics Research International Journal*, vol. 17, pp. 75-94.

Siniukhin, A.M. 1959, 'Nature of the variation of the bioelectric potentials in the regeneration process of plants', *Biophysics Russia*, vol. 2, pp. 53-69.

Squire, L.R. 1987, *Memory and Brain*. Oxford University Press, New York.

Stossel, T.P. 1994, 'The machinery of cell crawling', *Scientific American*, vol. 271, pp. 54-63.

Venables, P.H. & Christie, M.M. 1973, 'Mechanisms, instrumentation, recording techniques and quantification of responses', in *Electrodermal Activity in Psychological Research*, eds W.F. Prokasy & D.C. Raskin, Academic Press, New York, pp. 1-124.

Vogt, P.M., Thompson, S., Andree, C., Liu, P., Breuing, K., Hatzis, D., Brown, H., Mulligan, R.C. & Eriksson, E. 1994. 'Genetically modified keratinocytes transplanted to wounds reconstitute the epidermis', *Proceedings of the National Academy of Sciences USA*, vol. 91, pp. 9307-9311.

Vorontsova, M.A. & Liosner, L.D. 1960, *Asexual Reproduction and Regeneration*, Pergamon, London, UK.

Wiesner, J. 1892, *Die Elementarstructur und das Wachsthum der lebenden Substanz*, Wien.

Williamson, S.J. 1983, 'Biomagnetism. An interdisciplinary approach', in *Proceedings of a NATO Conference*, Plenum Press, New York.

Williamson, S.J. & Kaufman, L. 1981, 'Biomagnetism', *Journal of Magnetism and Magnetic Materials*, vol. 22, pp. 129-201.

Continuum in natural systems

It is an urgent task for the future to raise our sights, our thinking and our acting, from our preoccupation with segregated things, phenomena, and processes, to greater familiarity and concern with our natural connectedness, to the "total context." Weiss (1977)

Because of our intellectual endowments, simple ideas—clearly articulated—can profoundly influence the way we live and can change the course of history. *Continuum* is such a concept. Among other things, continuum describes the living matrix within and around us. Continuum refers to the state of continuous organic wholeness fundamental to the structure and behavior of the natural world. The most memorable and breath-taking moments in our lives are glimpses, of one sort or another, of a deep interconnectedness that is always present. Our inability to maintain this level of awareness and experience sustains the continuing dilemmas of our times.

In recent years, dramatic statements of this view have been brought to us by astronauts returning from their explorations. More than one of them, looking back at our planet, has trembled with the realization that the blue and green globe beneath them is a single living being.

> *Viewed from the distance of the moon, the astonishing thing about the Earth, catching the breath, is that it is alive.* LEWIS THOMAS (1973)

This perspective has been termed *the overview effect* in the book of the same title by Frank White (1987). Of course, native people the world over have always believed and functioned on the premise that the Earth and all of its inhabitants form a single living being.

Continuum is at the same time a philosophical, scientific, artistic, musical, poetic, and spiritual concept, a cosmology, and an advanced state of consciousness. When applied to our affairs, continuum leads us naturally to a saner and happier world. In other words, continuum as an experience is a direct involvement in the harmony and congruence of our inner and outer realities. It enables us to live the real lives of our bodies:

> *What is it that we want: To fully experience our aliveness.*
> *To feel in our bodies a streaming, like the rush of a river over stones.*
> *To be awake, alert, and responsive in our limbs and sensitive in our fingertips.*
> *To feel as if our inner and outer reality are congruent and that our efforts are rewarded by a sense of satisfaction.*
> *We aspire to have our private lives nestle within the valley of a public world which we can affirm.*
> *We long to feel connected with each other.*
> *We want to be able to embrace and be embraced.*
> *We want to live the life of our bodies and want our bodies to permit us to fully live our lives.* BEINFIELD AND KORNGOLD (1991)

As an anatomical and structural principle, continuum goes a long way toward explaining some of the miraculous consequences of "hands-on" bodywork and "hands-off" energy therapies, the martial arts, and other phenomena we have discussed. Those who use touch as therapy are leading-edge explorers of this realm. What you touch when you contact another person's body is the connective tissue and all it embodies and reaches. Continuum is one of the most profound descriptors of connective tissue. Training, experiences, and expertise in working and living from this point of view are a gift to all of us. We need to apply discoveries in this realm to persistent problems in biology, medicine, and society as a whole. Writing books such as this is a process of sorting through the underlying principles and their deeper meanings.

THE BOUNDARY OF THE BODY

We wish to document the value of defining connective tissue in a larger and more inclusive sense that encompasses all to which it is connected. The sum total of these elements we refer to as *the living matrix*. This continuum includes every organ, tissue, cell, molecule, atom, and subatomic particle within the body, as well as the energy fields and relations they engender and interact with. A vital aspect of this matrix, just as palpable as its measurable properties, is the way we think about it. What we refer to as the *living body* has no clearly definable boundary:

> *The recent results indicating an intimate spatiotemporal relationship of living things to their planetary field have raised to a significance the long-time academic problem concerning the limits of the individual organism, whether at the structural surface of the body or out in the area encompassed by the physical fields generated by the individual. The answer may be very different for the anatomist, on the one hand, and the physiologist, behaviorist and psychologist, on the other. The life of the individual organism cannot be fully described by its organized aggregate of molecules, nor can at this level the powerful tools and concepts of modern physics be brought to bear on it. Rather, it must be supplemented by the dynamic fields that are metabolically generated in the organism operating as a bioelectronic entity, and which continuously and mutually interacts with the counterpart fields of its ambient physical surroundings. The organism can only arbitrarily be defined as separate. The organism and the physical environment are mutually invasive.* (Brown 1980)

Schools and teachings that are unwilling to examine life from the continuum perspective perpetuate our predicaments in which our major and most expensive health problems are untreatable, intractable, and virtually unapproachable. There are some thinkers who still limit their definition of the organism to the visible and palpable phenomena lying at or under the surface of the skin. Environmental, social, and emotional aspects of disease and disorder are considered unapproachable and, therefore, outside of the domain of scientific medicine. The resulting myopia and inadequacy in the vital realm of health affect all of our other local and global affairs.

CONTINUUM PHILOSOPHY

Some approaches to the body are rarely able to tap into continuum methodology and consciousness because their philosophical base is advertised as rational and unemotional science, while the energetic continuum is repeatedly rejected on irrational and emotional grounds. The history, psychology, and biopolitics of this rejection is a fascinating subject. The point is that a medicine that fails to look at, and function from, life's most fundamental attributes is destined to become irrelevant and obsolete.

There are physicians and researchers who, by nature, successfully and compassionately operate from the state of mind we refer to as continuum. Some have written eloquently and thoughtfully about such matters in an effort to waken their colleagues to what has been missing in their training and practice (Heymann 1995; Pellegrino 1974). There also are scientists and scholars who have seen through the veil of confusion generated by our thinking minds and reinforced by our natural tendency to focus upon one piece of the world picture at a time. These visionary individuals have affirmed our systemic wholeness and have shown us how our world really works. Their message is timely and vital.

> *If Nature puts two things together she produces something new with new qualities, which cannot be expressed in terms of qualities of the components. When going from*

electrons and protons to atoms, from here to molecules, molecular aggregates, etc., up to the cell or the whole animal, at every level we find something new, a new breath-taking vista. Whenever we separate two things, we lose something, something which may have been the most essential feature.   SZENT-GYÖRGYI (1963)

Living matter seems to be a system of water and organic matter, which forms one single inseparable unit, a system, as the cogwheels do in a watch.   SZENT-GYÖRGYI (1957)

Dr. Ida P. Rolf was influenced by a man who was a leading continuum thinker, Ludwig von Bertalanffy (1901-1972). He introduced *General Systems Theory* in the 1930s and 1940s, and he helped found the Society for General Systems Research in 1956. He was convinced that "[a]ll physical, chemical, biological, social, and psychological systems operate in accord with the same few fundamental principles."

von Bertalanffy's last essay was published in Volume I of a series entitled, *A New Image of Man in Medicine* (Shaefer, Hensel & Brady 1977). The book includes presentations at a conference on "Man-Centered Physiological Science and Medicine" held in Herdecke, Germany, September 24 to 28, 1973. The gathering of distinguished scientists and physicians was largely supported by the Rudolf Steiner Foundation. The paper by von Bertalanffy was read posthumously at the meeting.

Weiss on connective tissue

The book contains a chapter that is a gem in the literature on continuum. It is written by Paul A. Weiss (1922–1980), at the time Professor Emeritus of Biology at the Rockefeller University in New York. We have long appreciated Weiss for his fundamental discoveries about the way connective tissue is laid down in healing wounds and in cell cultures (Weiss 1961). We have cited his experiments frequently in our writings about structural work (see Figure 22-3).

Weiss studied the fibrin filaments that form during wound repair. At first, the fibers are oriented randomly. As the clot dissolves, fibers that are not under tension dissolve first, leaving behind a web of oriented fibrin fibers. Fibroblast cells migrate into this web, become oriented along the fibers, and deposit collagen, primarily along tension lines. Any collagen fibers that are not oriented along tension lines are removed by a process similar to the readjustment that took place in the clot. The result is a tissue composed of fibers oriented in the direction that is appropriate to resist the tensional forces.

A comparable mechanism, operating on a larger scale, accounts for the way the musculoskeletal system adapts to tensions created by movement and gravity. The adaptation is accomplished by laying down extra collagen fibers where they are needed for support.

NEED FOR A SYSTEMS PERSPECTIVE

Before seeing the 1977 conference proceedings, we did not know that Paul Weiss was a contemporary of von Bertalanffy during their student days in Vienna. They met in coffeehouses and "milked" each other's minds about systems concepts they were working on from different perspectives.

Weiss needed a systems outlook because of discoveries he made during his Ph.D. research, which concerned the effects of gravity and light on the resting postures of butterflies. His experimental results did not fit with the mechanistic explanations of animal behavior that prevailed at the time. In order to develop a framework that reconciled classic behavior theory and his observations, Weiss turned to the new views of nature emerging in modern physics. He was in the right place at the right time for this, because it was during these years that great insights were leading to the synthesis of quantum theory. Two other young Viennese scientists, Wolfgang Pauli (Nobel Prize, 1945) and Erwin Schrödinger (Nobel Prize, 1933), were major contributors. Key papers on quantum mechanics were published in 1925 and 1926 by Pauli and Schrödinger, Werner Heisenberg (Nobel Prize, 1932), and Max Born (Nobel Prize, 1954).

Weiss wrote about his "holistic system-theory" in 1922, and his monograph was published, after some delay, in German in 1925 (Weiss 1925). An English translation was published in 1959 (Weiss 1959). Von Bertalanffy published his monograph on general systems theory in 1968.

The discoveries Weiss made and his thoughts about them have more than passing interest to those who are concerned with healing and human performance. During a lifetime of experimental laboratory work on development, Weiss observed from many levels and perspectives the way organisms assemble from their components. His system concept was a "silent intellectual guide and helper" during the studies for which he became well known and respected among biologists everywhere.

QUANTUM MECHANICS

Systems, continuum, and quantum concepts have guided the organization of information on the human body and the effects of various therapies. Quantum mechanics explains the fundamental structure of the universe on all scales, and our normal concepts, our language, and our whole way of thinking are inadequate to describe it. This fact created intellectual and emotional problems for physicists because it went to the heart of our conceptualization of the nature of our reality, of what we are taught about the way the world works from the first days of our lives. In the end, the crisis in physics and the soul searching and bewilderment were rewarded with deep and profound insights that have meaning for all of us.

A basic discovery in quantum physics is that, at the subatomic level, no particles exist except in relationship to others. What we call *objects* are in fact points of correlation in an unbroken and interconnected network of events, motions, relations, and energies—the continuum of nature. Subatomic particles and all matter made from them, including our cells, tissues, and bodies, are *patterns of activity* rather than *things*. There is *no thing* that exists by itself. Living nature and the universe as a whole form a seamless dynamic web of interrelated and intercalated parts and rhythmic processes. No fundamental unit or most important part can be isolated.

An earlier publication (Oschman & Oschman 1997) included a chapter on the rediscovery of relationship, quoting various scientists who participated in the great revolution in physics. One of them is Niels Bohr (Nobel Prize, 1922), the great Danish physicist who turned to the teachings of Buddha and Lao Tsu to find a foundation or grounding for the lessons that were emerging from the physics laboratory.

The crisis in perception

In *The Turning Point,* Capra (1982) describes how all of the great crises of our times—terrorism, war, pollution, crime, energy, unemployment, health care—are facets of one and the same crisis, *a crisis of perception.* We are applying concepts of an outdated mechanistic world view to a reality that is not understandable in those terms.

Capra takes the optimistic position that we already have been forced to acknowledge our interdependence and our position as but one component in the continuous fabric of nature. One experience that confirms Capra's position comes from watching the dramatic successes various scientists have when they look beyond their small pieces of the puzzle of nature and see glimpses of the broader context.

In the intellectual sphere, quantum physics has paved the way to a description of the deep interconnectedness of all things (although not all quantum physicists agree with this interpretation), while the various structural and energetic therapies can lead a person to a direct experience of the same natural truth. Modern biology and medicine lie somewhere in between, struggling to integrate reductionism, continuum, systems, and quantum thinking into their intellectual base.

UNSOLVED PROBLEMS OF DEVELOPMENT

Therapists have daily and intimate encounters with the intricate and wondrous forces that shape the human frame; therefore, they experience a personal relationship with the greatest unsolved mysteries of biology.

In *A New Image of Man in Medicine* (Shaefer, Hensel & Brady 1977), the chapter by Weiss follows that by von Bertalanffy. It is entitled "The System of Nature and the Nature of Systems: Empirical Holism and Practical Reductionism Harmonized."

Like other perceptive biologists (Strohman 1993; Wilson 1994), Weiss reiterates the basic fact that we simply do not know how an organism develops from a single cell or a seed into its mature form. There is an illusion prevalent in the halls of science that this is a solved problem, an answered question. The answer goes like this: Chemical processes, directed by deoxyribonucleic acid (DNA) in the chromosomes, build up the various molecules, cells, tissues, organs, and functions that comprise the adult organism. The inherited genetic material is a complete blueprint for the form and function of the adult body. All of life obeys the laws of chemistry and physics; therefore, life is reducible to a sequence of machinelike processes that involve reading the blueprint and translating it into structure. Life is the inevitable outcome of a sequence of linear cause-and-effect events. The chemist, and particularly the molecular biologist, will provide all of the answers to life's questions. Because life is chemistry, pharmacologists, pharmacists, and physicians collaborate to apply these answers to our medical problems.

There are simple and profoundly significant flaws in all of this reasoning and in the ways it is used in our medicine. It is true that physical and chemical processes are taking place in living systems; however, the laws of physics and chemistry do not include concepts of *the living plan* and *form*. In a physics book you can find details of the various *states* of matter: solid, liquid, and gas. You will not find a description of *the living state*.

To say that cells obey laws of chemistry and physics is correct but largely irrelevant. It adds nothing to our understanding, and it deletes the most significant attributes of "aliveness." Many of our medical problems, and their high cost in terms of human suffering and of dollars, persist because we think we have already answered fundamental questions when we have, in truth, not even asked them.

Yes, the DNA contains the hereditary code for the construction of protein molecules such as enzymes, collagen, hemoglobin, and neuropeptides. The mechanisms by which this code is conveyed from generation to generation and the way the code is read and translated into new proteins are among the most important discoveries of recent science. There is, however, a profound illusion that arises from careful observation of life at this microscopic level. In this parcel of the whole, one observes parts (proteins) being precisely and repeatedly created from existing patterns of order (base sequences in the DNA). The fundamental formula, the central dogma of molecular biology, is that DNA *replicates* itself to pass genetic information from one cell to another and from one generation to the next; DNA *transcribes* its information into another molecule, ribonucleic acid (RNA), which can migrate out of the cell nucleus; and the RNA *translates* the information to form specific proteins.

So impressive is the regularity, predictability, and reproducibility of protein synthesis that it seems obvious that the same mechanism must operate at, and thereby create, all levels of size and complexity. The logical mind has a natural tendency to seize upon processes occurring at one level and assume they operate similarly at other scales. It was logical for molecular biologists to predict that we soon would be able to understand how the entire plan of the organism is carried within the DNA and where the plan goes wrong to cause disease. The specific model is as follows:

DNA → RNA → protein → everything else, including disease

A similar point of view was expressed by James D Watson, codiscoverer with Francis Crick of the double helical structure of DNA:

> These successes (in understanding heredity) have created a firm belief that the current extension of our understanding of biological phenomena to the molecular level (molecular biology) will soon enable us to understand all the basic features of the living state.
> WATSON (1970)

Vast sums, many careers, and a whole molecular industry have been devoted to locating the "master plan," its defects, and ways to make repairs, all within the molecular machinery of the body. Although much was learned in the process and some interesting ideas have been tested, the "master plan" and the "causes" of major diseases, such as cancer, remain elusive.

> The quick harvest of applied science is the usable process, the medicine, the machine. The shy fruit of pure science is Understanding. BARNETT (1950)

To solve biological and medical problems, we must examine the basic aspects of life that mainstream biomedicine has been blind to, with chronically disastrous consequences. Viewed whole, the living organism displays properties that cannot be accounted for as a synthesis of the behaviors of parts. Mainstream science has had difficulty with these "whole systems" properties because of the current fashion of taking things apart, because the molecular approach has been enormously successful in some limited areas, because whole system inquiries are multidisciplinary, and because of an emotional fear that a close look at the whole might point to another "force" in the universe that is outside of our present set of beloved laws.

Forces of nature

The other "force" in nature has been given various names in both the scientific and metaphysical literature. Most of these names horrify many scientists: life force, form-giving force, vital substance, healing energy, archæus, Ch'i, K'i, orgone, wakan, puha, Kundalini, prana, bioplasma, odic force, entelechy, life field, guiding principle, love, great and holy spirit. The rhetorical method biomedicine has used to dismiss these concepts is to lump them together under a single concept, *vitalism,* and to brand them with the most dreaded of words: mystical, metaphysical, occult. In scientific circles, a vitalist is far worse than a Communist (Szent-Györgyi 1963).

The "other force" that science has been afraid to look for is not a force in the usual sense of physics (a force is an influence that changes the motion of an object). Instead, the concept of a *life force* encompasses the operation of whole systems properties, vital energy flows, and essential communications, all functioning in a continuum environment. This point of view was expressed more than a century ago by one of the fathers of modern physiology:

> *The genes create structures, but the genes do not control them; the vital force does not create structures, the vital force directs them.* CLAUDE BERNARD (1839)

In the words of Strohman (1993), the genes are important but not on top—just on tap! Genes undoubtedly are involved at every step of development, but this does not mean they deserve all of the credit for establishing order at every level. The most spectacular example to prove this point is provided by identical twins, two individuals who arise from a single egg and are endowed with identical complements of genes. If genetic determinism is the rule, identical twins should always develop into identical adults. Sometimes this happens, and sometimes it does not. "Identical" twins can look so different that it is hard to recognize that they are members of the same family, or they may be so similar that they display remarkable connections, such as two twins in England who speak in precise synchrony.

Scientists around the world now are discovering the nature of the whole systems processes that direct the formation of structures and their functional integration. Vitalism and vitality are being put back into living matter after centuries of being "anti-establishment" and banished from consideration.

The role of light

Light is a major component of the vital communications that lead to wholeness. Of course, mainstream biologists have been certain for a long time that light has no role whatsoever in communication within organisms and that those who profess to see light emanating from the human body are hallucinating. We now know that light is a major component of the vibratory communication system of the body, because such vibrations have been both predicted on the basis of fundamental biophysical theory and measured by sensitive devices (Oschman and Oschman 1994; Popp, Li & Gu 1992). The role of light in living processes has passed from the "hallucination" stage to current science, as documented in a new technical periodical entitled *Biophotonics International* (n.d.) and a number of books devoted to the subject.

We are very impressed with our technological achievements with light, such as fiberoptic systems that enable huge amounts of data to be sent quickly and reliably between continents. One key to the modern technology is the use of optical *solitons,* or *solitary waves.* A later chapter is devoted to explaining precisely what solitons are. Solitons are waves in perfect balance and coherence; the usual tendency to spread out or disperse is canceled because of a high degree of order. So effective are solitons in fiberoptic communications that NTT Labs in Japan found that soliton data can travel the equivalent of 4500 times around the Earth without any loss of information (Voss 1995). Of course, engineers have the audacity to believe they discovered these methods, without realizing that nature took advantage of them and refined them a long time ago. The obvious question is how light can be generated within the body. One answer has emerged from biophysical research on the living matrix. In the past, heat generated by metabolism has been thought of as useless and wasted energy; however, heat is a form of vibration that at the microscopic level involves the back-and-forth motions of atoms in the molecular lattice. Like any other vibration (mechanical waves or sound, electric, magnetic, and light waves) heat is conducted from place to place by the living matrix, and heat is converted into other forms of energy, including light. How this happens will be considered in detail in Chapters 14 and 21.

INTEGRATING KNOWLEDGE

Our exciting task has been to follow the new scientific information as it emerges and to fit it together with understandings and experiences of various therapeutic schools for the benefit of all concerned. This profoundly exhilarating process is not well supported in our culture. As pointed out by the editor of *Nature,* one of the world's most prestigious biomedical journals, grants for producing data are abundant, but there are hardly any awards for standing back in contemplation of the deeper meanings (Maddox 1988). Data proliferate at an astonishing pace, but there is a corresponding impoverishment in the subtle analysis needed to make the data intelligible and useful.

In his article, Weiss recounts the way the deep mystery of life presents itself in the development of the mature organism. He does so from the perspective of a lifetime of careful and distinguished laboratory investigation.

Embryology

The study of embryology has direct relevance to human health and performance. The same generative processes continue to operate in the adult for the constant "turnover" or replacement of parts. The structural, energetic, or movement practitioner can encourage and enhance the flow of information and new materials into the body's frame. This permits their clients to reconstruct, renew, or regenerate themselves along the ideal structural line or pattern that is their birthright (Rolf 1962). That such a regeneration can and does take place, on a large scale, is at the same time a marvel of bodywork, energetic, and movement therapies, and it is a concept that until recently has been virtually unknown in our health system.

To illustrate the central problem of biology in a simple manner, Weiss presents a diagram that we have redrawn as Figure 12-1. It shows a chick embryo that has been removed from its egg and suspended in a test tube in a balanced salt solution (Figure 12-1, *A*).

What happens next is an act of violence of the sort that takes place daily in many laboratories in the name of scientific analysis and medical progress. The hapless chick is homogenized or blenderized into a mush, a uniform suspension of its parts (Figure 12-1, *B*). These parts then are separated or fractionated by centrifugation. The enhanced gravity field in the centrifuge forces the heavier components to the bottom of the tube, while the lighter and soluble components stay on top (Figure 12-1, *C*).

The violent process just described is the fundamental method underlying modern biochemistry. The solution at the top of the last tube is still "alive" in the sense that it contains functional enzymes capable of carrying out chemical reactions essential to living matter. By studying these reactions in the test tube, free of annoying and dis- tracting and irrelevant *structures,* biochemists have been able to map out the metabolic pathways that convert sugars into energy and that, they think, construct living matter. Hence, if some sugar is added to the top of the tube in Figure 12-1, *C*, one can trace its stepwise and energy-producing breakdown into pyruvate, a process described in any biochemistry text. Sadly, though, the once sentient being we referred to as the chick is no longer in any condition to benefit from the process.

Can we resynthesize the chick?

Now Weiss poses the question, Can we reverse the process shown from left to right in Figure 12-1? Can we "synthesize" the chick from the mush? The problem is similar to the question of the origin of life. One popular theory is that living complexity arose from a nutrient-rich "primordial ooze" by the appropriate and more or less fortuitous combination of molecules.

The question Weiss poses may seem absurd, but anyone who spends his or her days thinking about and tinkering with the stuff in the middle and right parts of Figure 12-1 might say, "Why not?" The tubes on the right contain the same molecules as the tube on the left. All we need to know is how to assemble the scrambled parts in the proper order. Because this information is thought to be available in the DNA blueprint and the tube contains both the DNA and the enzymatic equipment needed to read the code and assemble proteins, we have all of the essentials needed to mesh the parts back together into a fully alive chick.

The sad fact that the process shown in Figure 12-1 cannot be reversed to recreate the chick is well documented. What is revealed by stating the question in this way is the central problem of biology: although we understand how the component parts are

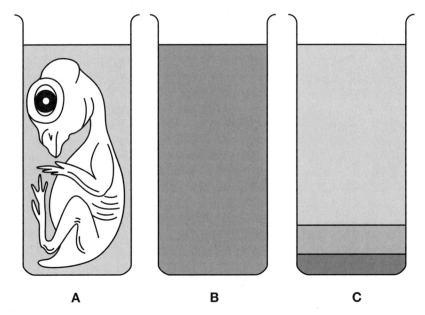

Figure 12-1 A chick embryo is taken from its egg and suspended in a salt solution (**A**). The contents of the tube are homogenized (**B**) and centrifuged to separate the heavier and lighter components (**C**). Does the tube on the right contain all of the information needed to reassemble the chick, that is, to reverse the process from right to left? (Modified from Weiss, P.A. 1973, 'The science of life: The living system—A system for living', in K.E. Shaefer, H. Hensel & R. Brady (eds), *A New Image of Man in Medicine, Volume I, Toward a Man-Centered Medical Science,* Futura Publishing Company, Mount Kisco, NY, Figure 1, p. 25. Used by permission of Blackwell Publishing, Oxford, UK.)

made, we do not know how these pieces assemble into larger structures and into the organism as a whole, with its myriad functions. Molecular and developmental biologists have devised a simple concept to explain all of this. Once the molecules of life are formed, they spontaneously or automatically *self-assemble,* much in the way atoms join together to form a crystal. What we refer to as *life* somehow arises in the process.

The appeal of self-assembly

Self-assembly has a simple appeal, and many examples can be found in the biological literature. As a concept, self-assembly makes an intricate question vanish. The biochemist can describe the various processes and energetics involved in the manufacture of a molecule, while the way this molecule manages to combine with others to produce larger structures is left to a spontaneous self-assembly processes that requires no further explanation. Of course, once the body is assembled, we are still faced with explaining how *functions* arise.

Self-assembly definitely is a piece of the puzzle, but it leaves out the areas of greatest fascination to us: how the whole organism is created and maintained, how it adapts during its lifetime, and how the whole arrangement evolved in the first place.

Self-assembly involves structures with certain charge distributions being attracted to other structures that have complementary charge distributions. They mesh together, similar to the way the bumpy surface of a key fits with the corresponding bumpy surface inside of a lock. Because of the way the charges are distributed along a molecule, a certain alignment will take place when two such molecules come close to each other. What happens in connective tissue involves very long tropocollagen molecules with an intricate distribution of charges that causes these molecules to assemble with their characteristic offset alignment, as shown in Figure 12-2.

Structures form hierarchically; each level is the construction unit for the next size level. One of the rules for such hierarchical ordering in connective tissue is the same as that followed in the manufacture of ropes. The direction of twist changes at each successive level in the hierarchy. The fibers pack more tightly together when the twist alternates. The greater strength achieved by adding more fibers is achieved with an economy of volume (Figure 12-3).

Emergent properties

At each level of complexity, new properties emerge that are not possessed by the parts. These have been called *synergistic, cooperative,* or *collective properties.* Some of these properties arise because dissimilar materials combine to form composites. Connective tissue is an example; it consists of a collagen network within a gel or ground substance. Some examples of properties that emerge in larger assemblies are given by Neville (1993):

- Contractility, as in actin-myosin complexes inside cells
- Infectivity, as in the self-assembled bacteriophage
- Piezoelectricity, or "pressure electricity"
- Colors
- Enzymatic activity

To these emergent properties we can add others:

- Strength
- Resiliency
- Stiffness
- Elasticity
- Transparency, as in the cornea
- Other optical properties, such as birefringence
- Ability to store and release energy
- Tensegrity, or tensional integrity
- Coherence in vibratory states
- Helicity at different levels of order
- Orthogonal cross-lamination (plywoodlike structure)

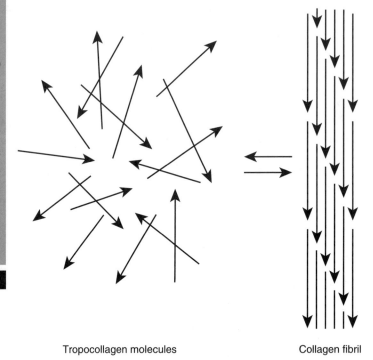

Tropocollagen molecules   Collagen fibril

Figure 12-2 Large numbers of tropocollagen molecules align and assemble to form stable fibrils that are virtually crystalline in nature. These fibrils then join to form bundles, and the bundles gather into larger arrays that form the fascia, tendons, ligaments, cartilage, and bone of connective tissue proper.

- Different arrangements, such as parallel, orthogonal, pseudo-orthogonal, helicoidal, and geodetic
- Liquid crystal properties
- Other electronic and solid-state properties

Each individual biomaterial has a particular combination of properties. When one material combines with another, the result is a composite with additional properties. The standard biochemical view is that essential features of the living state arise "spontaneously." These are the features that allow for the emergence of the higher forms of life. Those who have thought carefully about all of this think there has to be more to the story.

One aspect of the dilemma is that the assembly of connective tissue takes place outside of the cells where the tropocollagen molecules are manufactured. Hence, major portions of development, the formation of larger patterns, shapes, and forms, take place beyond the cell membrane, far away from the DNA "blueprint" and the various enzymes that read it. If DNA is, indeed, the blueprint, for the blueprint to be followed outside of the cell, some sort of remote control must be exerted over the assembly processes.

The collagen polypeptide chain is a right-handed helix.

The individual collagen molecule, called tropocollagen, consists of three collagen polypeptide molecules twisted together in a left-handed helix.

Many tropocollagen molecules twist together in a right-handed helix to form the collagen fibril.

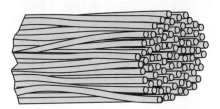

A collagen bundle is composed of many fibrils twisted together in a left-handed helix.

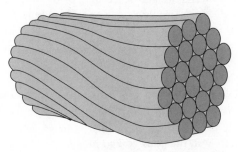

In a tendon, collagen bundles are twisted together in a right-handed helix.

Figure 12-3 The assembly of collagen protein to form larger fibers follows the same scheme used in rope manufacture. Individual strands are twisted together in one direction to make a larger fiber. Strands of these larger fibers are twisted together in the opposite direction to make yet larger fibers, and so on, until we have a tendon. (From *Gray's Anatomy*, 35th Edition, Warwick & Williams, Figure 1.51c, W B Saunders, Philadelphia, 1973, with permission from Elsevier Science.)

The simple answer to this problem is that the molecular structure, transcribed from the DNA blueprint, includes features that cause the molecules to spontaneously align and order themselves in particular ways. Another answer is that the DNA specifies the manufacture of specific enzymes that serve as the "hands" and "tools" of the genetic material, carrying out DNA's instructions by guiding assembly and gluing things together. Self-assembly is complemented with directed assembly.

The organism does not, in fact, assemble

Weiss has a surprising perspective on all of this. He tells us that the organism does not, in fact, assemble. "The basic premise that embryonic development is an assembly process is false." The orderly patterns within an organism and its integrated behaviors are not products of an assembly process in the usual sense of a set of linear cause-and-effect chains. This perspective demolishes Watson's ambitious plan for molecular biology to solve all of the riddles of nature. It also confirms that our current national research focus, the Human Genome Project, will not, as advertised, lead to cures for all diseases.

We can watch an automobile being put together on a linear assembly line. A large number of steps result in a complete functional car. The organism, however, is not the sum, the outcome of a comparable linear sequence of steps. In essence, the automobile is finished when it comes off the assembly line and deteriorates thereafter. In living systems the "assembly line" and the flow of components never cease. The organism possesses properties that the automobile lacks: it can repair and renew itself; it can change its structural organization, or adapt, according to the way it is used; it has consciousness, whatever that is. In contrast to the automobile, which wears out, the body always has the possibility not only of repairing but also of *improving* in its functional integration.

Something is missing

We have learned much about the mechanics and scheduling of the various processes taking place during embryonic development, but this is not the whole picture. That something is missing is revealed by careful study of the behavior of cells in the growing organism. Embryologists have long recognized that there are specific regions within the early embryo that are earmarked as forerunners of the various organs: heart, liver, kidney, brain, etc. The embryo is a busy place; cells wander about until they settle down and form relationships that create a specific organ. This process has been studied so thoroughly that there is no doubt that cells undergo a profound change when they switch from an early indeterminate or undifferentiated wandering state, with the capacity to form *any* organ in the body (called *totipotent*), to a determinate, differentiated, or *committed* state.

Breaking a law of biology

There is a biological law about this process: differentiation is a one-way street. The process is irreversible. Cells cannot resume their totipotent condition once they have gone beyond a certain point.

Like any statement of an absolute in biology or medicine, there are annoying exceptions that ruin the lovely law. For example, Gurdon (1968) showed that even highly differentiated intestinal epithelial cells in frogs retain the genetic information sufficient to produce the whole animal. He demonstrated this by transplanting nuclei from the highly differentiated intestinal cells into frog eggs that had undergone removal of their nuclei. These eggs developed into normal embryos and adults capable of reproducing. The conclusion of this and other experiments like it is that differentiation does not change the genome in a permanent way, but instead alters the *expression* of the genes so that only those needed in a particular tissue are active. Dedifferentiation can and does occur. This phenomenon is extremely relevant to regeneration.

Robert O. Becker has pointed out repeatedly that regeneration of damaged or missing body parts is a far more natural and effective medical technology than the use of prosthetic devices or transplants. The preponderance of biomedical research aims at developing mechanical substitutes for hearts and other organs, or transplantation, rather than determining how to induce the body to form its own replacements. The public is kept aware of this technology in adventures of various bionic creations (e.g., "Robocop").

Becker has shown that electrical stimulation can cause cultures of so-called *differentiated cells* to dedifferentiate into totipotent cells capable of forming all of the tissues needed to replace a lost or damaged part. His fascinating description of this breakthrough research can be found in his book written with Gary Sheldon entitled *The Body Electric* (Becker & Sheldon 1985, pp. 141-143).

Half a billionth of an ampere

The key to Becker's demonstration was reducing the strength of electrical stimulation. He had calculated that somewhere between a billionth and trillionth of an ampere would be effective. Following the usual way scientists tend to look at such matters, he assumed a larger current would be even more effective. This proved to be incorrect. In the experiments on frog red blood cells Becker conducted with a student named Frederick Brown, the test current was reduced one step at a time until they reached the lowest current the apparatus could produce, about half a billionth of an ampere. This stimulation produced a dramatic dedifferentiation.

The reason for mentioning these experiments is that they have profound medical implications, yet they have been virtually ignored by researchers in this country. Moreover, the extremely low levels of stimulation required to produce dedifferentiation and regeneration are comparable to the levels present in the human energy field. Hence it is not surprising to find accounts of acupuncture and other complementary approaches facilitating repair and even regeneration of tissue. This is an example of an important line of medical inquiry that is proceeding in the complementary medical community while it is virtually untouched by biomedical research.

NATURAL APPROACHES

These discoveries relate to our health crisis. Biomedicine continues to look toward the most difficult, painful, and expensive methods, such as artificial organs and organ transplants, instead of examining the more natural approach of regeneration:

All the circuitry and machinery is there; the problem is simply to discover how to turn on the right switches to activate the process. WEIL (1995)

The fact that researchers in the United States are not searching for the "switches" that control regeneration could provide good fodder for those who are fascinated by conspiracy theories based on who benefits the most and who suffers the consequences of the direction taken by medical research. This view must be tempered with the simple fact that basic biological myths, such as the irreversibility of differentiation, self-assembly, or the inability of mammals to regenerate limbs and organs, tend to take on a life of their own. They persist in the minds of scientists long after they have been found to be incorrect or incomplete. Whole avenues of beneficial research are closed because everyone agrees that certain phenomena cannot happen. Hence we wait for a new era of healing-oriented medicine that is possible if we are willing to look more deeply into regeneration and other phenomena that presently are considered impossible.

Embryological assembly

Now we return to the embryological assembly process. If we follow individual cells from one stage to the next, we find that they take different paths in different individuals. Weiss provides a drawing of this, which we have modified and show as Figure 12-4. Although the overall outcome is two adults with a similar set of similarly structured organs, the formative processes do not correspond to each other. Determination of an individual cell is a response to its position in the *orderly pattern of the future whole* rather than a result of an orderly assembly process.

Two complementary interpretations emerge from this discovery. First, there is far more order in the whole than would be expected from the seemingly capricious behaviors of the parts as they assemble; therefore, the ordered state in the adult cannot be the blind outcome of a defined sequence of microscopic cause-and-effect chain reactions in the embryo. Some other whole-systems processes are operating.

There are other ways of viewing the same principle. Despite their freedom to continuously move about and alter their shapes in an infinite variety of ways, the parts join together to form larger structures with definable organization—systems. There is a drive present within the embryo that leads toward a perfect outcome, even though there are many ways to get there. The story is far more exciting than an automobile assembly line.

Defining "system" and "systemic cooperation"

Now we come to a definition of *system*. A system is an entity, a whole, whose configuration and functioning are achieved and maintained. The form is not conserved *because of* a rigid linking together of constituents but despite the *absence* of tightly interlocked components. A system is a complex unit in space and time so constituted that its parts, by *systemic cooperation,* attain a certain configuration and function and preserve them in spite of constant changes in the environment. Despite their ability to change shape and position, the parts are coordinated by being enmeshed in a dynamic fabric—the continuum that is the subject of this chapter and that is considered throughout this book.

The pattern of the whole is a joint property of the interlinked components. This property was referred to earlier by Szent-Györgyi: *"Whenever we separate two things, we lose something, something which may have been the most essential feature."* It is the loss of this essential "something," the loss of "systemic cooperation," that prevents the reassembly of the chick from the mush shown in Figure 12-1.

Intractable diseases may not really have causes in the usual sense but instead may be the consequence of a loss of "systemic cooperation."

It is the restoration of Szent-Györgyi's "something" and Weiss's "systemic coopera-tion" to the whole organism that is the miracle of modern hands-on and energetic bodywork. Identifying precisely what it is that passes from therapist to patient and in the opposite direction is a subject worthy of investigation. Even more interesting are the emissions from diseased or disordered tissues. Indeed, this is one of the most excit-ing topics we could possibly research. This book opens up many opportunities for a study of energetic exchanges by describing the matrix that is the substrate for exchang-ing of information.

Wholism and reductionism are not separable

Now comes a simple revelation: wholism and reductionism are analytical procedures that are not separable. Neither can describe nature completely; therefore, they are *com-plementary* rather than *contradictory* representations. Here *complementary* is used in the same sense that Niels Bohr used *complementarity* to reconcile the wave-particle

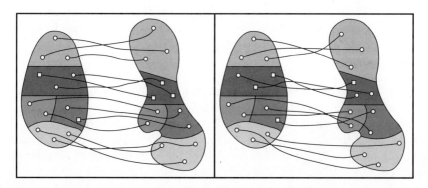

Figure 12-4 Two individual embryos of the same species are followed during the same time period of development. Specific regions, shown with different kinds of stippling, are earmarked as forerunners of specific organs such as heart, liver, kidney, brain, etc. Although these topographic regions are predictable, the cellular migrations are not identical. The lines connecting the early and later stages show what happens when we trace the movements of individual cells. Although the overall outcome is two adults with a similar set of similarly structured organs, the formative processes do not correspond to each other. Determination of an individual cell is a response to its position in the *orderly pattern of the future whole* rather than a result of a linear assembly process. (Modified from Weiss, P.A. 1973, 'The science of life: The living system—A system for living', in K.E. Shaefer, H. Hensel & R. Brady (eds), 1977, *A New Image of Man in Medicine, Volume I, Toward a Man-Centered Medical Science,* Futura Publishing Company, Mount Kisco, NY, Figure 2, p. 28. Used by permission of Blackwell Publishing, Oxford, UK.)

duality in physics. Matter can be viewed as a wave or as a particle. The phenomena are very different conceptually, and there is no way to reconcile the two ways of viewing nature, yet the two views complement each other in that the properties of matter cannot be fully understood without considering both of its aspects.

Notice how there is an evolution of our conceptualization of how nature works. For the physicists, the sequence of discovery involved particles, waves, their incompatibility, and resolution of the dilemma by complementarity. There now are signs of another step in this intellectual process because of a breakdown in the fundamental truth that the wave and particle views cannot be reconciled. Experiments have shown that atoms can have wavelike and particle-like behavior *at the same time*. For a review of this situation, see Gribbin (1995). The point is that complementarity is a useful point of view that, like most of our views, is giving way to a more comprehensive perspective. The irreconcilable is being reconciled.

If we analyze the whole by focusing on its parts (which in themselves are wholes of a smaller order), we acquire a certain microscopic precision in our descriptions. This precision is obtained at the expense of losing the information, the rules, Szent-Györgyi's "beautiful vistas," the "something," or the "systemic cooperation" that enables the higher levels, what we call *organization,* to exist.

Organization, precision, and information

Weiss presents an illustration to show this, which we have redrawn as Figure 12-5. We have modified his drawing to include the degree of organization in relation to precision and information. We also added our notation of the part of this scheme that is the focus of Structural Integration and other sophisticated hands-on somatic techniques. By operating on the level of principles of whole-body organization and information, integration, and systemic cooperation, significant results are obtained, without tinkering with or even knowing about details at the level of tissues, cells, molecules, atoms, and subatomic particles.

In physics, biology, and human lives, the transition from randomness to order is most remarkable and most visible at the level of the whole; however, all parts of the continuum participate. If the focus of a therapy is on the whole organism, the parts will take care of themselves. This is a predictable aspect, perhaps even a law of the natural living continuum. It is dramatically illustrated by the first part of the first session in Dr. Rolf's work. When breathing is liberated, by allowing the ribs and diaphragm to move more freely, the oxygen supply to every cell in the body is enhanced and metabolism is everywhere facilitated. When the musculoskeletal system is balanced, other systems inevitably will follow. Balance and flexibility are contagious. If the focus is on the parts, the whole may or may not respond in the manner desired. In this sense, cause and effect are more predictably linked in complementary wholistic approaches.

CLINICAL SIGNIFICANCE

There is an obvious clinical significance to all of this. Modern medicine tells us to wait for science to produce the analytical methods that will describe human life and enable us to solve medical problems. The public senses the fallacy of this approach and is justifiably impatient when obvious fundamental guiding principles are left out of the search. Complementary practices are growing in popularity because the public perceives that complementary techniques *complement* the wisdom of nature.

At this point, we can be specific about the importance of thinking about these ideas. The outcome of this line of inquiry will profoundly affect the future of all of us. This is so because our most debilitating, painful, and costly medical problems are breakdowns at the level of whole systems, which then lead to observable problems with the parts and not the other way around.

Mainstream medicine and mainstream politics continue to bumble around in the kindergarten intellect of linear cause and effect and in attempting to fix the broken parts. Occasional brilliant successes from this approach sustain the hope for predictable outcomes, expressed in our love for the concept of "a pill for every disease" that justifies a pharmaceutical approach to life and pathology. We mortgage our future to pay for ineffective and harmful drugs that often make matters worse. A child born today begins life with a share in the national debt of some $18,000, and a good por-

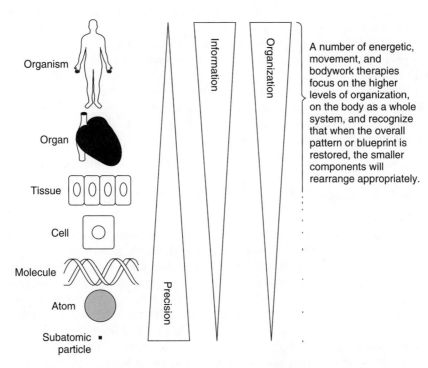

Figure 12-5 As our analysis descends from the whole to its smaller parts, we gain microscopic precision and detailed knowledge, but we lose the essential information that governs the structure and function of the whole, which we can refer to as *organization*. Structural Integration and other "hands-on" approaches operate at the levels of the whole organism and its larger, visible, and palpable components. The microscopic pieces participate, but they "take care of themselves" in the sense that they do not have to be given detailed attention.

tion of this has been given to the pharmaceutical industry for incompetent drugs. Real breakthroughs are available to our medicine once we go beyond our natural tendency to look at one piece of the world picture at a time. Our lack of success in solving fundamental biological and medical problems continues because we have focused our attention only on the linear sequences of chemical reactions and have left out the context or matrices in which these reactions take place, which include connective tissue, water, and the electromagnetic field.

Cause and effect

Weiss provides a perceptive analysis of our current mode of thinking. Yes, within the organism chains or sequences of chemical reactions can be mentally singled out and treated as though they were single tracks. On closer study, we see that these sequences are confined to their linear tracks because of ordering influences that are imposed from outside.

One of these influences is our choice of experimental methods and modes of perception. We look at nature and are struck by the phenomena that seem consistently and regularly coupled or correlated. We then discover that we can interfere with the connection between them: if we modify one, we observe a corresponding change in the other, and we delight in our demonstration of a cause-and-effect relationship.

In going through this process, we have not observed, ignored, or discounted as erratic or irrelevant the changes taking place in parts other than those we have focused on. Sometimes we note other events taking place but nullify them in order to "keep conditions constant." We fail to notice that in a dynamic continuum, changing one parameter *inevitably* changes many if not all of the others.

The distinguished regulatory physiologist Edward F. Adolph stated it this way:

The mature scientist knows that cause and effect are elusive because of the presence of multiple correlations. No properties are uncorrelated, all are demonstrably interlinked. And the links are not single chains, but a great number of criss-crossed pathways.

ADOLPH (1982)

In making correlations and finding cause-and-effect relations, we acquire some *data* about life and that is all; however, we seldom stop there. We go on to endow our correlations with a life of their own, separated from the context from which they were extracted. In this manner we erect simple laws from a small portion of the continuum of nature. Viewed from the whole, these laws are never so simple. Because of interconnectedness, no aspect is ever negligible. Only we, the observers and manipulators of nature, are able to *declare* an aspect to be negligible or relevant in relation to our particular problem, our level of study, or our vested interests.

The current health crisis and the increasing contrast between the results with conventional and complementary approaches can lead us toward a more sane, realistic, and balanced view of ourselves in all of our relations. Like it or not, the future of mainstream medicine depends on its willingness to look at the deep meaning of complementary medical practices and the natural wisdoms of the body they reveal. Ultimately the public will judge the entire biomedical research enterprise by its

ability to uncover methods that lead to improvements in curing, caring, and comfort of patients. Which innovations emerge from conventional versus unconventional sources, or from their integration, will be a matter of historical rather than practical importance.

Restoring systemic cooperation

We can get even more specific about the reasons for making these statements. Our serious, costly, and painful health problems arise when our inherent whole systems mechanisms for defense and repair are compromised. One of the accomplishments of complementary medicine is to slow or reverse the accumulation of subtle disorders, imbalances, or communication breakdowns that compromise our immune defenses and repair systems. Structural and movement approaches stimulate the body's repair systems to repair themselves, restoring "systemic cooperation," with many beneficial consequences. It is much easier to prevent than to cure, and prevention is definitely not a cure given in advance.

Wound healing and disease resistance are among the most remarkable of living processes involving the integrated and cooperative activities of many regulatory systems throughout the body. Level upon level of intricate control systems and feedback loops join together to accelerate and inhibit myriad activities, on a moment-to-moment basis. *Even the smallest pinprick alters the behavior of millions of cells and billions of molecules.* Whereas biology and medicine are aware of some aspects of this, what is truly remarkable is that therapists, with their "hands on," are able to interact with all of it in profoundly significant ways that have virtually escaped scientific investigation.

An invisible dynamic

The following summarizes technical details of the invisible dynamic that lies under the fingertips of the therapist using touch to enhance functioning by restoring structural, energetic, and kinetic balance to the living system:

- Any trauma sets off an intricate cascade of physiological activities and adjustments. If the injury is severe, both local and systemic responses are initiated, and all of the systems in the body can be involved.

- An injury or disease triggers the migration of a variety of kinds of cells toward the site of the problem. Platelets release clotting factors; waves of white cells move in to fight infection and to resorb nonself materials; epithelial cells, fibroblasts, and osteoblasts crawl into position to replace damaged tissues and to form scar tissue. These events are triggered by a variety of messages that radiate from a site of disorder. A wide range of stimulating and inhibiting factors activate and integrate wound repair or tumor absorption, and then wind down the processes when healing is complete.

- Cells migrating into a wound or a tumor must be replaced by cell division. If the problem is extensive, clotting, inflammatory reactions, or tumor resorption consume cells, which must be replaced by proliferation of stem cells in distant organs, such as the lymph nodes, bone marrow, liver, or even the brain, with its "brain marrow" (Suslov et al. 2000). Healing therefore involves an array of dynamic interactions between local and systemic processes. Fever, allergic reactions, and the fight-or-flight response are all examples of lifesaving systemic regulations.

- Many physiological responses are nonlinear because of the intricate web of whole-body feedback and feed-forward regulatory pathways that are involved. Some activities persist for weeks after an injury. Vital living processes must be maintained during repair. This may require temporary shifting of functions to other parts, systems, or pathways. Redundancy is the hallmark of vital functions.

Each injury and disease is different, and the body's response must be precisely appropriate if the organism is to be fully restored to its original condition. Appropriate responses *do not* depend on the competence of the parts to work together—they are poised all times to do this. Instead, the responses depend upon the context, the quality of the living matrix to act as a continuous medium to support essential communications and cellular migrations.

Healing references the blueprint

Within limits, defense and repair mechanisms operate to restore the body to its original form. This fact reveals a relationship between defense and repair mechanisms and the informational systems that define, organize, and maintain one's unique body structure and features, as well as the detailed architecture of every part. Moreover, damaged tissue is replaced with tissue of the same kind rather than tissue of a different kind. Bones are repaired with bone tissue, skin with skin of appropriate thickness, mucous membranes with mucous membranes, fascia with fascia, etc. It is obvious that wound repair activates the elusive and controversial form-giving forces that biologists have been debating and dismissing as vitalist for centuries.

More often than not, serious problems result from an accumulation of disorder rather than from a single isolatable cause. As an organism develops, ages, and acquires traumatic experiences, the network of physiological compensations and adjustments within the matrix becomes more intricate, leading to a continuous range in ability to respond to genetic problems or injury or disease. Each experience produces multiple rearrangements of the entire network. Eventually "systemic cooperation" itself is compromised.

The biologist or biomedical researcher recognizes that the entire body is extremely complicated, meaning it is composed of a great number of simplicities. The conventional approach is to study these simplicities one at a time and then add them to get the whole; yet organization, as we saw earlier, has the opposite property. Complex phenomena *cannot* be understood as the sum of the behaviors of isolated parts. Failing to notice this, conventional medicine has no measures of the properties of the connective tissue continuum. The living matrix is a given, a constant, something that everyone has, just as air and gravity are everywhere present, taken for granted. Connective tissue is the stuff the surgeon has to cut through or push aside to get to the target. Osteoporosis and range of motion are some of the few changes that conventional medicine recognizes and measures in the living matrix system.

INTERACTING WITH COMPLEXITY

What brings vitality and aliveness to all of the matrix are not the linear sequences of chemical reactions, messages, or physiological events but the ways they are regulated and integrated. We are dealing with a vast network of processes—proliferations, specializations, movements, differentiations, dedifferentiations, interactions, cross-linkages, feed-forwards and feedbacks of startling complexity and diversity—all directed at a goal of maintaining and restoring the *orderly pattern of the whole.*

Complementary medicine uses a variety of ways of interacting with this complexity. "Hands-on" and "hands-off" therapies focus on the physical and energetic matrix that supports essential communications and cell migrations instead of trying to deal with specific disorders. Systems and continuum thinking are ways of conceptualizing what is going on. Intuition and intention (which are systems and continuum properties) are keys to optimizing structure and function.

If there is one lesson to be learned from the broad spectrum of experiences of complementary practitioners, it is that detailed understandings of physiological and molecular processes, although of interest, are not essential for working at the level of the whole. There are simple reasons for this, articulated in a variety of ways by Weiss, von Bertalanffy, Szent-Györgyi, Capra, Rolf, and others. The whole is not the sum of the parts, and its behavior is not a synthesis of the behaviors of the parts. The whole is governed by certain natural processes, wisdoms, rules, or principles that are practical and discernible. *Continuum* is one of these principles. As we stated earlier, continuum refers to the state of continuous organic wholeness fundamental to the structure and behavior of the natural world. It is the willingness to operate at this level, which is at the edge of present scientific mystery, that distinguishes the results of complementary therapies from those of conventional medicine.

The central issue has been wonderfully articulated by Szent-Györgyi, and we repeat it here to lead us to a conclusion for this chapter:

> *If elementary particles are put together to form an atomic nucleus, something new is created which can no longer be described in terms of elementary particles. The same happens over again if you surround this nucleus by electrons and build an atom, when you put atoms together to form a molecule, etc. Inanimate nature stops at the low level or organization of simple molecules. But living systems go on and combine molecules to form macromolecules, macromolecules to form organelles (such as nuclei, mitochondria, chloroplasts, ribosomes or membranes) and eventually puts all these together to form the greatest wonder of creation, a cell, with its astounding inner regulations. Then it goes on putting cells together to form "higher organisms" and increasingly more complex individuals, of which you are an example. At every step new, more complex and subtle qualities are created, and so in the end we are faced with properties which have no parallel in the inanimate world, though the basic rules remain unchanged.*
>
> SZENT-GYÖRGYI (1974)

ENGENDERING NATURAL WISDOM

Here, then, is a central unsolved biological problem that we shall be returning to many times and about which we shall ask you for your insights: What are the "systems principles" that lead, at every level of the continuum of nature, to higher levels of order and harmony, and how can we live our lives from such a perspective? After half a century of looking into this question, Szent-Györgyi (1974) concluded that living matter has a "drive to perfect itself." The real nature of this drive provides an unanswered and even untouched question for the biologist. Where, how, and when this drive arose during evolution is an even greater mystery, but one that is worthy of exploration. After taking many directions, the search is beginning to focus on the most natural strategies for bringing forth natural wisdom, order, and systemic cooperation within the natural continuum.

REFERENCES

Adolph, E.F. 1982, 'Physiological integrations in action,' a supplement to *The Physiologist*, vol. 25, pp. 1-67, April.

Barnett, L. 1950, *Life*, 5 January.

Becker, R.O. & Sheldon, G. 1985, *The Body Electric: Electromagnetism and the Foundation of Life*, William Morrow and Company, New York.

Beinfield, H. & Korngold, E. 1991, *Between Heaven and Earth: A Guide to Chinese Medicine*, Ballantine Books, New York.

Bernard, C. 1839, *Des Liquides de l'Organisme*, Tome III. Bailliere, Paris.

Biophotonics International is a bimonthly publication of Laurin Publishing Co., Inc., P.O. Box 4949, Pittsfield, MA 01202-9933, telephone 413-499-0514, telefax 413-442-3180, e-mail: Photonics@MCIMail.com. It is distributed without charge to researchers, engineers, practitioners, technicians, and management personnel working in the fields of medicine or biotechnology.

Brown, F.A. 1980, Free-running rhythms and biological clocks: A 1980 perspective. Unpublished manuscript.

Capra, F. 1982, *The Turning Point: Science, Society and the Rising Culture*, Simon and Schuster, New York.

Gribbin, J. 1995, *Schrödinger's Kittens and the Search for Reality*, Little, Brown and Co., Boston.

Gurdon, J.B. 1968, 'Transplanted nuclei and cell differentiation: The nucleus of a cell from a frog's intestine is transplanted into a frog's egg and gives rise to a normal frog. Such experiments aid the study of how genes are controlled during embryonic development', *Scientific American*, vol. 219, pp. 24-35.

Heymann, J. 1995, *Equal Partners: A Physician's Call for a New Spirit of Medicine*, Little, Brown and Co., Boston.

Maddox, J. 1988, 'Finding wood among the trees', *Nature*, vol. 333, p. 11.

Neville, A.C. 1993, *Biology of Fibrous Composites: Development Beyond the Cell Membrane*, Cambridge University Press, Cambridge, UK.

Oschman, J.L. 1993, The natural science of healing. Manuscript available from N.O.R.A. P.O. Box 5101, Dover, NH 03821, tel. 603-742-3789.

Oschman, J.L. & Oschman, N.H. 1997, 'Book review and commentary', in *Biological Coherence and Response to External Stimuli*, ed. H. Herbert Fröhlich, Springer-Verlag, Berlin, 1988. Manuscript available upon request from N.O.R.A., P.O. Box 5101, Dover, NH 03821, telephone 603-742-3789.

Pellegrino, E.D. 1974, 'Educating the humanist physician: The resynthesis of an ancient ideal', *Journal of the American Medical Association*, vol. 227, pp. 1288-1294.

Popp, F.A., Li., K.H. & Gu, Q. (eds) 1992, *Recent Advances in Biophoton Research and Its Applications,* World Scientific, Singapore.

Rolf, I.P. 1962, 'Structural integration. Gravity: an unexplored factor in a more human use of human beings', *Journal of the Institute for the Comparative Study of History, Philosophy and the Sciences,* vol. 1, pp. 3-20.

Shaefer, K.E., Hensel, H. & Brady, R. (eds) 1977, *A New Image of Man in Medicine, Volume I, Toward a Man-Centered Medical Science,* Futura Publishing Company, Mount Kisco, NY.

Strohman, R.C. 1993, 'Ancient genomes, wise bodies, unhealthy people: Limits of a genetic paradigm in biology and medicine', *Perspectives in Biology and Medicine,* vol. 37, pp. 112-145.

Suslov, O.N., Kukekov, V.G., Laywell, E.D., Scheffler, B. & Steindler, D.A. 2000, 'RT-PCR amplification of mRNA from single brain neurospheres', *J Neurosci Methods,* vol. 96, pp. 57-61.

Szent-Györgyi, A. 1957, *Bioenergetics,* Academic Press, New York, pp. 38-39.

Szent-Györgyi, A. 1963, 'Lost in the twentieth century', *Annual Review of Biochemistry,* vol. 32, pp. 1-14.

Szent-Györgyi, A. 1974, 'Drive in living matter to perfect itself', *Synthesis,* vol. I, pp. 14-26.

Thomas, L. 1973, *Lives of a Cell: Notes of a Biology Watcher,* Bantam Books, Inc., Toronto

von Bertalanffy, L. 1968, *General System Theory. Foundations, Development, Applications,* Braziller, New York. Revised and enlarged edition: The Penguin Press, London, 1971.

Voss, D. 1995, 'You say you want more bandwidth? Solitons and the erbium gain factor', *Wired,* July, p. 4.

Watson, J.D. 1970, *Molecular Biology of the Gene,* 2nd ed, ed. W.A. Benjamin, New York.

Weil, A. 1995, *Spontaneous Healing. How to Discover and Enhance Your Body's Natural Ability to Maintain and Heal Itself,* Alfred A Knopf, New York.

Weiss, P. 1961, 'The biological foundation of wound repair', *Harvey Lectures,* vol. 55, pp. 13-42.

Weiss, P.A. 1925, 'Tierisches Verhalten als "Systemreaktion." Die Orientierung der Ruhestellungen von Schmetterlingen (Vanessa) gegen Licht und Schwerkraft', *Biologia Gen,* vol. 1, pp. 168-248.

Weiss, P.A. 1959, 'Animal behavior as system reaction: Orientation toward light and gravity in the resting postures of butterflies (Vanessa)', *General Systems: Yearbook of the Society for General Systems Research,* vol. IV, pp. 19-44.

Weiss, P.A. 1973, 'The science of life: The living system—A system for living', in Shaefer, K.E., Hensel H. & Brady R. (eds). 1977, *A New Image of Man in Medicine, Volume I, Toward a Man-Centered Medical Science,* Futura Publishing Company, Mount Kisco, NY.

White, F. 1987, *The Overview Effect,* Houghton-Mifflin, Boston.

Wilson, E.O. 1994, *Naturalist,* Island Press/Shearwater Books, Washington, DC.

13 A crisis in bioenergetics

Many of life's most remarkable experiences can begin to be understood by realizing that the body possesses a high-speed, solid-state, semiconducting electronic communication network that reaches into every nook and cranny of the organism. Of course, most biologists were not prepared for Prof's radical idea that electrons and protons might have something to do with life. To this day, many scholarly and knowledgeable academics remain blissfully unaware of the progress that has been made by scientists studying electronic and submolecular biology; however, Prof was convinced that life's activities are far too rapid and subtle to be accounted for solely by signaling in the nervous system and signal molecules randomly diffusing from tissue to tissue. The thesis of this book is that communications within the living matrix organize, integrate, and energize the myriad of functions that occur within us every instant of our lives.

DESIGNING AN ORGANISM

One way of looking at this is from the perspective of the biologist. Ask the question: What would be the best way of designing an organism so it can respond to its environment in a way that gives it the best chance of survival? Then give the organism millions of years of evolutionary trial and error to test various designs, using all of the laws of physics that are available in nature. What will the outcome?

I believe that the logical outcome will be an organism in which every tissue, cell, molecule, and atom is organized in a highly specific and precise manner that allows all parts to communicate as rapidly as possible with all others. This will enable the organism to find nourishment, protect itself, and respond to a crisis in a way that will provide the highest probability of survival.

INTRODUCING THE PROBLEMS OF ENERGY AND INFORMATION

A fundamental problem in biology is how energy and information move about within living systems. There is a standard textbook view implying that these are solved problems. Energy is conducted in the form of molecules, such as sugar or adenosine triphosphate (ATP), which have chemical bonds that can break and release their energy to power life's activities. Information is conducted by action potentials in the nervous system and by hormones and other factors circulating in the blood and other body fluids.

A CRISIS IN BIOENERGETICS

There are serious unsolved problems with current academic models of energy and information flow. There is, in fact, a continuing *crisis in bioenergetics*. The notion of a crisis comes from a symposium held in 1973 at the New York Academy of Sciences (Green 1974). This symposium was devoted to the mechanism of energy transport in biological systems. Leading experts in bioenergetics discussed the crisis, but no solution emerged. This crisis persists in the background of all of biochemistry.

A summary of the 1973 symposium was published in *Science* (Green 1973), and it makes such interesting reading that a copy is appended to this book (Appendix C).

One important aspect of the crisis in bioenergetics is stated simply in the illustration shown in Figure 13-1. Proteins transform chemical energy into mechanical energy, giving rise to all movements and cellular activities. Proteins are long polymeric molecules.

It is accepted that the universal unit of energy is the energy released by hydrolysis of ATP, which is synthesized in cellular organelles called *mitochondria,* during oxidation of food (e.g., glucose). The energy is stored in the phosphate bond, affectionately referred to as *squiggle-P* or ~P. When the energy is released, the phosphate is converted to inorganic phosphate, denoted as P_i. The ATP hydrolysis is catalyzed by enzymes, called *ATPases,* located at particular places on the protein molecule. Generally the energy released by the ATPase is used in some other part of the molecule. *How does this energy get from the site of ATP hydrolysis to the place where the energy is needed to produce an action?*

This problem has been solved by the standard model of muscle contraction. It has been discovered that the ATPase is located on the head of the myosin molecule, adja-

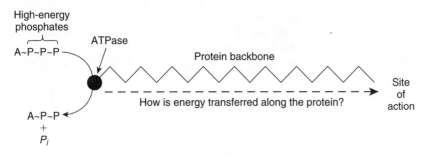

Figure 13-1 The crisis in bioenergetics. Energy released by the hydrolysis of adenosine triphosphate (ATP) by ATPase powers cellular activities. How does this energy get from the site of hydrolysis, shown at left on the protein, to the site where the energy is needed, shown at right?

cent to the place where the action takes place; therefore, the standard model does not require that energy travel along the myosin from one end to the other. The precise manner by which the energy released by ATP hydrolysis is converted into movement remains unknown. Moreover, the proximity of the ATPase to the site of action does not take away from the theoretical importance of the concepts that will be described next, which involve a mechanism for energy and information flow along proteins and thence through the entire matrix of the body.

There is another aspect of the crisis in bioenergetics. The principle of *conservation of energy* is a fundamental of physics. Energy cannot be created or destroyed; it can only be transformed from one form to another. If ATP and other high-energy phosphate compounds can supply the energy for muscle contraction, it should be possible to add up the ATP energy in a muscle before and after contraction and show that the energy the muscle is able to liberate is equal to the phosphate bond energy consumed. The problem is that the numbers do not add up right. This was recognized many years ago by researchers studying muscle energetics, but it has been all but forgotten. For two reviews of the subject, see Homsher and Kean (1978) and Curtin and Woledge (1978). The fact that these review articles are no longer cited by muscle researchers shows that the problem simply has been left behind.

This chapter has introduced a profound problem in biology. It is a problem that is being solved through the understandings that will be described later in the book. Before doing this, though, we need to take a close look at the textbook perspectives on how senses sense, nerves signal, and muscles produce actions. We then will propose that all of this information, although quite correct, is just part of the story. We will propose that there is a *primitive* (meaning evolutionarily ancient) system in the body that existed long before the nervous system evolved and that is capable of sensing and responding faster than the systems you will read about in the physiology texts. We will return to the crisis in bioenergetics and see how it can be resolved by this new understanding of the living matrix.

REFERENCES

Curtin, N.A. & Woledge, R.C. 1978, 'Energy changes in muscular contraction', *Physiological Review,* vol. 58, pp. 690-761.

Green, D.E. 1973, 'Mechanism of energy transduction in biological systems: New York Academy of Sciences Conference', *Science,* vol. 181, 583-584.

Green, D.E. (ed.) 1974, 'The mechanism of energy transduction in biological systems', *Annals of the New York Academy of Sciences,* vol. 227, pp. 1-675.

Homsher, E. & Kean, C.J. 1978, 'Skeletal muscle energetics and metabolism', *Annual Review of Physiology,* vol. 40, pp. 93-131.

14 Introducing biological coherence

The existence of coherent excitations in active biological systems has been established in recent years. . . . This does not mean that a "theory of biology" has been established but it implies that such a theory will make use of such excitations.   Herbert Fröhlich (1988)

Science is built on the premise that Nature answers intelligent questions intelligently; so if no answer exists, there must be something wrong with the question.   Szent-Györgyi (1972)

WHAT IS COHERENCE?

As a foundation for the chapters ahead, the reader will benefit from insight into the nature of *biological coherence*. We will define coherence, describe the origins of the biological aspects of the subject, and provide references to the literature for those readers who wish to delve more deeply.

A valuable definition of coherence can be found in Mae Wan Ho's book (1993), *The Rainbow and the Worm*.

A key notion in the new perspective of living organization . . . is "coherence." Coherence in ordinary language means correlation, a sticking together, or connectedness; also, a consistency in the system. So we refer to people's speech or thought as coherent, if the parts fit together well, and incoherent if they are uttering meaningless nonsense, or presenting ideas that don't make sense as a whole. Thus, coherence always refers to wholeness. However, in order to appreciate its full meaning and more importantly, the implications for the living system it is necessary to make incursions into its quantum physical description, which gives insights that are otherwise not accessible to us.

Ho (1993)

Stated differently, if this book is put together logically, coherently, and with respect to wholeness, the information within it will resonate with the readers' experience of themselves and enable them to enhance the coherence of their life.

- Coherence is apparent in celestial rhythms
- Coherence can arise in body structures
- Coherence can arise in the body's energy fields
- Coherence can arise in thought and action

HERBERT FRÖHLICH

For those interested in the human body as an energetic system, it is important to study the research of Herbert Fröhlich and his many colleagues who have continued this important line of inquiry. One of the foremost physicists in the study of crystalline or highly ordered nonliving materials, Fröhlich became fascinated with the cooperative or coherent phenomena displayed by biological systems. What drew him into biology was the realization that the cell membrane, although extremely thin, has an enormous voltage across it, amounting to some millions of volts per meter. The molecules in a liquid crystal with that voltage across it should vibrate strongly and emit signals.

In 1988, Fröhlich edited a book that provides a deep and rich source of insight and stimulation. Unfortunately the book is no longer in print and is difficult to obtain. For this reason, we have written a short review and commentary on the book to make its key points available to those who are interested (Oschman & Oschman 1994). This chapter contains excerpts from that publication.

Each reading of the literature related to biological coherence enhances one's understanding of an important model that is emerging in the field of biophysics. This is a model that eventually may validate many of the extraordinary energetic experiences in bodywork, the martial arts, and human performance in general.

A line of thought that began in the late 1960s led to the realization by Fröhlich that molecular systems must produce giant coherent or laserlike oscillations that will move about within the organism and be radiated into the environment. Fröhlich predicted that these signals would have important roles in the regulations that lead to unity of function, that is, wholeness, in the organism.

The ability of molecules to vibrate and produce strong oscillations at visible and near visible light frequencies goes hand in hand with the great sensitivity of living systems to external energy fields of all kinds.

Study of these phenomena is likely to enlighten us about a number of seemingly esoteric observations made on a regular basis by bodyworkers and movement and energetic therapists. For those therapists who are sensitive to radiations (light, colors, sounds, and electric and magnetlike sensations) in the spaces around their clients, the material reviewed here is worthy of careful study.

Coherent vibrations recognize no boundaries, at the surface of a molecule, cell, or organism. They are collective properties of the whole, and they radiate their messages into the environment in various ways.

We have attempted to keep to a minimum the use of technical terms and mathematics and have defined those concepts that may be new to you and that are essential for understanding the subjects covered. We interpret the main points of each chapter in Fröhlich's book from the perspectives being discussed here. Although the concepts apply to a variety of molecular arrays in living systems, such as those illustrated in Figure 9-2, our comments are mainly in reference to collagen, connective tissue, and fascia, as well as the cells that produce and nourish them.

Nonlinearity

Nonlinear systems are mentioned frequently. Nonlinearity, chaos theory, and fractals are related phenomena that are attracting attention in scientific circles at the present time. They form the basis of a "new science" that reveals order and pattern in phenomena that previously seemed to be random, erratic, and unpredictable. The radiations and communications discussed in Fröhlich's book are primarily of the nonlinear type.

In the summary that follows you will find some sections in parentheses, denoting that the author has added material to clarify the subject. Each section heading includes the name of the author(s) of that chapter of the book edited by Fröhlich.

Theoretical physics and biology (H. Fröhlich)

The two approaches, *reductionist,* or working with the properties of individual parts, and *holistic,* working with properties of large systems, supplement each other. An understanding of systems involves different and new principles that are not simply extensions of the behavior of parts. Study of energetics in living systems has elucidated some striking effects, but sometimes they are not reproducible. This is due to the nonlinear or chaotic properties of such systems. A very small change in an initial condition may lead to a large change in the outcome.

Theory of nonlinear excitations (F. Kaiser)

Nonlinear excitations can result in a single frequency becoming strongly excited and then stabilized in the excited condition because of the high degree of order present in the surrounding matrix. In fact, an entire system (such as a particular fascial layer or an entire tendon or bone) may react as a unit and transform as a whole. When this happens the system may emit a coherent signal, much like a laser beam.

This approach helps us understand the great sensitivity of living systems to extremely weak environmental electromagnetic fields. Energy applied from the outside of the organism can couple to coherent signals inside that are involved in whole-body regulations. The existence of coherent signaling systems in organisms has important consequences for the establishment of temporal (time) and spatial (structural) order.

One important type of coherent wave is the *soliton* or *solitary wave*. (In Chapter 19 we discuss the origin and significance of the soliton concept in detail.) The soliton is a wave that preserves its shape and speed, even when it collides with another soliton. Solitons are capable of loss-free energy transport—they do not dissipate or lose their energy as they move through tissues. (Because the living matrix is an excitable medium, a soliton can *gain* energy as it spreads through tissues.)

Solitons can carry a variety of messages, but they are nonlinear. This means that a very weak external field, containing very little energy, can destabilize a soliton. When this happens, the soliton can collapse and release a large amount of energy. In this way a strong effect can be produced by a minute external field.

(We shall see later in the chapter by Smith that such effects may be involved in allergic reactions. Oschman [1993] has suggested that energetic bodywork sometimes may stimulate the production of palpable soliton waves that flow through a client's tissues. This wave is in the form of a ripple that spreads away from the place where you are working. The soliton wave provides a good example of the difference between a linear and nonlinear system. Linear waves spread out and disperse. Their energy becomes disorganized and lost. In contrast, the nonlinear soliton wave remains self-sufficient— it holds itself together. The velocity of a soliton can be slow compared with the speed of sound [Hyman, McLaughlin & Scott 1981], although ultrasonic solitons have been predicted by reference.)

Structures, correlations, and electromagnetic interactions in living matter (E. Del Guidice and colleagues)

Quantum mechanics provides a method for predicting the properties of complex systems. In essence it is a statistical study of assemblies of parts. Order can arise in sets of seemingly nonordered microscopic components. Physicists have accounted for the behaviors of crystals, ferromagnets, superconductors, and plasmas using these methods. (With some outstanding exceptions, physicists generally shy away from living systems because they think cells and tissues are too complicated and therefore too messy for the application of basic physical principles. Few physicists seem to realize that living systems consist primarily of regions, such as the connective tissue, fascia, bone, cell membranes, and cytoskeleton, that are highly ordered because of the crystalline packing of the constituent molecules. "Living crystals" therefore can be described by many of the well-established approaches used in solid-state physics.)

Fröhlich (1968a, 1969b) made a big advance in the analysis of living systems when he recognized that the electrical polarity of a molecular component can be taken as a basic "order parameter." (A *parameter* is a numerical value that remains constant for the members of a population.) Once this step was taken, it became possible to study and predict the effects of a wide range of influences, all of which modify the basic parameter. (The electrically polarized collagen molecule is the most abundant of the building blocks of living matter. Fröhlich's studies provide a basis for understanding the behavior of many such units when they are arrayed together to form a larger structure, such as a collagen fibril, collagen bundle, tendon, ligament, cartilage, fascial plane, or bone. Each collagen molecule has a permanent electric polarity along its axis, as shown in Figure 14-1, *A*. As a result, an array of collagen molecules will have a net electric charge, as shown in Figure 14-1, *B*.)

Del Guidice and colleagues present a detailed mathematical application of quantum field theory (QFT) to living systems. They then develop a scheme for living systems that is based on the idea that life is the final step in an evolutionary process involving the assembly of electrically polarized molecules.

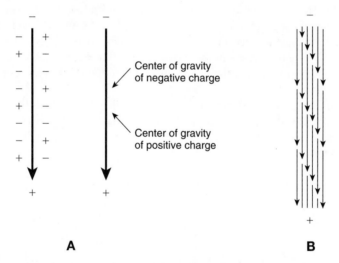

A   **B**

Figure 14-1 A, The collagen molecule (tropocollagen) is a stiff rod, shaped like a knitting needle. Positively and negatively charged groups are distributed along the length of the molecule *(left)*. Because charges are grouped asymmetrically, gravitational centers of positive and negative charge do not coincide but are instead a certain distance apart along molecular axis *(right)*. As a result, collagen has a permanent electric dipole moment, or polarity, along its axis. For simplicity, the dipole is represented as an arrow pointing toward the positive end. **B,** An array of collagen dipoles forming a collagen fibril. The molecules are not packed one on top of the other; instead they are staggered or offset from each other. The polarities of the individual molecules add together so that the fibril as a whole has an electrical polarity. It is an electret, that is, it has a long-lasting electric polarization along a specific direction. Electrets usually are piezoelectric, that is, they generate electric fields when they are compressed or stretched. The formation of the electret is the first step in the self-organization of the living system. Because the bulk of the collagen fibrils in the vertebrate body are oriented vertically, the organism as a whole has an overall electrical polarity, with the head negative with respect to the tail or feet (Athenstaedt 1974).

There is a detailed treatment of filamentary units (such as collagen molecules in connective tissues and microtubules in cells) that can orient water molecules. We shall see that the relationship between water and collagen is a key element in understanding biological quantum coherence.

Figure 14-2, *A*, shows that the water molecule is electrically polarized because the oxygen end is strongly electronegative. The water molecule can oscillate or vibrate in various planes (Figure 14-2, *B*). It also can spin or tumble in different planes (Figure 14-2, *C*).

(Previous work [Berendsen 1962] showed that the spacing of charges along the collagen molecule is ideal for orienting nearby water molecules, which associate with the collagen in hydrogen-bonded chains, as shown in Figure 9-5. The water molecules form hydrogen-bonded filaments that stabilize the three-dimensional structure of the collagen fibrils.)

Filamentary structures and their associated water molecules can confine and channel electromagnetic signals, and they can protect or shield them from outside influences (Figure 14-3).

(The structure of these systems includes water that is closely associated with the protein and more distant layers of water molecules whose orientation also is affected by the presence of the protein. When a wave of energy in the form of an electric, magnetic, or electromagnetic pulse passes along the protein, it will cause adjacent water molecules to vibrate or spin about their axes (detailed in Chapter 25). The vibrations or spins of water molecules a distance away from the protein will be less influenced by the wave.)

Of great importance is the fact that such systems will have interfaces between highly ordered structures and associated layers of water molecules (such as occur around collagen fibrils and cell membranes) and less organized regions (such as the extracellular

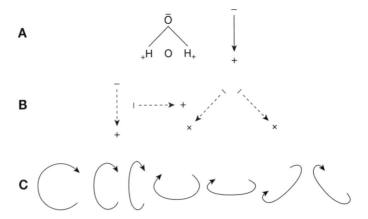

Figure 14-2 A, The water molecule is electrically polarized because the oxygen is strongly electronegative with respect to the hydrogen atoms. As with the collagen molecule, the water electric dipole can be represented as an arrow. **B,** The water molecule can vibrate or oscillate in various planes. **C,** The water molecule can spin or tumble in different planes.

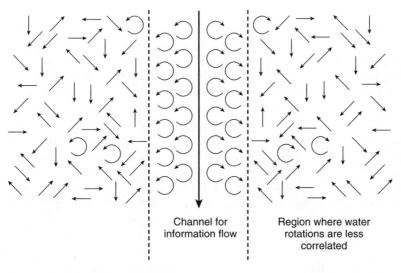

| Channel for information flow | Region where water rotations are less correlated |

Figure 14-3 When a wave of energy passes along a protein such as collagen, water molecules surrounding and closely associated with the protein will spin about their axes. The protein and coherently spinning water molecules provide an information channel that confines electromagnetic signals. The spins of water molecules some distance away from the protein will be less influenced by signals passing through the channel. The interface between ordered and disordered water phases, shown with the *dashed lines,* has a steep electrical gradient across it. The disordered or uncorrelated water layer protects, shields, or insulates the information channel from outside influences. The extent of this shielding depends on the degree of spin correlation, which can be calculated on the basis of quantum field theory (this is known as the Goldstone correlation among dipoles).

fluid) where the rotations of water molecules are far less correlated. These interfaces have very steep electrical gradients across them. "Life then appears to be possible only on the interface between order and disorder or between differently ordered domains."

Detailed mathematical treatment of this system leads to a startling result in relation to cancer. Cancer cells and cells infected with viruses have a higher than normal microscopic order. The cytoskeleton of the cancer cell contains a few thick filaments instead of the distributed fibrillar network found in healthy cells. The excessively strongly correlated structure in the cancer or virus-infected cell would prevent external electromagnetic fields from penetrating into the system. In essence, the cells would be shut off or insulated from environmental stimuli. (We find this to be a most intriguing concept because of its relation to the concept in acupuncture that cancer is caused by "stagnant Ch'i" and Albert Szent-Györgyi's concept that cancer is caused by a breakdown in electronic communications.)

Living systems are dominated by coherent vibrations. The coherence or structural order, "the matter field," triggers the coherence of the electromagnetic field. In the technical language of physics, the coherent field arises because of spatial variation of the condensation of Goldstone bosons. Again, detailed mathematical treatment of this system leads to the conclusion that biological systems will exhibit phenomena that usually are associated with superconductors, including influences of weak magnetic fields and Josephson effects. (Brian D. Josephson [Nobel Prize, 1971] predicted that certain kinds of currents, called *supercurrents*, would pass between two superconductors separated by a thin insulating layer. The effect was demonstrated experimentally in 1963. Josephson junctions now are widely used in electronics and computers because they can detect and amplify weak fields, switch signals from one circuit to another at extremely high speed, and store information, all with extremely low power dissipation and in very small spaces.)

(The Josephson effect is particularly interesting because it provides the basis for the ultrasensitive magnetometer known as the *SQUID [superconducting quantum interference device]*. A SQUID instrument is capable of detecting the magnetic fields produced in the spaces around the body as a consequence of physiological processes such as muscle contractions, glandular secretions, and brain activity, as discussed in Chapter 1.)

The SQUID consists of two superconductors separated by a thin insulating barrier. Josephson effects have already been predicted and investigated by others. (This discussion is pointing toward a mechanism by which living systems may be able to detect biomagnetic fields, which is, in fact, the basis for certain unconventional approaches such as aura balancing and magnetic healing.) As biocircuit elements, Josephson junctions and extended arrays of such junctions could be used for information storage and processing. They also provide another means by which living systems could sense exceedingly weak electromagnetic fields present in the environment.

An event such as a deformation or chemical reaction at a particular site along a molecular chain will trigger solitons, which will propagate along the chain. The lengths of molecular chains in living systems and the velocities of soliton propagation imply soliton lifetimes up to a few seconds.

A soliton can trap an electric charge and carry it along. Once the soliton is formed, this charge transfer does not require further input of energy; thus, a sort of supercurrent is created. As the soliton wave passes along the molecular chain, electromagnetic fields may be radiated into the space around the chain. Dipolar water molecules close to the molecule and in the surrounding space will cooperate by oscillating (or spinning) coherently.

Resonant cellular effects of low-intensity microwaves (W. Grundler and colleagues)

This chapter discusses the effects of microwaves on living systems. Resonant interactions between microwaves and living systems are well documented, but the results are not always reproducible. There are other, as yet unidentified, parameters that need to be taken into consideration.

The influence of low-intensity millimeter waves on biological systems (F. Kremer and colleagues)

This study documents both thermal and nonthermal influences of low-intensity millimeter waves on selected biological systems. The authors provide a detailed set of criteria for establishing such effects.

Metastable states of biopolymers (J.B. Hasted)

(Dictionary definition of *metastable:* marked by only a slight margin of stability.) Consider a sphere in a bowl that has a convex spherical bump in the center. The stable configuration is with the sphere in the lowest part of the valley. If the sphere is placed on the side of the bump, it is unstable—it will roll down into the valley. If the sphere is balanced on top of the bump, it is metastable—it is temporarily stable but will roll off if its balance is even slightly disturbed (Figure 14-4). Physicists refer to such a disturbance as a *perturbation.*

(Collagen is an example of a biopolymer. A *biopolymer* is a chainlike molecule formed by the polymerization or chemical linking of small biomolecules, each of which is called a *monomer.*)

In the case of vibrations within biopolymers, a metastable state can occur for a limited period of time if sufficient energy is supplied to the system to get it to a balance point. A perturbation or input of extra energy can destroy the metastable state and release the energy that was used to create the metastable state in the first place.

The ability of the metastable state to withstand slight perturbations depends on its environment. For example, in nature proteins are wet. The water molecules associated with a protein can absorb a certain amount of the energy of a perturbation and therefore protect the metastable state.

Halsted presents a detailed technical discussion of the relationship between a protein and its aqueous film in relation to the absorption of perturbations of different kinds (heat, vibration, electromagnetic fields) and at different frequencies. An

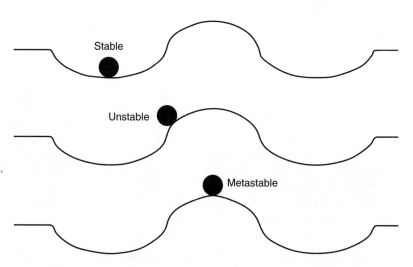

Figure 14-4 Types of stability. Consider a sphere in a bowl that has a convex spherical bump in the center. The *stable* configuration is with the sphere in the lowest part of the valley. If the sphere is placed on the side of the bump, it is *unstable*—it will roll down into the valley. If the sphere is balanced on top of the bump, it is *metastable*—it is temporarily stable but will roll off if its balance is even slightly disturbed.

assembly of electric dipoles with a permanent or semipermanent polarization, such as the collagen fibril, will be a stable or metastable *electret*. (An electret has a long-lasting electric polarization along a specific direction and usually is piezoelectric.) In an electret, polar molecules are oriented in a way that enables the dipoles to reinforce each other.

Electromagnetic fields interact with biomolecules by both resonant and nonresonant processes. (Resonance occurs when a relatively strong vibration is produced in an electrical or mechanical system by applying a much smaller periodic stimulus that is at or near the natural frequency of the larger system.) Radiofrequency interactions are dominated by nonresonant interactions because the wavelengths are quite long with respect to the molecular lengths. Only at gigahertz (giga = billion = 10^9 Hz), frequencies with millimeter and less wavelengths (microwaves, infrared, etc.) do resonance processes begin to dominate the absorption of electromagnetic signals. Some preliminary work on such resonances has been reported.

Physical aspects of plant photosynthesis (F. Drissler)

The primary physical event in photosynthesis is of great interest because it is the first step in the production of all biological energy from solar radiation. Remarkable transport mechanisms convey energy trapped by chlorophyll molecules to reaction centers, where the energy is converted into chemical bonds. If the excitation does not reach a reaction center within a short period, it is re-radiated (this is called *delayed fluorescence*). This study examines the possibility that coherent vibrational states in the chloroplast membrane may be involved in these energy transfers.

Emission of radiation by active cells (J.K. Pollock and D.G. Pohl)

Fröhlich's original suggestion was that the highly ordered molecular arrays in living systems could cooperate under certain conditions to set up high-frequency electrical oscillations in the range from 10^{11} to 10^{12} cycles per second. This article summarizes evidence that such signals exist and are radiated into the cellular environment. Particular emphasis is given to the studies of Pohl using microdielectrophoresis. This is a method in which cells are suspended in a solution containing tiny charged particles that are attracted to areas where there is a strong electric field. It was repeatedly found that cells produced the strongest fields at or near mitosis (the stage in cell division in which the chromosomes and nuclei divide). They studied bacteria, fungi, algae, and mammalian cells. The method revealed patterns around the cells, suggesting that the radiation is from certain parts of the cell surface.

Physiological signaling across cell membranes and cooperative influences of extremely-low-frequency electromagnetic fields (W.R. Adey)

Originally the cell membrane was regarded as a boundary between the cytoplasm and the environment. Emphasis has shifted to the role of the membrane as a window through which the living material senses its environment. Of great interest are studies showing that there are proteins that span the membrane, joining the inside and outside environments. These proteins are energetic and signaling pathways, conveying external stimuli to the cell interior (and vice versa).

The huge voltage across the cell membrane should prevent weak, very-low-frequency electromagnetic fields from influencing cellular processes; however, we now know that minute low-frequency signals have a profound influence on cellular events. This implies some form of cooperative communication processes along the lines that Fröhlich proposed some years ago.

Adey and colleagues have used low-frequency signals to identify the steps involved in coupling information flows from the outside to the inside of the cell. Cell membranes seem to act as powerful amplifiers, boosting minute electromagnetic fields as the first step in a series of long-range quantum processes. Adey likens the cell surface, with its protruding proteins, to "a field of waving corn, responding to an infinite variety of faint electrochemical breezes that blow along the membrane surface."

The inward and outward flows of signals are related to pathological problems, including cancer, and there is the prospect of distinguishing between normal and abnormal signal streams.

The way calcium binds with brain tissue provides an example of an amplifying effect such as just described (and alluded to in the chapter by Kaiser, in which a small external signal produces a much stronger internal influence). Very weak, low-frequency electromagnetic fields at strengths comparable to those produced by normal brain waves (as measured with the electroencephalogram) have profound cooperative effects on calcium and hormone binding in brain tissues. The effects depend on signal strength and frequency.

The pineal gland is sensitive to the orientation of the head with respect to the earth's geomagnetic field. The earth's field also influences calcium fluxes in response to applied electromagnetic fields.

There is a discussion of cellular events involved in bone healing and the ways external fields can be used to stimulate fracture therapy. Much attention also is given to epidermal growth factor (EGF), nerve growth factor (NGF), and their receptors in cell membranes. Both factors extend across cell membranes and have a similar sequence of 23 amino acids that reside inside the membrane. This segment is not well adapted as a mechanical linkage or as an ion transfer system. Adey suggests that transmembrane signaling instead may be accomplished by Davydov-Scott soliton waves (Hyman, McLaughlin & Scott 1981).

Adey and colleagues discovered three examples of hormonal responses inside cells that can be triggered with electromagnetic fields only, independent of the presence of the hormones.

Adey discusses the possibility that external electromagnetic fields can initiate cancers by mechanisms that do not involve direct effects on the genetic material. Instead, harmful fields may distort inward signal streams directed toward the nucleus and other cell organelles. Several examples are given.

Adey then discusses models of organization in physiological systems with respect to cooperative phenomena.

Assemblies of many components can develop dynamic patterns as a result of complex flow patterns. "These flow patterns can undergo sudden transitions to new self-maintaining arrangements that will be relatively stable over time."

The flow patterns referred to are energy flows. Two or more different interactions can give rise to the same dynamic pattern. What makes this possible is *cooperativity,* which is defined as the way the components of a system act together to switch from one stable state to another. These transitions can involve phenomena that physicists refer to as *phase transitions, hysteresis,* and *avalanche effects.* (A *phase* is a state, condition, or aspect of a system. The conversion of water to ice is an example of a *phase change,* as is a change from disorganized or diseased to organized or healthy. *Hysteresis* is a situation in which the relationship between two quantities is not linear but depends on their prior history. *Avalanche effects,* as the name suggests, occur when a small disturbance becomes multiplied to cause a rapid and regenerative flow of something such as a current.) Very weak signals can trigger such transformations. When this happens, we refer to the situation as an *amplification effect.* Classic linear systems respond slowly to large stimuli. In cooperative systems, large transitions can take place sharply and rapidly in response to minute inputs of energy. The extreme sensitivity of living systems, and the biological responses to remarkably minute signals we already know about, remind us of how much we have yet to learn about the couplings of which cooperative systems are capable. (We must be careful of the artificial energy fields we create in our environment.)

Sustained oscillations are an essential component of living systems. Every process in the organism that can be measured shows rhythmic variations. Adey discusses the theory of how environmental rhythms may entrain internal oscillations. He then considers chaotic models and how they may interact with strongly structured rhythms to the advantage of the organism.

Again, the possible role of solitons in signaling and control of molecular states is discussed. When the energy applied to a system is raised, a certain sharp threshold is reached and soliton waves are formed. If the applied energy is increased further, another threshold is reached above which solitons will not form.

The focus is on the cell membrane, which acts as an amplifier of natural or imposed environmental fields in the extremely-low-frequency (ELF, below 3000 Hz) range, which is referred to as the *biological spectrum.* The low-frequency oscillations can readily couple to rhythms in the polarization state of macromolecular arrays. (A *macromolecule* is a large molecule such as a biopolymer. A *macromolecular array* is a crystalline assembly of such large molecules. Tendons, bones, and sheets of fascia are examples of macromolecular arrays in the human body.) The macromolecular array would swing from a highly excited and strongly polarized state to a weakly polar ground state. Slow chemical oscillations could be coupled to these slow rhythms of electrical polarization. The strongly polarized state would oscillate at 10^{11} Hz. (Hz is the abbreviation for *Hertz,* a measure of frequency equivalent to cycles per second. A frequency of 10^{11} Hz is in the microwave portion of the electromagnetic spectrum.) Slow chemical and electrical oscillations would have frequencies around 10 Hz (this is an interesting frequency for healers because it approximates the average frequency of the Schumann resonance in the earth's electric and magnetic field. In separate articles, we have discussed evidence that various types of traditional healers may couple their brain waves and their bodily biomagnetic fields to the Schumann resonance when they are in their healing state).

Also relevant to the ELF biological spectrum is the fact that the strength of the geomagnetic field is ideal for inducing cyclotron resonances in free calcium ions and gyrofrequencies in both singly and doubly charged ions of biological significance. (*Cyclotron resonances* are situations in which charged particles develop a collective helical motion in a magnetic field. At resonance, large amounts of energy can spiral through a conductor or a semiconductor.) In other words, the large magnetic field of the planet could transfer large amounts of energy into the various ions found in living tissues.

(All of this evidence points to a wide spectrum of previously unsuspected energetic relations between the components within the organism and between the components and the environment.) There even is some evidence that biological systems may exhibit superconductivity.

In conclusion, Adey discusses the various hierarchies of order in living systems and the new concepts of communication that are emerging from physical studies at the atomic and molecular levels. Electromagnetic fields applied to cells can be used to probe the sequences of events involved in hormonal regulations. A dramatic component of these sequences is the amplification that takes place at the cell surface. This amplification enables fields that are millions of times weaker than the membrane potential to modulate cellular activities.

The interaction of living red blood cells (S. Rowlands)

It has been recognized for more than a century that clotting red blood cells form stacks. This is called *rouleau formation*. In 1972, Rowlands observed that when cells and rouleaux were a short distance apart, they seemed to move rapidly toward each other. At the time he considered this to be an illusion, particularly because the outer surface of the membrane of each red cell is mainly negatively charged and erythrocytes should repel each other.

A review of coherent excitations published by Fröhlich in 1980 theorized that long-range attractive forces would be exerted between cells as a consequence of coherent oscillations of polarized membrane molecules. Upon reading Fröhlich's work, Rowlands immediately recognized that red blood cells provide an ideal system to test the long-range attraction idea. Red blood cells are simple in structure and have no intrinsic motility. A series of studies confirmed Fröhlich's hypothesis. Figure 14-5 shows the microscopic appearance of rouleaux.

Rouleaux do not form in normally circulating blood in healthy people, although in some diseases, particularly connective tissue diseases, rouleau formation can be observed in the small blood vessels in the retina and conjunctiva of the eye.

According to Fröhlich's theory, long-range interactions require an intact cell membrane, an electrical potential across the membrane that polarizes the membrane macromolecules, and a supply of metabolic energy to maintain coherent waves of membrane polarization. Rowlands and colleagues tested and confirmed each of the postulates of Fröhlich's theory for the interaction of human erythrocytes.

Polarized macromolecules in cell membranes may polarize nearby water molecules. Multiple layers of these ordered water molecules may provide an ideal medium for the transmission of Fröhlich's coherent polar waves from cell to cell.

In conclusion, Rowlands points out that the usual biochemical view of molecules wandering about inside cells by random motion until they happen to bump into the correct enzyme is inconceivable. Much more logical is the idea that long-range coherent Fröhlich interactions bring about the orderly and efficient movements and actions of molecules and cells that constitute life.

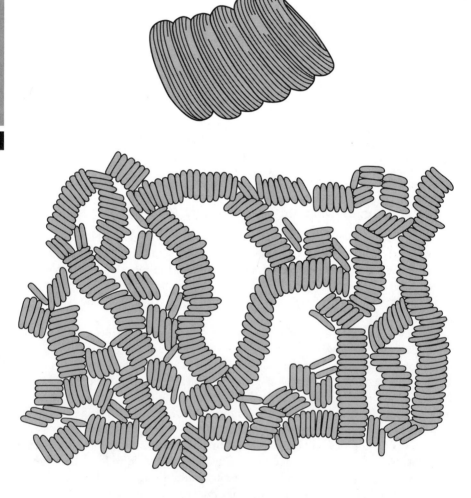

Figure 14-5 Rouleau formation. In 1980, Fröhlich predicted that cells would attract each other as a consequence of coherent oscillations of polarized membrane molecules. Rowlands recognized that the stacking of red blood cells would provide an ideal system to test Fröhlich's ideas, and Rowlands was successful in demonstrating this. (The illustration is Figure 177 from *Anatomy and Physiology*, 15th ed, by Kimber/Gray/Stackpole, © Reprinted by permission of Pearson Education, Inc., Upper Saddle River, NJ.)

The genetic code as language (H. Fröhlich)

Geneticists have identified a significant portion of deoxyribonucleic acid (DNA) that does not code for the amino acid sequence in proteins. This *nonsense*, imprecisely sequenced component of the genetic material has been called *selfish* or *junk* DNA. The existence of such DNA presents a paradox in that the evolutionary forces of natural selection and "survival of the fittest" should have weeded out any random or functionless DNA long ago.

In this exceptionally fascinating chapter, Fröhlich suggests broadening the genetic code to include more of the functions of language. Specifically, he suggests the junk or selfish DNA may function to regulate higher levels of complexity.

The genetic code as it usually is viewed does not account for the collective aspects of chromosomal functioning, the *planetary genome,* that might have analogies with language. Specifically, Fröhlich contrasts the inactive, one-dimensional structure of a protein, which is determined by its amino acid sequence, and its active, three-dimensional form. This three-dimensional form gives proteins their remarkable specificity for interacting with particular substrates. In order to become active, a protein must assume a single correct shape out of the infinite variety of configurations that are possible. Such specificity usually is considered to arise spontaneously as a consequence of self-organization. Fröhlich regards these as concepts that "mask many difficulties."

One of the pieces of information that may reside in the junk DNA is involved in the chromosomal replication process. A chromosome is a huge molecule that, during mitosis, uncoils, separates, finds its appropriate partner with which to pair, and then folds up again without becoming tangled up. All of this happens in a very small space. The process illustrates the essential difference between the concept of a code and the concept of a language. Corresponding chromosomes may have the same resonant frequency that enables them to communicate over some distance, drawing like to like.

At higher levels of order, developing cells take on particular forms and functions, sometimes moving considerable distances to join with specific neighbors. Perhaps the extra DNA is used for organizing such collective behavior. Fröhlich suggests that collective coherent oscillations in the junk region of the DNA may produce resonant vibrations that help the cell select which proteins will be produced at particular times during development so that they may take up their functional positions in tissues and organs. This model would explain why a young kidney cell, when transplanted into a liver, adapts to that organ because it is overwhelmed by the resonance of the surrounding liver cells. When an organ becomes large enough, the collective conversations reaches a level that would stop further proliferation. Without such control, a cancer would result.

Fröhlich goes on to discuss autoimmune diseases and the problem of self-recognition. The conventional dogma is that all cells of an organism have self-markers that enable cells to recognize invading, nonself cells, which are attacked by the immune system. What causes this, and how does it come about? The answer must involve some collective property of the entire genome. (Of great interest in relation to the ideas summarized in this chapter is the work of the Institute of HeartMath on heart/DNA resonances.)

Electromagnetic effects in humans (C.W. Smith)

C.W. Smith has developed some of the most important applications of biological coherence to human health. He begins with an historical review of coherence and electromagnetic effects in humans. He then discusses a fascinating range of related topics.

Our reliance on clocks and calendars shows that we order ourselves in a coherent fashion with the coherent motions of the celestial bodies. In spite of the obvious manner in which we link our behavioral and physiological rhythms with natural and astronomical cycles, the subjects of biological rhythms and electrical phenomena in living systems have been vigorously and acrimoniously debated by scientists.

Tesla, who invented much of our modern electrical technology, described the first well-documented case of electromagnetic hypersensitivity. In his experiments he exposed himself to massive electrical and magnetic fields, which led to an illness similar to an allergy. (There now are many people who have mysterious electromagnetic illnesses that conventional medicine is unable to diagnose. Those who are hypersensitive to 60-cycle electricity may become dizzy or nauseous or develop migraine headaches that are triggered by walking past a hidden transformer or by standing next to an appliance such as a toaster. Sometimes these patients are given drugs to treat their symptoms, the drugs produce side effects, and more drugs are given to treat those side effects.)

Lakhovsky used high-frequency fields to treat cancer. His theory (1939) was that health involved a balance or equilibrium in the electrical oscillations in living cells, and disease arose from oscillatory disequilibrium. (This idea not only coincides with the concepts that now are emerging from Fröhlich's work, but they fit well with the ancient concepts that form the basis of acupuncture. Health involves the harmonious balance between pulsating energy flows and communications within the various meridians.)

Piccardi studied the correlations between rates of chemical and biological processes and extraterrestrial influences. Gauguelin reviewed biological rhythms that are highly coherent and become "phase locked" to environmental fluctuations.

There are many advantages to using coherent signals for biological communications. Smith discusses the advantages of serial versus parallel data processing channels and the ways coherence can reduce interference from natural and artificial fields in our environment. In the same way that harmful chemicals can disrupt the genetic material and cause disease, harmful coherent signals can sensitize and disrupt electronic signals within the organism. (Our long-term survival as a species may depend upon how well our internal communication systems are able to carry out essential functions in an environment that we are increasingly polluting with artificial coherent energy fields of various kinds. Robert O. Becker [1985] believes electromagnetic pollution eventually may have more significant health effects than chemical pollution of our air and water.)

Smith also makes a profoundly important but rarely appreciated point that the widely studied chemical signaling processes (e.g., hormonal regulations) are really interchangeable with electrical signals. The whole field of chemical analysis by spectroscopy shows that there is a "fundamental duality between chemical structure and coherent oscillations." (Spectroscopy is the main method used by physicists, chemists, and even astrophysicists to study the structure of matter of all kinds. When energy is applied to atoms or molecules, they vibrate and produce luminescence or radiation. Spectroscopists study these radiations to determine the precise structure of matter that is too small or far away to be examined directly.)

Smith makes the additional point that survival in predator-prey situations has forced living systems to evolve sensors that operate at the limits of the laws of physics. These sensors enable biological rhythms to be entrained with variations in the geomagnetic field. The geomagnetic field, in turn, varies from moment to moment in relation to an intricate fabric of extraterrestrial cycles, which include solar and lunar rhythms, sunspots, the rotation of the sun about its axis, solar winds, the auroras, oscillating currents in the upper atmosphere, magnetic storms, planetary positions, and the interstellar wind.

Important work of Wever is reviewed next. Influences of environmental signals on human physiological rhythms were studied by isolating subjects underground in shielded rooms. Twenty years of German research involved 325 subjects who were placed in isolation for 1- to 3-month periods. There were 52 experiments in which bodily rhythms were altered or desynchronized by a constant environment. Applying a 10-Hz field via electrodes hidden in the walls of the isolation chamber could prevent or reduce rhythm desynchronization. The pineal gland is sensitive to both light and minute magnetic field variations. A number of investigators think the pineal is the time-keeping organ in humans.

Homing pigeons have large pineals that may serve as their geomagnetic compass. Honeybees also have a compass sense, but insects do not have pineal glands. It appears that the whole insect body may be cooperatively involved in detecting variations of magnetic field strength and in navigation.

Experiments regarding the possible involvement of magnetic fields in dowsing are discussed, but at this stage we are not sure of how to interpret the findings.

Submarine crews experience physiological changes that have been attributed to the fact that they are shielded from ELF geomagnetic variations by the steel hull and surrounding salt water. Astronauts orbiting the earth may have different responses because they experience each day's geomagnetic cycle during a much shorter time period. There are indications from studies of cultured cells that effects of shielding from the geomagnetic field may not show up until many generations after the experiment is performed.

The frequency range from 1 to 30 Hz is particularly important physiologically. It also is the frequency range of normal variations in the geomagnetic field. Of particular importance is the 8- to 12-Hz range of the brain alpha rhythm. The geophysical rhythms are strong enough to influence living tissues. Of particular interest are the ELF frequencies that are present in all human subjects and that resonate with homeopathic remedies. Ludwig (1987) has reported on the resonant frequencies of particular physiological functions in man.

Smith turns next to puzzling aspects of enzyme chemistry. Enzymes are catalysts of chemical reactions. They are not consumed during the reaction, but they stimulate the reaction to progress at a very high rate. Enzyme-catalyzed reactions can be compared to electronic devices such as high-gain amplifiers in which the gain is regulated by feedback mechanisms.

The great specificity of enzyme action (the enzyme will act upon one and only one substrate) has been compared to a mechanical "lock and key" arrangement. However, what chemists do not usually consider is the manner by which the key finds its way to the keyhole. (The problem is comparable to the attraction of the human immunodeficiency virus to its receptor on a lymphocyte.) Fröhlich developed a model for this type of situation in which strong attractions arise because of giant dipole oscillations of the two molecules involved. The appropriate frequency for such attractions is around 10^{13} Hz, which corresponds closely to the frequency of cell membrane electrical oscillations at body temperature.

Other studies have shown that water is important in determining the way proteins behave, and the way the water is added to a protein powder is important. Water can be added as a vapor or as a liquid, and different results are obtained. This was the first indication that water might maintain a sort of memory of its recent history, and this effect turned up later in connection with studies of allergies.

Biomolecules are not supposed to have magnetic interactions with water, but nevertheless they show effects 10,000 times larger than expected. There are a number of detailed technical matters that must be taken into consideration for these studies to be repeatable. Enzyme activities may be sensitive to the light beam used in a spectrophotometer and even to the geomagnetic field strength at the time an experiment is performed. (This is an important point that shows that experiments may not be repeatable unless one takes into consideration factors that previously were considered inconsequential.) Solutions of enzymes seem to "remember" for a long time the frequencies to which they are exposed. Chemists often use magnetic stirrers to mix solutions being studied, and Smith points out that this could mask magnetic influences that might be present.

Enzymes and other biological processes show responses to weak magnetic fields that decrease or disappear if stronger fields are used. Biological systems are nonlinear. The idea that the absence of a biological response to a strong field means that a weaker field will surely have an even smaller effect is not valid.

There are indications that Josephson effects may take place in biological systems. If so, the various electronic and computational tricks (switching, amplification, information processing) that are possible with Josephson junctions probably are available to living cells and used by them for communication and information processing.

Studies on bacteria and yeasts indicate that specific coherent frequencies can trigger proliferation. This may be a part of the *Candida* problem. For example, the clocks in computers operate at high frequencies that may be biologically active. One of Smith's allergy patients had attacks of colitis that seemed to be triggered by working next to a computer with an 8-MHz (eight million cycles per second) clock frequency. Smith and others have shown that this frequency affects the growth of yeasts.

All of this information provides a basis for examining allergies, which now are recognized as disturbances of regulatory systems. About 15% of the population is debilitated to some degree by allergies. In extreme cases, an individual may respond to as many as 100 different allergens.

Since 1982, Smith has studied more than 100 electrically sensitive multiple-allergy patients. The sensitivities can be triggered by specific electromagnetic frequencies in the range of a few thousandths of a Hertz to a gigahertz. New allergic responses can be acquired when a patient is exposed to a previously innocuous substance while reacting to an allergen. A coherent electromagnetic signal at a particular frequency can become an allergen and trigger a specific set of symptoms. ". . . [T]he pattern of allergic responses is the same whether the trigger is chemical, environmental, nutritional or electrical."

Traditional therapy for allergies has involved pricking the skin with a diluted allergen. The result is a wheal on the skin. As the allergen is taken through a series of dilutions, the wheal gets smaller, then larger, then smaller, etc., until a dilution is reached in which no wheal is produced. If the allergen was taken through a further series of dilutions, the wheal would again cycle through the pattern of response and no response. Eventually a dilution is reached at which no response occurs. This is known as the patient's *neutralizing dilution,* which could be injected to protect the patient from subsequent exposure.

Work by Monro and colleagues showed that extremely sensitive allergic patients only needed to hold a glass tube containing a dilution of the allergen to show symptoms or neutralizing effects. The most sensitive patients could even distinguish tubes of allergen that were merely brought into the same room with them. On the basis of this finding it was discovered that the pinprick was unnecessary. Dilutions could simply be dropped on the patient's skin and then wiped off.

Smith and colleagues used a similar method for testing and treating electrically hypersensitive allergy patients. Increasing the frequency has the same effect as diluting the allergen. Eventually a frequency is reached that has the same effect as the neutralizing dilution.

In conducting these studies, care must be taken to keep the strength of the field low. If the field is too strong at a frequency that causes an allergic reaction, the system becomes overstimulated and saturated, and further testing must be put off for some hours or days. The studies are begun with the signal generator in the adjacent room to see if the patient can tell whether the signals are on or off.

For sensitive patients, the signal generator does not even need to have an antenna attached to its output terminals. For less sensitive patients, a short length of wire is adequate. In no case is the wire physically connected to the patient. The amounts of radiation are no larger than that leaking from a television set or home computer.

Not only can patients be extremely sensitive to electromagnetic fields, but they can emit signals during their reactions. These signals can be large enough to produce malfunctions in electronic equipment. Computers, factory robotic systems, electronic ignition systems, and facsimile links have been interrupted by reacting allergy patients. (This phenomenon has been known for a long time and is called the *Pauli effect,* after the Nobel Prize [1945] physicist Wolfgang Pauli. Pauli was a theoretical physicist who was "allergic" to laboratory apparatus and experimentation. It was said that if Pauli walked into a laboratory, equipment would break. Although such phenomena usually are regarded as coincidences, we now know from Smith's work that there is a valid biophysical explanation for it [see Peat 1987 for a description of the Pauli effect].)

Electronic equipment has been developed that detects signals from allergic patients, phase inverts them, and feeds them back into the body for therapeutic purposes. The equipment shows that it is possible to connect via an acupuncture meridian to organ-specific regulatory systems.

A "personal oscillator" set to a patient's neutralizing frequency might seem to be an effective treatment for the individual's allergy, but the correct frequency for one person can be another's allergic trigger. Because of this reaction, a better method is to expose a vial of mineral water to the patient's neutralizing frequency. The patient can merely hold the vial of water in his or her hand to neutralize an allergic reaction. The water retains its effectiveness for at least 1 to 2 months; however, if the patient undergoes a strong allergic reaction, the water seems to lose its effectiveness. Smith thinks this happens because such patients emit signals that overwrite the signal stored in the water.

Homeopathic techniques, such as allergy treatments with neutralizing dilutions, involve diluting solutions to the point where none of the original tincture molecules remain. Water seems to retain a memory of all coherent electromagnetic signals to which it has been exposed since it was last distilled. Many scientists consider it improbable that water molecules can store frequency information, but others are investigating the mechanisms by which this is accomplished. Smith and colleagues have published such a mechanism in which the hydrogen bonds hold water molecules together in a helical structure that acts as a solenoid. A magnetic flux would induce a current flow through the helix, which would regenerate the magnetic flux. Once this metastable state is achieved, current flows and resonances would persist after the original magnetic flux was removed.

In conclusion, Smith points out that acoustic vibrations and electrical vibrations in polarized biomolecules are equivalent and interchangeable. DNA has highly coherent microwave acoustic modes. Living systems are thought to be able to respond to close to a single quantum of light, and Smith believes they also are able to respond to single magnetic quanta. Josephson devices may be involved in capturing such signals. Coherent vibrations in living systems are as fundamental as chemical bonds.

Coherent properties of energy-coupling membrane systems (D.B. Kell)

The ultimate problem in bioenergetics is to locate the final and precise mechanism by which metabolism gives rise to the high-energy phosphate bond that provides a ready source of energy for biological processes such as muscle contraction. Metabolism gives rise to a flow of electrons through a chain of molecules located on the membranes of mitochondria, and this electron flow somehow is coupled to the actions of the enzyme that attaches phosphate bonds to adenosine to form adenosine triphosphate (ATP). Because Fröhlich has developed a mechanism by which the vibrations of one molecule can be linked to those of another molecule some distance away, the possibility exists that the ultimate step in energy metabolism involves such coherent transfers.

There is a detailed discussion of how enzymes, which are like chemical machines, change their shapes as they carry out their catalytic functions. The possibility arises that collective or coherent motions may take place *within* individual enzyme molecules.

One suggested mechanism for energy flow involves highly cooperative channeling of protons along an enzyme's surface. Solitons may be involved. We have seen that soliton waves, although very robust, can be destroyed by very weak signals from the outside if these weak signals are of exactly the right strength and frequency. Kell presents evidence for such effects and discusses their interpretation in terms of the thermodynamics of molecular motions in membranes. Although much remains to be learned, the direction of the research is toward the discovery of specific electromagnetic signals that can influence biochemical processes with a very high degree of specificity without the side effects of current chemotherapies.

Coherence in the cytoskeleton: Implications for biological information processing (S.R. Hameroff)

Living cells are organized and regulated by elaborate networks of dynamic protein filaments. The whole cellular fabric is called the *cytoskeleton* because of its obvious structural role. Hameroff has been a leader in recognizing that the cellular matrix functions in communication and information processing *(intelligence)*. It is the nervous system of all cells, including nerve cells. The main components are microtubules, actin and intermediate filaments, and microtrabeculae, all of which are composed of electrically polarized subunits (electrets) capable of Fröhlich-style interactions and excitations. Coherent interactions may be involved in a wide variety of cellular processes, including holographic information storage. (This subject is discussed in more detail later, beginning with Chapter 23.

After a brief history of the discovery of the cytoskeleton, Hameroff details what is known of microtubules, "the most visible and widely studied cytoskeletal elements." These long tubes (which can be several meters long in a nerve cell) are composed of tiny protein subunits, called *tubulin,* that assemble to form filaments and can quickly disassemble. There are different kinds of tubulin and, therefore, the possibility of different sorts of microtubules, even within the same cell. Tubulin contains a high proportion of charged amino acids, so microtubules can attract and structure a layer of charged ions and water molecules around them.

The hollow core of the microtubule is a mystery. Del Guidice and colleagues have speculated that superconductivity and electromagnetic focusing may take place in the cores, but we really do not know.

Microtubules assemble at "microtubule organizing centers" (see Figure 22-1, *B*) and remain attached to these structures. Other proteins, called *microtubule-associated proteins,* attach to the surfaces of the microtubules at specific points. These proteins are involved in various processes, including generation of cytoplasmic movements, sensation of the environment, and transport of substances.

Taken together, the microtubules and other filamentous components create a network with a huge surface area that has been estimated at 69,000 to 91,000 μm^2 per cell. Motions of water and ions attracted to this surface are dipole coupled to coherent vibrations passing through the cytoskeleton.

Hameroff builds up a dynamic image of cytoskeletal activities, including wiggling of proteins, interactions with hormones, cellular movements, changes in consistency, metabolic processes, electron flows, and even the actions of anesthetics. Because the components are piezoelectric, electric fields change their shape and mechanical stimuli change their electrical condition.

Fröhlich's analysis of cooperative systems such as the cytoskeleton has far-reaching biological implications. Energy flow through living matter, as envisioned by quantum physicists, is quite different compared to the way we have described it in the past. "Solitons, massless bosons, and Fröhlich's coherent polarization waves may be synonymous." According to the Milan group, what we see as structure is, in fact, a *consequence* of coherent focusing of polarized waves of energy. These waves can be confined and propagated as beams to be about 15 nm in diameter, which also is the inner diameter of the microtubules. What we observed as an ordered filamentous network arises, in part, because of the alignment of *rotating* components such as water.

Finally, Hameroff considers information processing in the cytoskeleton, a concept that has been suggested by at least a dozen groups of scientists. These ideas have implications for our understanding of consciousness and memory. Hameroff reviews 13 models of cytoskeletal information processing, the experimental evidence for them, and the ways they are strengthened by involving biological coherence.

Of particular interest is the cited work of DeBrabander and colleagues, who developed a double staining technique using immunogold particles that bind to specific tubulin subunits. This technique made it possible to use the electron microscope to visualize patterns of protein deposition on microtubules. These modifiable patterns could provide a mechanism for coding information, or *cellular memory.*

Hameroff's own model for information storage involves the use of coherent waves to create interference patterns or holographic images that are stored within the microtubular lattice. The cytoskeleton then can be viewed as a programmed electronic computer with the capability of storing, recalling, and processing information of all sorts. Soliton waves traveling through neural and nonneural cells would leave in their wakes memories in the form of patterns of cytoskeletal structure and/or vibrations. This information subsequently could be read and used to make informed decisions about the appropriate timing for activation of cellular processes. Within the brain, the collective functioning of cytoskeletons could give rise to phenomena we refer to as *images, thoughts,* and *ideas.*

In the past we have thought of these as properties of neuronal networks. We now can look deeper for the contributions made by the interconnected cytoplasmic fabrics within all types of cells scattered throughout the organism.

REFERENCES

Athenstaedt, H. 1974, 'Pyroelectric and piezoelectric properties of vertebrates', *Annals of the New York Academy of Sciences,* vol. 238, pp. 68-94.

Becker, R.O. 1985, *Cross Currents,* JP Tarcher, Los Angeles, CA.

Berendsen, H.J.C. 1962, 'Nuclear magnetic resonance study of collagen hydration', *Journal of Chemical Physics,* vol. 36, pp. 3297-3305.

Fröhlich, H. 1968a, 'Long-range coherence and energy storage in biological systems', *International Journal of Quantum Chemistry,* vol. II, pp. 641-649.

Fröhlich, H. 1968b, 'Bose condensation of strongly excited longitudinal electric modes', *Physics Letters*, vol. 26A, pp. 402-403.

Fröhlich, H. (ed.) 1988, *Biological Coherence and Response to External Stimuli*, Springer-Verlag, Berlin.

Ho, M.-W. 1993, *The Rainbow and the Worm: The Physics of Organisms*, World Scientific, Singapore.

Hyman, J.M., McLaughlin, D.W. & and Scott, A.C. 1981, 'On Davydov's alpha-helix solitons', *Physica*, vol. 3D, pp. 23-44.

Ludwig, H.W. 1987, 'Electromagnetic multiresonance: The base of homeopathy and biophysical therapy', *Proceedings of the 42nd Congress International Homeopathic Medical League*, Arlington, Texas, 29 March–April, 1987, American Institute of Homeopathy, Washington, DC.

Oschman, J.L. 1993, 'Sensing solitons in soft tissues', *Guild News (The newsletter for the Guild for Structural Integration, Boulder, Colorado)*, vol. 3, pp. 22-25.

Oschman, J.L. & Oschman, N.H. 1994, 'Book review and commentary: Biological coherence and response to external stimuli', Edited by H. Fröhlich, Springer-Verlag, Berlin.

Peat, F.D. 1987, *Synchronicity: The Bridge Between Matter and Mind*, Bantam Books, Toronto, p. 21.

Szent-Györgyi, A. 1972, *The Living State*, Academic Press, New York.

Limitations of the neuron doctrine

If you want to model the brain you don't model it, you make a baby!

Sir John Eccles

NERVE, BRAIN, AND CONSCIOUSNESS

The 1963 Nobel Prize in Physiology or Medicine was awarded to Sir John Carew Eccles, Alan Lloyd Hodgkin, and Andrew Fielding Huxley for fundamental discoveries about how nerves conduct impulses and communicate with each other. Hodgkin and Huxley described the ionic basis for nerve impulse conduction. Eccles, an Australian, made fundamental discoveries about how one nerve can excite or inhibit another. He worked "under the enchantment of the synapses."

As you read these words, the electrical events these distinguished scientists elucidated are taking place within your very own body. The research recognized by the 1963 Prize brought an unprecedented level of clarity to these processes.

The model Sir John Eccles (1903-1997) developed began to explain how the brain makes the myriad of decisions to process and integrate sensory information and memories to give rise to the conscious moment. The process involves thousands of millions of nerve cells in the cortex.

A small portion of the cortex is shown in Figure 15-1, *A*. One of the key pyramidal cells is illustrated in part in Figure 15-1, *B*. The illustration is partial because it can show only a few of the thousands of synaptic boutons on the cell. Each bouton, in turn, contains numerous synaptic vesicles filled with some 5,000 to 10,000 molecules of a neurotransmitter substance. A close look at one of these synapses is shown in Figure 15-1, *C*.

One of Sir John's achievements was the recognition that a cell such as the one shown in Figure 15-1, *B*, will interact with countless similar cells through a myriad of connections at the boutons. Some of these nerves will be excitatory; others will be inhibitory. Signal processing takes place when the cell "polls" all of the incoming

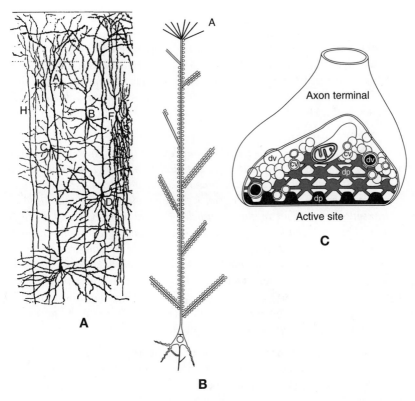

Figure 15-1 Neurons and their synaptic connections. **A,** The preparation shows eight neurons in the three superficial layers of the frontal cortex from a 1-month-old child. The cells have many dendrites covered with spines. **B,** A pyramidal cell with its apical dendrite showing the side branches and the terminal tuft studded with spine synapses (not all are shown). **C,** Details of the axon terminal or bouton showing dense projections from the active site with cross linkages forming in the presynaptic vesicular grid. (From Eccles keynote in Pribram KH (1993). Rethinking neural networks: quantum fields and biological data, with a keynote by Sir John Eccles. Erlbaum Hillsdale NJ. These illustrations are Sir John's Figure 1a on page 19, Figure 3a on page 21, and Figure 6 on page 24, respectively.)

signals and determines which is dominant: excitation or inhibition. The results of the polling process will be a decision as to whether to fire and send a signal to other neurons or to remain dormant, until the next polling takes place a small fraction of a second later. This is called an *integrate-and-fire model*. The phenomenon of consciousness, then, arises from the sum total of the decisions made in thousands of millions of such neurons.

TRANSCENDENT PROPERTIES

Generations of neurophysiologists have followed up on those fundamental discoveries. Their findings continue to be impressive, but we still are at a complete loss with regard to some fundamental questions relating to the origins of what Eccles referred to as the *nonphysical* and *transcendent properties*, such as feelings, thoughts, memories, intentions, and emotions. As he approached the end of his distinguished career, Eccles concluded in 1993 that the methods and theories he had pioneered, and for which he had received the Nobel Prize, were inadequate to the task he had set for himself. It would be necessary, as Szent-Györgyi also had decided, to take his investigations to a smaller level of scale, to the ultramicroscopic study of the synapse and its quantum properties (Eccles 1993). Here he would find the ultimate connection between mind and brain.

> . . . it is necessary to move to a higher level of complexity, the ultramicrosite structure and function of the synapse. The boutons of chemical transmitting synapses have a presynaptic ultrastructure of a paracrystalline arrangement of dense projections and synaptic vesicles, a presynaptic vesicular grid. Its manner of operation in controlling chemical transmission opens up an important field of neural complexity that is still at its conception. The key activity of a synapse concerns a synaptic vesicle that liberates into the synaptic cleft its content of transmitter substance, an exocytosis. A nerve impulse invading a bouton causes an input of thousands of calcium ions, 4 being necessary to trigger an exocytosis. The fundamental discovery is that at all types of chemical synapses an impulse invading a single presynaptic vesicular grid causes at the most a single exocytosis. There is conservation of the synaptic transmitter by an as yet unknown process of higher complexity. ECCLES (1993)

Yasue, Jibu, and Pribram (1991) used holonomic brain theory to derive a linear time-dependent Schrödinger equation to describe the interaction between the ionic bio-plasma density distribution and the phase patterns of neural membrane potential oscillations. Dawes (1993) calls this a *bottom-up perspective* on the ultimate principles of neural dynamics. Independently, Dawes found that the same Schrödinger equation is key to describing the large-scale properties of the parallel distributed neurocomputing architecture (for references, see Dawes 1993). This Dawes refers to as a *top-down perspective*.

NEURONS AS CELLS

Recent thinking about the nervous system is taking us in some new directions. One of these is that the nervous system is not the only communication system in the body and that some of the properties ascribed to the brain, such as memory and consciousness, actually may be whole-body phenomena. For those who work daily in therapeutic relationships this is obvious; for neuroscientists, it is not. It is remarkable how insights that one group has to offer, another group cannot tolerate.

A second direction, mentioned by Eccles, is that neurons are cells with parts that do things. We need to look at the ultramicroscopic scale to understand what really is going on. This productive perspective has been brilliantly and humorously articulated by Stuart Hameroff, of the University of Arizona:

> As presently implemented, the neuron doctrine portrays the brain's neurons and chemical synapses as fundamental components in a computer-like switching circuit,

supporting a view of brain = mind = computer. However, close examination reveals individual neurons to be far more complex than simple switches, with enormous capacity for intracellular information processing (e.g. in the internal cytoskeleton). The neuron doctrine, currently in vogue, is too watered-down to explain how the brain gives rise to mental life. Neuroscience is not being applied deeply enough. The neuron doctrine considers only certain activity at neuronal surfaces, ignoring internal features, including the fact that neurons are living cells. Each neuron is treated as a "black box," ignoring internal activities. The present characterization of the neuron is a cartoon, a skin-deep portrayal that simulates a real neuron much as an inflatable doll simulates a real person. HAMEROFF (1999)

Hameroff continues (Figure 15-2):

Consider a single-cell paramecium, which swims gracefully, avoids predators, finds food, mates, and has sex, all without a single synapse. Remarking on the complex behavior of motile protozoa, C.S. Sherrington (1951) said, "of nerve there is no trace. But the cell framework, the cytoskeleton might serve." If the cytoskeleton can be so useful in protozoa, what might it be doing in massive parallel arrays within neurons? Are neurons stupid in comparison to protozoa? HAMEROFF (1999)

Hameroff, recognizing that neurons contain very long arrays of microtubules, has researched the possibility that the microtubules can store information *(memory);* however, microtubules in neurons are not different from those found in other cells throughout the body. We suspect that the elegant work Eccles began on the structure of the synapse and Hameroff's work on microtubules apply to all cells throughout the body.

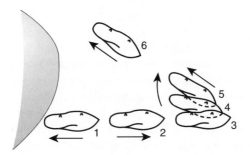

Figure 15-2 "Consider a single-cell paramecium, which swims gracefully, avoids predators, finds food, mates, and has sex, all without a single synapse. Remarking on the complex behavior of motile protozoa, C.S. Sherrington (1951) said, 'of nerve there is no trace. But the cell framework, the cytoskeleton might serve.' If the cytoskeleton can be so useful in protozoa, what might it be doing in massive parallel arrays within neurons? Are neurons stupid in comparison to protozoa?" (Modified with permission from the website of Stuart Hameroff, http://www.consciousness.arizona.edu/hameroff/slide%20show/slideshow_1.htm)

NEURONS, GLIA, AND THEIR SYSTEMIC INTERACTIONS

For decades, scientists thought that all of the missing secrets of brain function resided in neurons. However, a wave of new findings indicates that glial cells, formerly considered mere supports and subordinate to neurons, participate actively in synaptic integration and processing of information in the brain.

VESCE, BEZZI, AND VOLTERRA (2001)

The past decade of studies has changed our view of the integrative capacities and roles of glia. A picture is emerging in which neurons and astrocytes, a subtype of glial cell, are in a continuous regulatory dialogue. . . . It is likely that the results of these recent studies will signal a new way of thinking about the nervous system, in which the glial cell comes to the forefront of our attention.   MAZZANTI, SUL, AND HAYDON (2001)

Glial cells are active partners of neurons in processing information and synaptic integration. The active properties of glia, including long-range signaling and regulated transmitter release, are beginning to be elucidated. Recent insights suggest that the active brain should no longer be regarded as a circuitry of neuronal contacts, but as an integrated network of interactive neurons and glia.   BEZZI & VOLTERRA (2001)

Development of microelectrode and sensitive electrophysiological recording devices has made neurophysiology the dominant paradigm for the study of biological communication. Robert O. Becker (1990, 1991) attempted to draw attention to the importance of the perineural system in a pair of classic papers. A decade later, researchers in a number of laboratories began to make a series of discoveries that confirmed and expanded upon Becker's insights.

The nervous system consists of an astronomical number of neurons, but there are some five to ten times more supporting cells, collectively called the *neuroglia* (from the Greek, meaning "nerve glue"). These cells include the astrocytes, oligodendroglia, and microglia. Outside the brain, in the peripheral nervous system, there are other cells that can be included in this supportive category: Schwann cells, satellite cells of peripheral ganglia, and ependymal cells. The latter are epithelial cells that line the ventricles, choroid plexus, and central canal of the spinal cord.

Nerve cells have always seemed to be the primary actors in exchanging information. The relatively tiny glial cells (in contrast to the very long neurons) were never considered to have any major or direct role in communication, although they have been studied extensively for their roles in development and repair of nerve damage (Matsas & Tsacopoulos 1999; Vernadakis & Roots 1995). For a description of the structure of these fascinating cells, see a histology textbook such as Fawcett (1994).

The recent revolutionary discovery is that the perineural connective tissue is also a dynamic communication system with multiple interactions with the neurons. Researchers are exploring the roles of these cells in neuronal-glial, neuronal-neuronal, and glial-glial interactions (McGrath et al. 2001; Miykata & Hatton 2002; Morale et al. 2001; Vesce, Bezzi & Volterra 2001; Watanabe 2002). Neuron-glial signaling cascades are being described (Morale et al. 2001).

Figure 15-3 shows a scheme for nonsynaptic communication within the brain through the astrocytic syncytium. The dotted line denotes a calcium wave that can propagate through this network (Cornell-Bell et al. 1990; Ventura & Harris 1999). Astrocytes have a repertoire of neurotransmitter receptors that mirror those of the neighboring neural synapses (Vesce, Bezzi & Bolterra 1999).

Glia now appear to play a key role in the pathophysiology of pain (Watkins, Milligan & Maier 2001) and in Parkinson's disease, mental illnesses, the action of psychotropic drugs, and learning, to name a few (Hertz, Hanson & Ronnback 2001).

It even has been proposed that fully differentiated glial cells can serve as stem cells that can give rise to cortical neurons (Alvarez-Buylla, Garcia-Verdugo & Tramontin 2001).

Glial cells are emerging from the background to become more prominent in our thinking about integration in the nervous system. Given that glial cells associated with synapses integrate neuronal inputs and can release transmitters that modulate synaptic activity, it is time to rethink our understanding of the wiring diagram of the nervous system. It is no longer appropriate to consider solely neuron-neuron connections; we also need to develop a view of the intricate web of active connections among glial cells, and between glia and neurons. Without such a view, it might be impossible to decode the language of the brain. HAYDON (2001)

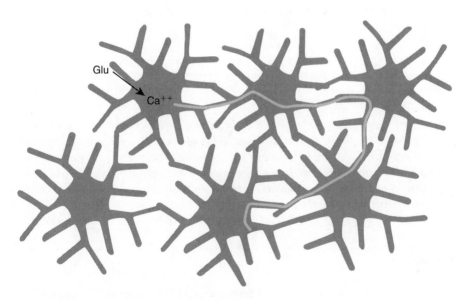

Figure 15-3 Astrocytes form an interconnected system called a *syncytium*. The astrocytic syncytium allows a nonsynaptic means of communication within the brain. (From Ventura, R.E. 'Astrocytes in the hippocampus', Synapse Web, Boston University, *http://synapses.bu.edu/*)

WAVE ASPECT OF CONSCIOUSNESS

Memory and consciousness cannot be understood without a better picture of the energetics of cells and tissues. As an interesting example, Charman (1997) logically views what we refer to as *mind* as a brain-generated neuromagnetic field. To paraphrase:

> *When the mosaic of neurons resonates at preferred frequencies, so will their associated microfields. These will interact with each other to form a complex neuromagnetic whole that permeates through the magnetically transparent physical structure of the brain as if it was not there.*

Add to this concept of mind the biomagnetic fields of the peripheral neurons and the cell and tissue structures associated with them, and we begin to see a dynamic picture of the biomagnetic body as a whole.

Freeman described the situation as follows:

> *Pulses coming into a set of neurons are converted into synaptic currents which we call waves. These currents are filtered and integrated over time and space in the wave mode. The wave activity reaching trigger zones is converted back into the pulse mode. The pulse is then transmitted from one place to another across the synapse. This involves further delay, dispersion in time, etc.*
>
> FREEMAN, QUOTED BY PRIBRAM IN BULSARA AND MAREN (1993)

That a deeper level of analysis would be required for understanding consciousness was recognized long ago by Max Planck:

> *We must assume behind this force (in the atom) the existence of a conscious and intelligent mind. This mind is the matrix of all matter.*   MAX PLANCK

The idea has been developed formally by Pearson:

> *Mind is a property of the "nuether," a sub-quantum level of reality. . . . the nuether (is) structured like a neural network.*   PEARSON (1997)

Romijn also has elaborated on these ideas:

> *. . . [T]he hypothesis is put forward that the fleeting, highly ordered patterns of electric and/or magnetic fields, generated by assemblies of dendritic trees of specialized neuronal networks . . . encode for subjective (conscious) experiences such as pain and pleasure, or perceiving colors. Because by quantum mechanical definition virtual photons are the theoretical constituents of electric and magnetic fields . . . it is the highly ordered patterns of virtual photons that encode for subjective (conscious) experiences.*
>
> ROMIJN (2002)

REFERENCES

Alvarez-Buylla, A., Garcia-Verdugo, J.M. & Tramontin, A.D. 2001, 'A unified hypothesis on the lineage of neural stem cells', *Nature Reviews Neuroscience*, vol. 2, pp. 287-293.

Becker, R.O. 1990, 'The machine brain and properties of the mind', *Subtle Energies*, vol. 1, pp. 79-97.

Becker, R.O. 1991, 'Evidence for a primitive DC electrical analog system controlling brain functions', *Subtle Energies,* vol. 2, pp. 71-88.

Bezzi, P. & Volterra, A. 2001, 'A neuron-glia signaling network in the active brain', *Current Opinion in Neurobiology,* vol. 11, pp. 387-394.

Bulsara, A.R. & Maren, A.J. 1993, 'Coupled neural-dendritic processes: Cooperative stochastic effects and the analysis of spike trains', in *Rethinking Neural Networks: Quantum Fields and Biological Data,* ed. K.H. Pribram, Lawrence Erlbaum Associates, Hillsdale, NJ, p. 96.

Charman, R.A. 1997, 'The field substance of mind: A hypothesis', *Network,* vol. 63, pp. 11-13.

Cornell-Bell, A.H., Finkbeiner, S.M., Cooper, M.S., Smith, S.J. 1990, 'Glutamate induces calcium waves in cultured astrocytes: long-range glial signaling', *Science,* vol. 247, pp. 470-473.

Dawes, R.L. 1993, 'Advances in the theory of quantum neurodynamics', in *Rethinking Neural Networks: Quantum Fields and Biological Data,* ed. K.H. Pribram, Lawrence Erlbaum Associates, Hillsdale, NJ, pp. 149-159.

Eccles, J. 1993, 'Evolution of complexity of the brain with the emergence of consciousness', keynote in *Rethinking Neural Networks: Quantum Fields and Biological Data,* ed. K.H. Pribram, Lawrence Erlbaum Associates, Hillsdale, NJ, pp. 1-28.

Fawcett, D.W. 1994, *A Textbook of Histology,* 12th ed, Chapman & Hall, New York. Hameroff, S. 1999. 'The neuron doctrine is an insult to neurons', *Behavioral and Brain Sciences,* vol. 22, pp. 838-839.

Haydon, P.G. 2001, 'GLIA: listening and talking to the synapse', *Nature Reviews Neuroscience,* vol. 2, pp. 185-193.

Matsas, R. & Tsacopoulos, M. (eds) 1999, 'The functional roles of glia cells in health and disease: Dialogue between glia and neurons', in *Advances in Experimental Medicine and Biology,* Kluwer Academic/Plenum Publishers, New York.

Mazzanti, M., Sul, J.Y., Haydon, P.G. 2001, 'Glutamate on demand: astrocytes as a ready source', *Neuroscientist,* vol. 7, pp. 396-405.

McGrath, B., McCann, C., Eisenhuth, S. & Anton, E.S. 2001, 'Molecular mechanisms of interactions between radial glia and neurons', *Progress in Brain Research,* vol. 132, pp. 197-202.

Miykata, S. & Hatton, G.I. 2002, 'Activity-related, dynamic neuron-glial interactions in the hypothalamo-neurohypophysial system', *Microscopy Research and Technique,* vol. 56, pp. 143-157.

Morale, M.C., Gallo, F., Tirolo, C., Testa, N., Caniglia, S., Marletta, N., Spina-Purrello, V., Avola, R., Caucci, F., Tomasi, P., Delitala, G., Barden, N. & Marchetti, B. 2001, 'Neuroendocrine-immune (NEI) circuitry from neuron-glial interactions to function: Focus on gender and HPA-HPG interactions on early programming of the NEI system', *Immunology and Cell Biology,* vol. 79, pp. 400-417.

Pearson, R.D. 1997, 'Consciousness as a sub-quantum phenomenon', *Frontier Perspectives,* vol. 6, pp. 70-78.

Romijn, H. 2002, 'Are virtual photons the elementary carriers of consciousness?' *Journal of Consciousness Studies,* vol. 9, pp. 61-81.

Ventura, R. & Harris, K.M. 1999, 'Three-dimensional relationships between hippocampal synapses and astrocytes', *Journal of Neuroscience,* vol. 19, pp. 6897-6906.

Vernadakis, A. & Roots, B.I. (eds) 1995, *Neuron-Glia Interrelations During Phylogeny: II: Plasticity and Regeneration,* Humana Press, Totowa, NJ.

Vesce, S., Bezzi, P. & Volterra, A. 1999, 'The active role of astrocytes in synaptic transmission', *Cellular and Molecular Life Sciences,* vol. 56, pp. 991-1000.

Vesce, S., Bezzi, P. & Volterra, A. 2001, 'Synaptic transmission with the glia', *News in Physiological Sciences,* vol. 16, pp. 178-184.

Watanabe, M. 2002, 'Glial processes are glued to synapses via Ca(2+)-permeable glutamate receptors', *Trends in Neurosciences*, vol. 25, p. 7.

Watkins, L.R., Milligan, E.D., Maier, S.F. 2001, 'Spinal cord glial: new players in pain', *Pain*, vol. 93, pp. 201-205.

Yasue, K., Jibu, M. & Pribram, K. 1991, 'Appendix', in *Brain and Perception, Holonomy and Structure in Figural Processing*, ed. K.H. Pribram, Lawrence Erlbaum Associates, Hillsdale, NJ, pp. 275-330.

Sensation and movement

Sensation may not be what we think it is. Oschman (2001)

In the previous chapter we learned that a single-cell paramecium swims gracefully, avoids predators, finds food, mates, and has sex, all without a single neuron or synapse. How can this happen, and what does this mean for other cells, such as those that make up our bodies?

AN IMPORTANT THEORY

Jelle Atema is a dedicated flute player and an accomplished sensory neurophysiologist and marine behaviorist. After years of research at the Woods Hole Oceanographic Institution, he became the director of the Boston University Marine Program (BUMP) at the Marine Biological Laboratory in Woods Hole.

In 1973, Atema published an important theoretical paper entitled *Microtubule Theory of Sensory Transduction*. In his article, Jelle cites earlier work of Lowenstein, Osborne, and Wersäll (1964) suggesting the remarkable concept that *sensory systems may be movement systems working in reverse.*

The remarkable concept that the same cellular structures can be responsible for both sensation and movement has many implications for the properties of the living matrix and for our exploration of continuum movement.

To be specific, Lowenstein and his colleagues suggested that the sensory *cilium* found in many receptors may act as a motile structure in reverse, responding to deformation in a certain direction with the initiation of electric changes. The idea originally came from earlier work of Gray and Pumphrey (1958) and Gray (1960).

Atema's discussion supports the idea that the same molecular mechanisms that convey sensory information in cells can produce movement and vice versa. This is a remarkable concept and has many implications. We shall see that it represents quantum coherence in action.

Bacteria as an example

A dramatic example is provided by simple and evolutionarily ancient organisms such as bacteria, which are able to sense their environments and move toward or away from particular physical stimuli such as light, temperature, pH, oxygen levels, magnetic fields, attractant/repellent chemicals, and gradients in osmolarity.

Bacteria are composed of only a few thousand molecules and are entirely lacking in what we usually refer to as *sense organs, nerves,* or *muscles.* Thus they are extremely efficient in sensing their environment, processing the information, and responding through movement. They are able to integrate their responses to several simultaneous sensory stimuli in order to balance the overall response and maintain themselves in the optimal conditions for growth and reproduction. Bacteria can move toward or away from a stimulus at a rate of several millimeters per hour (Armitage 1992).

Perhaps Valerie Hunt's observations on the dancers imply that the evolutionarily ancient sensory/movement systems found in lower organisms, even bacteria, still exist in the human body and can be accessed under appropriate conditions. Perhaps this is the same system Albert Szent-Györgyi was describing in his research on electronic biology.

One colleague, Deborah Stucker, refers to this ancient system as *amoeba mind.*

Bacterial flagella

Motile bacteria have movable levers called *flagella* (Figure 16-1, *A* and *D*). In the seventeenth century, Leeuwenhoek observed the movement of bacteria and suggested that they had "legs" a millionth of the thickness of a hair from his beard, an estimate that has proved to be reasonably accurate.

The bacterial flagellum is a single filament composed of the protein flagellin. Each filament is a polymer composed of about 20,000 flagellin monomers. In protozoa and higher organisms, the flagella are composed of larger structures called *microtubules.* The protein making up microtubules is called *tubulin.*

Protein sensory receptor molecules are located on the surfaces of flagella or on the microtubules within them. These sensors are capable of perceiving gravity as a mechanical stress, as well as light, chemicals, sound, and electricity. Figure 16-1, *B* and *C,* documents that bacteria can be repelled or attracted by particular chemicals in their environment.

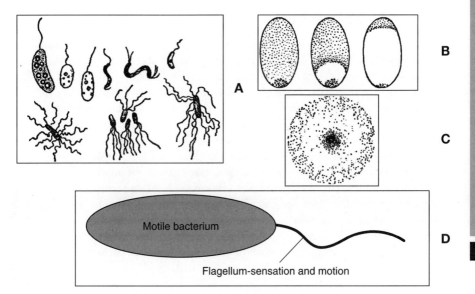

B

C

Motile bacterium

D

Flagellum-sensation and motion

Figure 16-1 Bacteria are extremely simple organisms, having the most rudimentary mechanisms for sensing and responding to their environment. Despite their simplicity, they are capable of moving toward or away from specific stimuli, finding the optimal location in their environment to maximize survival. Some bacteria are simple particles, others are capable of rapid movements produced by the whiplike protoplasmic processes called *flagella* or *cilia*. **A**, Different species of flagellated bacteria, some of which have a single flagellum, and others with large numbers, either at one end of the body or scattered over the surface. **B**, Response of bacteria *(Spirilla)*, which are repelled by sodium chloride (NaCl) crystals. **C**, Response of bacteria (again, *Spirilla*) to oxygen produced by an alga. The bacteria quickly gathered closely to an alga in light. **D**, Atema and others have suggested that the same organelle, the flagellum, is capable of both sensing the environment and acting upon it. (**A, B**, and **C**, from Jennings, H.S. (ed) 1906, *Behavior of the Lower Organisms,* New York, Columbia University Press, The Macmillan Company, Figures 23 (p. 26), 24B (p. 28), and 25B (p. 30), respectively. Reprinted with permission of the publisher.)

A locomotor-sensory system

The Russian scientist Y.A. Vinnikov refers to this as the *locomotor-sensory system* (LMSS). He has researched the evolutionary origins of sensory organs in the context of locomotion. The various planetary fields of gravity, light, sound, electricity, motion, smell, and taste have propelled the evolutionary selection of mechanisms for both sensing and responding to gravity and other environmental factors, that is, for exercising directed locomotion.

In other words, sensitivity to the environment, that is, the *reception* of energy/information, evolved in close functional relationship with locomotor mechanisms, that is, the *production* of movement or kinetic energy (Vinnikov 1995).

Sensory/motile cilia in higher animals

Jelle Atema points out that cilia and flagella are components of many mammalian and invertebrate sensory cells. The basic structure of the cilium is shown in Figure 16-2, *A*, and some examples of sensory systems using cilia are shown in Figure 16-2, *B* and *C*. Cilia occur in olfactory receptors; the acoustico-lateralis system, which senses gravity and sound; and common photoreceptors. Figure 16-3 shows the most thoroughly studied sensory system of them all, the retinal rod cell in the human eye. Each of these sensory/movement systems can be traced from an evolutionary perspective back to the primitive motile bacteria, although the structure of the bacterial flagellum is much simpler.

Cilia and flagella in mammalian and invertebrate sensory endings have in common a 9 + 2 core arrangement of microtubules (Figure 16-2, *A*). In some cases, the tip of the sensory cilium thins down and contains fewer microtubules (*inset,* Figure 16-2, *C*). In sensory cells of higher animals, the microtubules connect within the cell to the cytoskeletal matrix, which is composed of additional microtubules, microtrabeculae, and microfilaments. This is none other than the living matrix pathway described in earlier chapters.

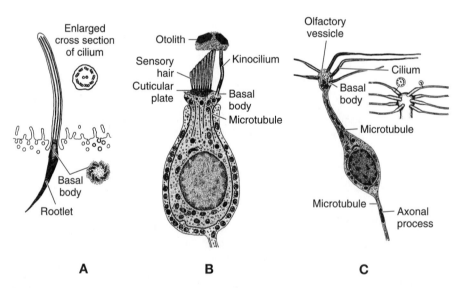

A   **B**   **C**

Figure 16-2 **A,** The sensory/motor cilium found in invertebrate and vertebrate organisms contains a core of microtubules with nine peripheral doublets and a pair at the center. **B,** Gravity or acceleration detection cell in the inner ear or vestibular labyrinth. Motions of the crystalline otolith granules are detected by the kinocilia. **C,** An olfactory cell. Note in the *inset* that the cilia become thinner at their tips, and the 9 + 2 core microtubule arrangement changes to a smaller number. (From Lentz, T.L. 1971, *Cell Fine Structure. An Atlas of Drawings of Whole-Cell Structure,* WB Saunders Company, Philadelphia, Figure 3, p. 4, Figure 179, p. 409, Figure 86, p. 201, with permission from Elsevier Science.)

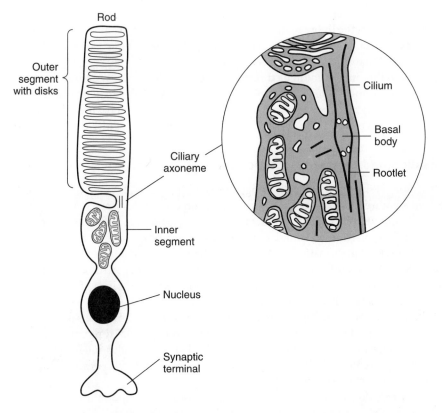

Figure 16-3 In both rods and cones, the outer segment is connected to the inner segment by a narrow stalk containing a ciliary axoneme. Developmental studies have shown that the photoreceptor lamellae are derived from the ciliary membranes. (Illustration of the rod cell is modified from Kleinsmith, J. & Kish, V.M. 1995, *Principles of Cell and Molecular Biology,* Harper Collins, New York, p. 768. Inset is modified from Lentz, T.L. 1971, *Cell Fine Structure: An Atlas of Drawings of Whole-Cell Structure,* WB Saunders Company, Philadelphia, p. 381.)

CONDUCTION OF SENSATIONS THROUGH THE LIVING MATRIX

Jelle Atema proposed that the ciliary microtubular apparatus in sensory cells receives environmental information and transmits it via *propagated conformational changes* in the filamentous microtubule proteins. The microtubules thus are active functional units in the reception and transmission of sensory information. Waves of conformational change in the microtubules could be conducted through the cytoskeleton, across the cell surface via integrin molecules, and then into neighboring cells or into the fibrous system of the connective tissue. We will show some drawings of the structures involved later.

The main point we are approaching is that there is a theoretical basis for sensory energy inputs at any receptor being conveyed rapidly and without loss into the living matrix system and then throughout the body.

Sensory systems can respond to one or a few energetic quanta. Because it is an excitable medium, the living matrix can amplify an incoming stimulus to produce a large or even a huge response. The classic example is the carelessly discarded cigarette or match that starts a raging forest fire.

One place this energy could be directed is into the contractile system in muscles. Hence sensory information could bypass or be conveyed parallel to neuronal signals. This would provide for a very rapid response to the environment. It could initiate motor responses prior to perception by the normal neuronal cognitive processes.

Sensation and generator potentials

Each sensory system in the body is designed to detect a particular kind of energy. We speak of photoreceptors, thermoreceptors, chemoreceptors, and mechanoreceptors. The mechanoreceptors come in different sorts, each responding preferentially to particular kinds of mechanical stimulation, such as stretch of muscles or tendons, rotation of joints, light deformation of the skin, bending of hairs, and distention of hollow structures such as blood vessels or abdominal viscera. Receptors are thought to be the "pass filters" of the body, admitting to the nervous system some kinds of information about the external world and rejecting other kinds.

Although the specialization of particular receptors for particular kinds of stimuli is obvious, it is important to point out that even the most specialized receptors may be excited by more than one kind of stimulus, provided the stimulus intensity is sufficiently high. Receptor selectivity is relative. For example, the retina can respond to both light and to magnetic fields (see Appendix A).

Sensory receptors differ from conventional neurons in an important way. Neurons have an *all-or-none* response (to be discussed in more detail later). Under adequate stimulation, there is a virtually explosive change in the electric field across the nerve membrane that is self-perpetuated and conducted along the neuron. Neurons either conduct an action potential or they do not. They are quantized or digital—there are only two states available for the action potential, on or off.

Sensory receptors are different. They are not normally activated by changes in electrical fields; they are activated by a variety of energy forms such as light, heat, mechanical distortion, and chemicals. Electrical fields develop across the membranes of receptor cells in a direct relationship to stimulus intensity. They are *analog* systems; the bigger the stimulus the bigger the response. These fields are called *generator* or *receptor potentials*. These are stationary potentials in that they are confined to the receptor cells and are not propagated along the attached afferent nerves unless the stimulus reaches a critical intensity that is sufficient to trigger a depolarization of the nerve. In other words, the generator potential is *graded* and *stationary;* the action potential is *all or nothing* and *conducted* (for a discussion of these points see standard physiology texts, such as Ruch and Patton 1965).

Transducers

Receptors are transducers that convert one kind of energy into another. All receptors convert particular kinds of energy into electrical information. What happens at the interface between a sensory receptor and an afferent neuron is that the magnitude of the generator potential is converted into a repetitive discharge of the neuron. The frequency or some other attribute of the discharge encodes the amplitude of the generator potential. For example, the neural spike frequency may be directly related to the stimulus intensity.

Stated differently, the receptor is an *analog* system, the output of which is in definite relation or proportionality to the intensity of the stimulus. In terms of electronics, this is called *amplitude modulation*. In contrast, the nerve net is digital.

The transduction process varies in different receptor systems and can involve intermediate events. For example, vision involves an intermediate chemical event, the breakdown of a visual pigment.

Evidence for a microtubule theory of sensory transduction

The microtubule theory of sensory transduction provides an answer to a question posed by Beidler (1970). He was discussing the mechanism of taste receptor operation:

> ... *[A]lmost all chemical stimuli tested showed an increase in response as the stimulus concentration is increased. ... From this observation one must conclude that a conformational change in the receptor molecule must produce an effect that is transmitted to areas of the cell distant from the receptor sites by means other than electrical depolarization associated with changes in ionic fluxes. The process by which such effects can be propagated is not yet known.* BEIDLER (1970)

Evidence for a role of microtubules in sensation comes from studies showing that treatment of sensory cells with drugs that cause microtubules to disassemble also inhibit the responses to stimuli (Moran & Varela 1971). The authors of that study speculate that "the 350-1000 parallel microtubules of the sensory process function as *mechanochemical engines driven backwards by the force of the stimulus,* creating conditions favorable to formation of generator current. ... The vast proliferation of microtubules in the sensory process may serve to increase the 'gain' of the mechanoreceptor."

Evidence for cilia being used simultaneously for sensation and locomotion comes from studies of the cilia on the gill of the mussel *Mytilus*. These cilia are motile and they are mechanosensors (Thurm 1968). Atema also mentions a model system that demonstrates the underlying principle. It is a synthetic polymer that shortens when calcium is added and expels calcium when it is mechanically stretched (Kuhn 1949).

More clues

At the time Jelle Atema published these ideas (1973), little was known about the way conformational changes actually could be conducted through microtubules. However, this was the period, in the early 1970s, when rapid advances began to take place in our understanding of conformational and vibrational properties of the protein lattice. We are referring to the works of Fröhlich and Davydov, and the more recent biophysical research on quantum coherence by Ho, Popp, del Guidice, Smith, and others. Our investigations into the crystalline and semiconductor properties of the connective tissue and cytoskeleton play into this as well.

It may be important that the long, thin peripheral nerves that connect the brain to sensory receptors and muscles are packed with microtubules. These microtubules are thought to be involved in the fast transport of various chemical factors back and forth between the cell body and the synaptic terminals, but they could have an additional role in conveying waves of conformational change related to sensation and movement.

In other words, the nervous system itself could have two parallel and distinct mechanisms for the transmission of sensory and movement information: a fast mechanism, involving waves of conformational changes in the microtubules, and a slower, classic mechanism, involving ionic currents and action potentials.

Another clue comes from the research of Andrew Packard at the Stazione Zoologica in Naples. Packard studied color changes in squids. Squids and other cephalopods change color with muscle-operated pigment cells called *chromatophores*. Orderly patterns of color change are controlled by the brain. When the brain is damaged or the nerve supply to the skin is cut, the color patterns become disorganized. Packard nonetheless observed wavelike excitations that continue to flow through the chromatophore system of squids some days after the nerve to one side of the body is cut. Without the nerve supply, the chromatophores develop supersensitivity. They continue to communicate with each other via some non-neural mechanism. This autonomous behavior highlights the general principle that non-neural control of a muscular tissue is possible (Packard 1995).

The emerging picture of how sensory information can be transduced *directly* into movement will be presented later, but first we will summarize the textbook explanations of sensation, nerve conduction, and muscle contraction. The intention is not to prove that the conventional texts are wrong but to show that the explanations they offer are a part of the story and that there are other possibilities. From a variety of perspectives it appears that Emilie Conrad, Valerie Hunt, Albert Szent-Györgyi, and Saotome all were correct. There is, indeed, a way of moving that is different from the classic neuromuscular processes that are so well known to physiologists and neurophysiologists.

A final clue from phototherapy

There exists a relationship which is largely predictable between light frequency, environment, and the restoration of health following departures from normal, which are still within the physiologic limits. . . . H.R. SPITLER

Dr. Spitler's observation that, within limits, health can be restored with specific light frequencies poses a profound biological question: How can light "jump start" the healing process for a wide range of clinical problems involving tissues throughout the body? An answer to this question would have much medical significance.

Gottlieb (2000), discussing the research of Tiina Karu in Russia, states the problem this way:

How does light find the right places to work to heal the body? Normal tissue is much less affected by light than out-of-balance tissue. Starving cells are far more sensitive than well-fed ones. GOTTLIEB (2000)

The story being documented in this book is beginning to answer this question.

Can we follow a light stimulus through the living matrix to the places where structure and function have been compromised? The goal is a logical explanation of how the appropriate application of light can reach and benefit any part and process in the organism.

Following a photon. The retina is the most remarkable and thoroughly studied sensor in nature. The standard picture of photoreception begins with the interaction of a photon with the visual pigment rhodopsin. The 1967 Nobel Prize research of George Wald and others showed precisely how rhodopsin in the retinal rod cell absorbs the energy of one or a few photons (Wald 1967). A conformational change in the rhodopsin molecule initiates a cascade of chemical reactions, the flow of millions of sodium ions across the rod cell membrane, and an electrical signal that is transmitted by the optic nerves to the brain. In essence, the energy contained in a single photon is amplified many times to produce a nerve impulse (Stryer 1987). From many such impulses the brain constructs our image of the world around us.

There is, however, another story to be told. This other story is not meant to replace the standard textbook description of photoreception and visual image formation. The neurobiology texts are not wrong, but when we look inside the rod cells, the nerves, and the other cells associated with them, we find another pathway by which the body becomes aware of the photon. This pathway *includes* the nervous system but is not limited to it. This pathway conducts information far faster than nerves. This pathway is not slowed by synaptic delays. The pathway is inside the cellular "black boxes" mentioned by Hameroff and connects to the connective tissue in which the cells are embedded.

We will look at this pathway to see if it can help us understand how phototherapies have their remarkable system-wide effects and to see if it can confirm Atema's theory in action.

How phototherapy affects the whole body. The main point is that there is a theoretical basis for sensory information being conveyed rapidly and *directly* into the body, bypassing or running parallel to the neuronal circuitry. This provides for a very rapid response to light, *a response that reaches into every part of the body.*

In terms of phototherapy, the visual perception of a color may not be the most important part of the process. We propose that healing responses are initiated by mechanisms that involve the *entire* living matrix, not just the neuromatrix. Light transduced by the retina into nerve impulses is processed by specific optic fibers in the nervous system, whereas light-induced waves of conformational change entering the living matrix are, in principle, capable of being rapidly conducted to every nook and cranny of the body, even to the nuclear matrices and genes located in *every* cell, not just nerve cells. It is a way for light to generate a message that reaches everywhere, not just where nerves go. This makes sense because, as we have seen, the most important cells involved in any repair process, such as fibroblasts or osteoblasts, are not in direct communication with neurons.

A dramatic application of phototherapy is in brain injury. This gives rise to a question: How can the application of light to the retina revitalize or regenerate neural pathways? To answer this question, it is important to recognize that the organization of the neuromatrix is primarily determined by activities of the perineural connective tissues, which are composed of astrocytes, glia, and oligodendrocytes. It was discovered that these supporting cells actually form a communication system of their own, with synapselike connections to the neurons proper (Lo Turco 2000). Light produces signals that reach all of these cells—it affects all parts of the matrix.

A clue from the anatomy of the retina. A look at the anatomy of the retina reveals a likely site for photoreceptor signals to connect to the perineural connective tissue system and from there to all parts of the body. Light microscopists have identified a dense-staining line called the *outer limiting membrane* that lies between the photoreceptor layer and the outer nuclear layer of the retina (Figure 16-4, *A*). Electron microscopy has revealed that this is not a membrane at all; instead it is a precisely aligned planar array of densely spaced plaque-bearing junctions with bundles of actin filaments attached to them. This distinctive row of adhering junctions, now called *the outer limiting zone,* attaches the photoreceptor cells to the Müller cells, which are neuroglial connective tissue cells (Figure 16-4, *B*). The junctions have an obvious architectural role in keeping the photoreceptors in position. The junctions are composed of a particular set of proteins that form an intricate and novel kind of cell-to-cell junction (Paffenholz et al. 1999).

The anatomical arrangement is such that a signal set up as a conformational wave in the cytoskeleton of a retinal rod or cone cell would reach the outer limiting zone and Müller cells first and then travel to the synaptic zone (Oschman 2001). It is reasonable, on the basis of Atema's hypothetical sensory conduction mechanism, that the signal could reach the outer limiting zone, be conducted by cytoskeletal fibers, through the novel cell-to-cell junctions just described, into the Müller cell cytoskeleton, and then throughout the entire connective tissue system. All of this could happen extremely rapidly, before a nerve impulse is initiated in the bipolar cells of the internal nuclear lamina that carry the signal to the brain via the optic nerves (see Figure 2-3).

The mechanism by which the conduction takes place will be discussed in detail in later chapters, but first we need to introduce the anatomical and physiological terminology neuroscientists use to describe sensation, nerve conduction, and muscle contraction. This terminology will be used again when we look at alternative pathways of sensation and action.

Sensation

The primary sensory systems in the body have been carefully researched, although this does not mean that we know everything about them. It is often stated that there are five basic senses, although some scholars list many more. For example, Murchie (1978) lists 32 senses. The traditional senses are sight, hearing, smell, taste, and touch.

We have seen that one of the important aspects of the structure of sensory endings is that many of them have tiny hairs on them, called *cilia*. The cilia, in turn, are composed of arrays of microtubules. Sensory neurophysiologists have suggested that the

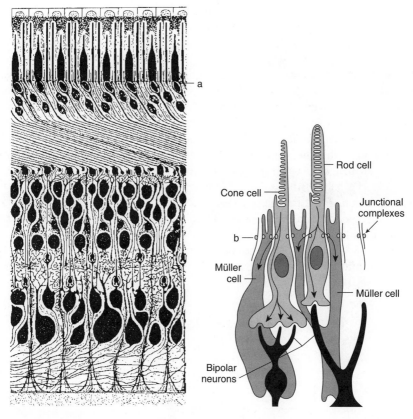

Figure 16-4 Light microscopists have identified a dense-staining line called the *outer limiting membrane (a)* lying between the photoreceptor layer and the outer nuclear layer of the retina. Under the electron microscope, this membrane shows up as a precisely aligned planar array of densely spaced plaque-bearing junctions with bundles of actin filaments attached to them *(b)*. This row of adhering junctions attaches the photoreceptor cells to the Müller cells, which are neuroglial connective tissue cells. It is suggested that the waves of conformational change developed in the rod and cone outer segments are conducted through the cytoskeletons of the photoreceptor cells and that the signal splits into two pathways *(arrows)*. One pathway takes the signal across the junctional complexes and into the Müller cells and thence throughout the living matrix of the body. The other pathway takes the signal to the synapses at the base of the photoreceptor cells and thence to the bipolar and other neurons, through the optic nerve, and into the brain. (Diagram on left is Figure 34-27, page 895, in Bloom & Fawcett 1994, *A Textbook of Histology*, 12th Edition, by permission of Hodder Arnold, London.)

molecules responsible for sensation actually are located on the cilia or in the microtubules themselves.

Sensation begins with a particular kind of energy (sound, light, pressure) interacting with a receptor molecule. The receptor molecule undergoes a conformational or shape change. Atema's hypothesis is that the change in the receptor induces a wave of conformational changes in the microtubules. These waves are propagated along the length of the tubule and then to other structures in the cell and beyond.

It is important to mention that molecules such as those producing sensations of taste or smell need not physically touch receptor molecules. Odor and taste molecules vibrate and emit electromagnetic fields that travel a certain distance through space. This phenomenon is the basis for spectroscopy, the technique scientists use to learn about the structure of atoms and molecules from the vibrations they emit and absorb (Oschman 2000, Chapter 9). It is probably the electromagnetic field, and not the molecule itself, that is detected by the receptor molecule. This *distance* or *noncontact interaction* is discussed by Callahan (1975) and Oschman (2000). Callahan's discovery arose from studies of pheromones, chemical attractant molecules, in insects. In the past it had been thought that the pheromone is an odor molecule that is *smelled* by the male insect. The challenge in explaining pheromone effects is that a male moth, for example, can detect a female who is emitting the sex attractant *downwind* from the male. Callahan concluded that the male cannot be responding to the pheromone molecule itself but is responding instead to a radio signal it emits when energized by infrared light from the night sky.

Figure 16-5 compares radio communication with molecular communication. A tuned transmitter circuit (*a*) generates an oscillating electrical field that flows into an

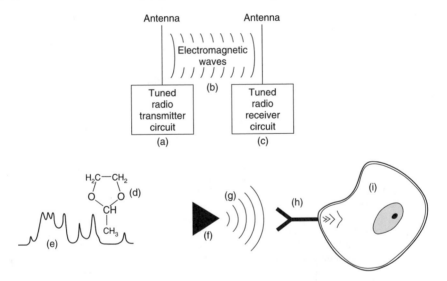

Figure 16-5 A tuned transmitter circuit (*a*) generates an oscillating electrical field that flows into an antenna wire. As electrons move back and forth in the antenna, an electromagnetic field (*b*) is set up in the surrounding space. This field can travel a very long distance to a receiving antenna connected to a tuned receiver circuit (*c*). Familiar radio and television technology uses this kind of setup. Low-power radio transmitters have been sent far into space and are continuing to send radio signals billions of miles back to Earth. Molecules (*d*) contain charged parts that vibrate because of thermal agitation. Like a radio transmitter, these moving charges produce electromagnetic fields with a spectral fingerprint (*e*) that can be used to characterize the molecule's structure. A hormone or other signal molecule (*f*) also vibrates and emits an electromagnetic field (*g*) that resonates with a receptor molecule (*h*), which acts like a receiving antenna. Vibrations of the receptor molecule can activate processes within the cell (*i*). (*d* and *e* are from Whiffen 1996, Fig. 8.2a, p. 102, used with permission from Pearson Education Limited.)

antenna wire. As electrons move back and forth in the antenna, an electromagnetic field *(b)* is set up in the surrounding space. This field can travel a very long distance to a receiving antenna connected to a tuned receiver circuit *(c)*. Familiar radio and television technology uses this kind of setup. Low-power radio transmitters have been sent far into space and are continuing to send radio signals billions of miles back to Earth.

Molecules *(d)* contain charged parts that vibrate because of thermal agitation. Like a radio transmitter, these moving charges produce electromagnetic fields with a spectral fingerprint *(e)* that can be used to characterize the molecule's structure. A hormone or other signal molecule *(f)* also vibrates and emits an electromagnetic field *(g)* that resonates with a receptor molecule *(h)*, which acts like a receiving antenna. Vibrations of the receptor molecule can activate processes within the cell *(i)*.

REFERENCES

Armitage, J.P. 1992, 'Bacterial motility and chemotaxis', *Scientific Progress Oxford,* vol. 76, pp. 451-477.

Atema, J. 1973, 'Microtubule theory of sensory transduction', *Journal of Theoretical Biology,* vol. 38, pp. 181-190.

Beidler, L.M. 1970, 'Physiological properties of mammalian taste receptors', in *Taste and Smell in Vertebrates,* eds G.E.W. Wolstenholme & J. Knight, Churchill, London, pp. 51-70.

Callahan, P.S. 1975, *Tuning in to Nature,* Devin-Adair, Old Greenwich, Connecticut.

Gottlieb, R. 2000, 'Scientific findings about light's impact on biology', *Journal of Optometric Phototherapy,* Apr., pp. 1-4.

Gray, E.G. 1960, 'The fine structure of the insect ear', *Philosophical Transactions of the Royal Society B,* vol. 243, p. 75.

Gray, E.G. & Pumphrey, R.J. 1958, 'Ultrastructure of the insect ear', *Nature,* vol. 181, p. 618.

Kuhn, W. 1949, 'Reversible Dehnung und Kontraktion bei Änderung der Ionisation eines Netzwerks polyvalenter FadenmolekÜlionen', *Experientia,* vol. 5, p. 318.

Lo Turco, J.J. 2000, 'Neural circuits in the 21st century: Synaptic networks of neurons and glia', *Proceedings of the National Academy of Sciences USA,* vol. 97, pp. 8196-8197.

Lowenstein, O., Osborne, M.P. & Wersäll, J. 1964, 'Structure and innervation of the sensory epithelia of the labyrinth in the Thornback Ray *(Raja clavata)*', *Proceedings of the Royal Society of London Series B,* vol. 160, pp. 1-12.

Moran, D.T. & Varela, F.G. 1971, 'Microtubules and sensory transduction', *Proceedings of the National Academy of Sciences USA,* vol. 68, p. 757.

Murchie, G. 1978, *The Seven Mysteries of Life,* Houghton Mifflin Company, Boston, pp. 178-180.

Oschman, J.L. 2000, *Energy Medicine: The Scientific Basis,* Harcourt Brace/Churchill Livingstone, Edinburgh.

Oschman, J.L. 2001, 'Exploring the biology of phototherapy', *Journal of Optometric Phototherapy,* Apr., pp. 1-9.

Packard, A. 1995, 'Organization of cephalopod chromatophore systems: A neuromuscular image-generator', in *Cephalopod Neurobiology,* eds N.J. Abbott, R. Williamson & L. Maddock, Oxford University Press, London. pp. 331-367. Videos strips of the color changes can be seen at: http://www.gfai.de/www_open/perspg/heinz.htm.

Paffenholz, R.C., Kuhn, C., Grund, C., Stehr, S. & Franke, W.W. 1999, 'The *arm*-repeat protein NPRAP (Neurojungin) is a constituent of the plaques of the outer limiting zone in the retina, defining a novel type of adhering junction', *Experimental Cell Research,* vol. 250, pp. 452-464.

Ruch, T.C. & Patton, H.D. 1965, *Physiology and Biophysics*, WB Saunders Company, Philadelphia. See Chapter 4 for a discussion of receptors and generator potentials.

Stryer, L. 1987, 'The molecules of visual excitation', *Scientific American*, July, pp. 42-50.

Thurm, U. 1968, 'Steps in the transducer process of mechanoreceptors', in *Invertebrate Receptors*, eds J.D. Carthy & G.E. Newell, Academic Press, New York, pp. 199-216.

Vinnikov, Y.A. 1995, 'Gravitational mechanisms of interactions of sensory systems in invertebrates in the evolutionary aspect [article in Russian]', *Aviakosm Ekolog Med*, vol. 29, pp. 4-19.

Wald, G. 1967, *Les Prix Nobel en 1967. The Molecular Basis of Visual Excitation: Nobel Lecture*, The Nobel Foundation, Stockholm, Sweden.

17 Neural communication

Sensory endings communicate their responses to environmental stimuli to the brain via nerve impulses, which carry information from place to place within the body. Physiologists agree that the nervous system is the basic control system in the body, although this is an oversimplification because it leaves out communication we are describing within the connective tissue, cytoskeletons, and nuclear matrices, collectively termed *the living matrix*.

Nerves are long, thin cells that are said to be *excitable*. The initiation and propagation of nerve impulses has been the subject of intense research, following the development of a technique called *voltage clamping* by K.S. Cole in the late 1940s and the research of Hodgkin and Huxley at Cambridge University, who received the Nobel Prize in 1963.

In essence, it has been discovered that all cells develop and maintain a large and steady transmembrane electrical potential, with the cell interior negative with respect to the outside. Some cells, such as nerves, have another distinctive property of being *excitable*. For such cells, a change in the environment, generally called a *stimulus*, brings about a temporary depolarization of the cell membrane, followed by spontaneous recovery or repolarization. The permeability change rapidly depolarizes nearby parts of the membrane, in a regenerative manner, and the excitation spreads as a wave over the membrane of the entire cell. The wave is called an *action potential*. Nerve cells are said to have *conductivity* or *self-propagation*, which enables information to be transmitted from one part of the body to another in the form of *action potentials*.

Much is known about action potentials from the study of giant axons in the squid. The Marine Biological Laboratory in Woods Hole, described in Chapter 3, is a major world center for studies on the squid giant axon.

The action potential involves a stereotyped sequence or cycle of permeability changes that cause ions to move in and out of the cell. The potential across the membrane decreases rapidly toward zero and then overshoots, so for a brief period the membrane potential is reversed, that is, the inside of the nerve fiber becomes positive with respect to the outside. The potential then returns back to the resting level. The whole process occurs in 0.5 to 0.6 ms in a large myelinated fiber. These fibers conduct at speeds of up to 120 m/sec. In contrast, sound travels at **330** m/sec in tissues. The wavelength of a nerve impulse is about 4 inches.

A classic example of nerve and muscle excitability is the simple reflex arc, shown in Figure 7-2. If you touch a hot object, your muscles quickly contract to pull your hand away to avoid or reduce burning. The nervous system detects a change in the environment and rapidly transmits messages to the muscles, which make an appropriate response.

Important aspects of the action potential are its *threshold* and all-or-nothing properties. The threshold is the minimal strength of a stimulus required to evoke a propagating action potential. Once the threshold has been reached, an action potential results. A stronger stimulus does not produce a stronger action potential. Nerve impulses are said to be *nondecremental*—they do not diminish with distance. The response is fixed in size, shape, duration, and conduction velocity no matter where it is recorded along the fiber. It is a *digital* signal, in that it has only two values, zero or one. This is termed *all-or-none behavior.*

Over the years, a series of thoughtful biologists have searched for more rapid and subtle processes that must be occurring to enable the sentient organism to function as an integrated, coordinated, and conscious whole. From time to time phenomena are encountered that do not fit with the classic sequence:

sensation→communication→decision making→action

Valerie Hunt's observations quoted in Chapter 7 document one such phenomenon. Other examples were mentioned in Chapter 5 and 6.

SYNAPTIC DELAY

There is a significant delay in the transfer of the signal from nerve to nerve or nerve to muscle. This is called the *synaptic delay,* and it arises because the nerve terminal secretes a neurotransmitter, such as acetylcholine, which must diffuse across the gap between the nerve and the muscle membrane and then activate processes in the muscle membrane. This diffusion-activation process takes a long time.

Synaptic delay is an important phenomenon because it places a limit on the speed of a response to a stimulus. Each synapse in a neural circuit, such as the reflex arc described earlier, adds to the delay. Indeed, neurophysiologists use information on the speed of conduction through a neural circuit to determine the number of synapses involved. Each synapse delays the signal by about 0.5 ms. The most rapid responses are the simple reflexes such as the one shown in Figure 7-2, which is called a *monosynaptic reflex.*

The synaptic delay arises because of the time consumed in each of the following processes:

A. Discharge of transmitter from the presynaptic terminal

B. Diffusion of transmitter across the synaptic cleft

C. Action of the transmitter on the receptor molecules in the membrane of the nerve or muscle

D. Inward diffusion of ions across the membrane to raise the membrane potential to the threshold for depolarization

An advantage of a non-neural mechanism for signal propagation is that it would avoid the synaptic delays.

18 Muscle contraction

The doctrines which best repay critical examination are those which for the longest period have remained unquestioned. **A.N. Whitehead**

In the illustration of a reflex arc, we see that the action potential is delivered to a motor end plate on the muscle. Figure 18-1 details the structure of the motor end plate where it contacts a mammalian muscle. When a nerve impulse arrives at the motor end plate, an electrical field arises in the surrounding tissues. Recordings of this field are called *electromyograms*, and this is the field that was measured by Valerie Hunt in the passage quoted in Chapter 7.

The muscle membrane is excitable. When acetylcholine interacts with the muscle membrane, a wave of depolarization spreads over the muscle surface and activates the contraction process by triggering the release of calcium ions inside the muscle. The electrical wave traveling along the surface of the muscle gives rise to a magnetic field that can be measured by magnetomyography.

Discovery of the role of calcium in muscle contraction was a major step in the history of physiology and was accomplished by a neighbor in Woods Hole, Floyd Wiercinski, in collaboration with L.V. Heilbrunn. In 1949, they published a paper describing the triggering of muscle contraction by injection of calcium ions into a muscle (Heilbrunn & Wiercinski 1949).

Muscle has been researched extensively by generations of physiologists and biochemists. The most widely accepted model of muscle contraction is called the *sliding*

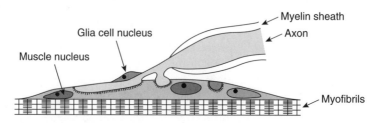

Figure 18-1 The motor end-plate. (Based on Fig. 10-31A, p. 288 in Bloom & Fawcett's *Textbook of Histology*, 12th edition, 1994, with permission from Hodder Arnold.)

filament hypothesis. The name is derived from the electron microscope observation that muscle is composed of microscopic fibers that slide over each other during contraction (Figure 18-2).

Let us take a close look at this conventional model of muscle contraction. The story is one of dynamic relationships between various kinds of fibers. The illustration of the myoneural junction (see Figure 18-1) gives a glimpse of the myofibrils, the contractile system in the muscle. Now we take a closer look.

Striated muscle consists of a virtually crystalline array of fibers. The organization is so regular that muscle structure can be studied by the same x-ray diffraction techniques that are used by crystallographers to determine the structure of mineral and protein crystals.

There are two types of fibers, the thin filaments, composed of *actin;* and the thick filaments, composed of *myosin* (Figure 18-3).

The thin actin filaments have several other proteins associated with them (bottom of Figure 18-4). The key regulatory proteins are tropomyosin and troponin. They are thought to strengthen the actin filament and to regulate the calcium-sensitive interactions with myosin.

The thick myosin filament contains a third type of filament that was discovered recently. It is composed of an elastic protein called *titin.* It may function as a template

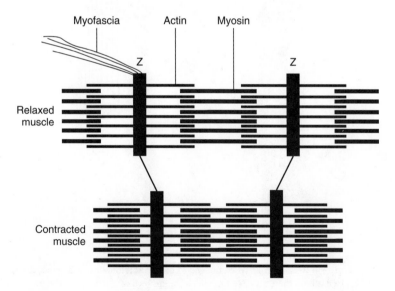

Figure 18-2 The sliding filament model of muscle contraction. In contraction, the thin filaments are displaced relative to the thick filaments. This causes the thin filaments to tug on the Z lines, which are attached to the weave network of collagen that makes up the endomysium. This network in turn is continuous with the larger tendonlike bundles of interwoven collagen forming the perimysium, which is continuous with the tendon proper. In cardiac muscle, and possibly in skeletal muscle, there is a preferential insertion of the endomysial collagen near the Z lines (Borg & Caulfield, 1979, 1980).

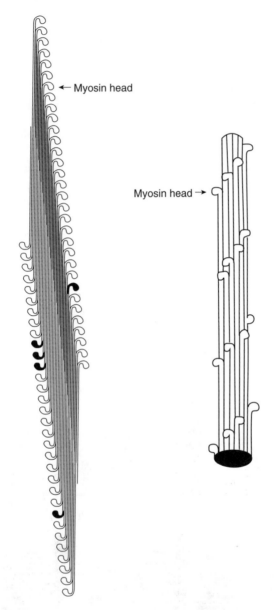

← Myosin head

Myosin head →

Figure 18-3 The myosin thick filament is composed of many individual myosin molecules stacked together in a way that allows their heads to line up along the surface *(left)*. The tilting movement of these heads is thought to provide the force developed in muscle contraction. The two-dimensional illustration to the left does not reveal the fact that the heads actually protrude in a spiral manner along the length of the filament *(right)*. (Diagram on left is Fig. 3, page 18, in Cross, R.A., Hodge, T.P., and Kendrick-Jones, J. 1991, 'Self-assembly pathway of nonsarcomeric myosin II', In Cross, R.A., Kendrick-Jones, J. (editors): *Motor Proteins: A volume based on the EMBO workshop,* Cambridge, September 1990. Supplement 14, *Journal of Cell Science* Supplement 14, used by permission of The Company of Biologists Limited.)

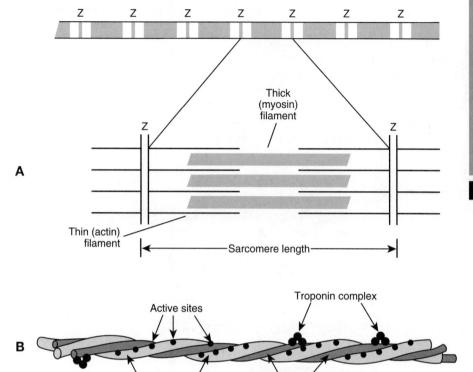

Figure 18-4 A, Sarcomere, which is the basic unit of the muscle fiber. B, Detail of the thin filament showing the major protein components, filamentous or F-actin, tropomyosin, and troponin. The active sites are places where the myosin heads are thought to attach for the "power stroke" in the contraction process. (A is from *International Review of Cytology*, Vol 106: 183-225, Davydov, Excitons and solitons in molecular systems, Fig 8, page 204, 1987, Academic Press, with permission from Elsevier Science.) B is Figure 4, page 578 in Sharma PH 1995 Muscle, molecular genetics of human, in Molecular Biology and Biotechnology, Myers RA (editor) VCH Publishers, Inc., 220 East 23rd Street, New York, NY.)

for controlling the assembly of the myosin filaments. It also connects the myosin with the Z line and therefore may function to hold the myosin in its position relative to actin.

The thick and thin filaments slide past each other during contraction, pulling on the Z lines. The Z lines, in turn, attach to collagen fibers in the myofascia, which then convey the tension to tendons and bones to create movement of the body (Borg & Caulfield 1979, 1980; Caulfield & Borg 1979).

To understand the anatomical arrangement, we need to take a close look at the way myosin molecules join together to form the thick filament. This is shown in Figure 18-3. Here we see that the myosin filament is composed of many myosin molecules, each of which has a protrusion or *head* at one end.

The motive force for muscular movement is thought to arise as these heads attach to actin, tilt, and then release. Details of the structure are shown in Figure 18-4, and details of the movements of the myosin heads are shown in Figure 18-5 (Murray & Weber 1974). Each myosin tilt moves the myosin relative to actin by some 4 to 10 nm (1 nanometer = 1 billionth of a meter).

To get an even better picture of this process, visualize it in three dimensions. The flat image of the myosin filament shown to the left in Figure 18-3 does not reveal that the filament is actually a cylinder, with the heads spiraling along the length of the fiber. This is shown more clearly in the illustration to the right in Figure 18-3.

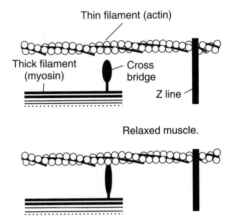

Relaxed muscle.

Myosin head elongates and bonds with actin filament.

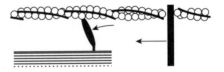

Myosin head rotates, displacing actin filament
and tugging on Z line.

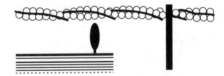

Myosin head separates from actin filament.

Figure 18-5 The standard model of striated muscle contraction showing details of how the myosin heads interact with the actin filaments. (Based on Murray JM, Weber A 1974 *Scientific American* 230(2):59–71. © George V. Kelvin.)

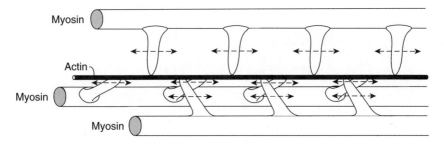

Figure 18-6 Close look at the theory for the sliding of filaments in skeletal muscle. Cross bridges or hooks on the thick filaments (myosin) are able to hook onto the thin actin filament, shown as a *black line*, at specific active sites. The arrows indicate that the bridges oscillate and thus produce muscular contraction by hooking onto an active site, pulling the thin filament a short distance, releasing it, and then hooking onto the next active site. (From Huxley H 1958. The contraction of muscle. © *Scientific American*, May issue, page 66 et. seq. with permission from Eric Mose, Jr., Executor of the estate of Eric Mose.)

The smoothness of muscle contraction arises because of the way multiple heads interact with actin molecules in three dimensions. This is shown in Figure 18-6.

This completes the posing of questions and our summary of the conventional picture of the way environmental stimuli lead to actions. The textbook stimulus→response scheme was summarized in Figure 7-2. Now we look at another view that has emerged from the work of Davydov, Fröhlich, and others on the way information can flow through tissues as conformational waves in the living fabric. We shall see that the work of Davydov (1973) led to an alternative theory for muscle contraction that connects Atema's theories of sensory transduction with the phenomena that interest us. Together, these theories have enormous implications. What has been missing in the past are relevant clinical observations. We suggest that the observations of Valerie Hunt and Emilie Conrad provide direct support for these theories. In fact, the importance of the observations of Hunt and Conrad provides a strong motivation for looking for such theories.

REFERENCES

Borg, T.K. & Caulfield, J.B. 1979, 'Collagen in the heart', *Texas Reports on Biology and Medicine*, vol. 39, pp. 321-333.

Borg, T.K. & Caulfield, J.B. 1980, 'Morphology of connective tissue in skeletal muscle', *Tissue and Cell*, vol. 12, pp. 197-207.

Caulfield, J.B. & Borg, T.K. 1979. 'The collagen network of the heart', *Laboratory Investigation*, vol. 40, pp. 364-372.

Davydov, A.S. 1973, 'The theory of contraction of proteins under their excitation', *Journal of Theoretical Biology*, vol. 38, pp. 559-569.

Heilbrunn, L.V. & Wiercinski, F.J. 1949, 'The action of various cations on muscle protoplasm', *Journal of Cellular and Comparative Physiology*, vol. 29, pp. 15-32.

Murray, J.M. & Weber, A. 1974, 'The cooperative action of muscle proteins', *Scientific American*, vol. 230, pp. 59-71.

19 Biological coherence: the Davydov soliton

It accumulated round the prow of the vessel in a state of violent agitation, then suddenly leaving it (the vessel) behind, rolled forward with great velocity, assuming the form of a large solitary elevation, a rounded, smooth and well defined heap of water, which continued its course along the channel apparently without change of form or dimension of speed. Scott Russell (1844)

Recall the 1973 meeting on the crisis in energetics discussed in Chapter 13. A.S. Davydov was one of the presenters at that meeting. In a series of important papers beginning in 1973, Davydov and his colleagues, at the Institute for Theoretical Physics, Academy of Sciences of the Ukrainian SSR, Kiev, developed an explanation of how the energy released by adenosine triphosphate (ATP) hydrolysis can be transferred along α-helical protein molecules as a special collective excited state called a *soliton* (reviewed in Davydov 1987). This is a singular or individual wave with remarkable properties (see box).

The accuracy and importance of Davydov's work is widely recognized by physicists the world over, as evidenced by a week-long international symposium held near Göteborg in 1978 and published in *Physica Scripta* (Wilhelmsson 1979). Solitons have been used to explain many phenomena in different fields of nonlinear optics, the physics of the condensed state, field theory, gravitation theory, plasma physics, and other sciences. Fascinating research on solitons continues in laboratories around the world. A search of the National Library of Medicine's *PubMed* website showed 628 papers on solitons. The point is that solitons are a subject that is receiving increasing attention.

Solitons have been the key to the development of modern high-speed multiwave-length optical networks used in the telecommunications industry (Stern & Bala 1999).

Solitons have the advantages of being able to carry a signal through a medium that ordinarily would tend to distort and broaden a propagating pulse. The soliton has

Solitons

The *soliton wave* is an ideal means of transferring energy in living systems. Usual waves, as on water, are periodic repetitions of elevations and depressions. Waves in a three-dimensional medium, such as sound waves moving through air, can be described as periodic densities and rarefactions. Such waves gradually disperse and lose their strength over distance. The *soliton wave*, in contrast, is an extremely stable solitary self-organizing wave that can transfer a large amount of energy with a constant velocity and without loss or dispersion or dissipation. Tidal waves and tsunamis are examples of oceanic soliton waves. Davydov and his colleagues in Kiev demonstrated that solitons can be propagated along α-helical proteins such as those found in living matter.

remarkable self-focusing properties that enable the propagation of a narrow stable pulse over long distances without any distortion or loss. Moreover, solitons propagating in opposite directions pass through each other transparently, which enables simultaneous communication in both directions. In an experiment conducted in Japan, soliton data were transmitted at a rate of ten billion bits per second, the equivalent distance of 4500 times around the Earth, with no data degradation (Voss 1995).

All of the properties of solitons have advantages for the design of living systems, and it would be astonishing if evolutionary processes had left them out of the equation of life.

From our present perspective, it would appear that Jelle Atema's ideas about sensory information transfer, Valerie Hunt's observations on the dancers, Emilie Conrad's work with paralysis and Continuum, and Albert Szent-Györgyi's work on electronic biology have a common denominator. They appear anomalous from the logical scientific perspective until one begins to explore Davydov's work on the soliton. Atema's paper on sensory transduction in microtubules sets the stage for the application of the Davydov soliton in sensory motor physiology.

According to Davydov, the importance of the soliton is as follows. The bulk of the living organism is composed of proteins. Energy and information are propagated through the protein matrix in the form of vibrations and electrons. The soliton is a method of vibratory energy transfer. It is a coherent wave. Biological media are nonlinear and dispersive, which are ideal conditions for the propagation of solitons.

SOLITONS IN NATURE

A soliton is a solitary or singular wave that can occur on the ocean or in any other medium. An example is the *tsunami*, or *tidal wave*, produced by a submarine earthquake or volcanic eruption.

The first qualitative description of solitary waves on water occurred in 1834, on a narrow channel connecting Edinburgh with Glasgow, in Scotland. A naval engineer, J. Scott Russell, was trying find a design for the shape of a ship's hull that would offer the least resistance to the water. He was carefully studying a boat being drawn along the channel by a pair of horses. The boat suddenly stopped. His description of what took place in the water is quoted at the beginning of this chapter. This was the first published account of a soliton. Scott Russel was struck by the exceptional stability and self-organizing characteristics.

This is the first published account of a soliton. Scott Russell was struck by the exceptional stability and self-organizing characteristics.

About 50 years later, two Dutch scientists, Korteweg and de Vries (1895) developed a mathematical description of solitary waves on water, using a nonlinear differential equation. Their equation accounted for the lack of dispersion, which would occur with ordinary linear waves. In a nonlinear situation there is extensive interaction between the monochromatic constituents of the wave, causing redistribution of energies between them. The energies of any more rapidly moving waves are transmitted to retarded waves so that the excitation becomes very stable and propagates as a unit without spreading.

SOLITONS IN BIOLOGICAL SYSTEMS

In his review article, Davydov describes the history of the mathematical approaches to waves in general and to solitons in particular. In the past, much thought went into the application of equilibrium thermodynamics and mechanics based on linear equations, but this work has been of little relevance in biology because it cannot explain energy transfer and transformation in nonlinear biological media.

Solitons can carry a large amount of energy over a long distance without loss. In other words, in contrast to normal waves, solitons do not disperse or dissipate their energy by spreading out. This means that a conformational wave produced by a sensory receptor, or by any other means, could carry energy directly to the contractile apparatus (such as a muscle) without loss. In the corresponding neurophysiological terminology, the soliton is *nondecremental,* like a nerve impulse.

Various therapists have noted that waves resembling solitons appear from time to time in the body, and they appear to have beneficial effects, including the release and/or resolution of traumatic memories (see Chapter 21). This is a fascinating phenomenon, often experienced in the therapeutic setting but seldom discussed in scientific circles. At least three schools of bodywork use this phenomenon (two schools of bodywork based on somatic recall and emotional release are *Holographic Repatterning* and *Holographic Memory Release.* Upledger cranial-sacral techniques also have a component that is referred to as *somato-emotional release*).

Subsequent chapters will give a detailed model of how traumatic memories may be stored in tissues, accessed by coherent vibrations, and released by solitons (Oschman & Oschman 1995a, 1995b). This point is mentioned here because of the possibility that the coherent motions described by Hunt and Conrad may access traumatic and emotional memories stored within the connective tissue and cellular matrices.

CONFORMATIONAL WAVES

Now we summarize Davydov's explanation of how conformational waves are propagated along a protein such as a microtubule. A thorough presentation can be found in

Davydov's review entitled "Excitons and Solitons in Molecular Systems" (Davydov 1987).

We begin with a close look at the protein backbone. There are two basic ways conformational waves can travel through a protein. One involves vibrations and rotations of the peptide groups, and the other involves vibrations of the side chains, called *amide I groups*. These two types of conformational changes interact with each other to produce two different kinds of wave motions, *excitons* and *solitons*, which will be described later.

Figure 19-1, *A*, shows a segment of a protein chain consisting of two peptide groups (shaded areas). The curved arrows show the bonds that are free to rotate. Figure 19-1, *B*, shows the amide I groups that interact resonantly with each other. All of these bonds resemble springs and are characterized in the mathematical equations as having particular "spring constants."

A special property of peptide groups is their ability to form hydrogen bonds with each other, and with water and other molecules in the environment. Figures 19-2, *A*, and 19-2, *B*, show the chains of hydrogen bonds between adjacent peptide groups. It is the tension in these bonds that cause the polypeptide to coil into a helix. This is called an α-*helix*. The rotation of the bonds shown in Figure 19-1 allows the protein to form into the α-helix shown in Figures 19-2, *A*, and 19-2, *B*. This phenomenon was discovered by Pauling and Corey (1951).

Davydov points out that there are millions of different kinds of proteins in nature, but they all are constructed along similar lines so that the theories he presents apply widely. The α-helix is present in a large number of different kinds of proteins, such as myosin, tropomyosin, and troponin in muscle; collagen in connective tissue; and globular proteins. For example, hemoglobin has 32 α-helical sections.

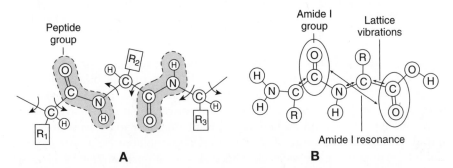

Figure 19-1 **A,** Segment of a protein chain consisting of two peptide groups *(shaded)*. Curved arrows show some of the bonds that can rotate and vibrate, like springs. These are called *lattice vibrations.* **B,** Two kinds of vibrational transfer in an α-helical protein. The amide I vibrations (C=O) are the strongest vibrations. They show up in the infrared absorption spectrum in the region 1650 to 1660 cm^{-1}. They last 10^{-13} second. Resonant interactions take place between adjacent amide I groups *(long arrow)* and transfer energy along the chain. The lattice vibrations *(short arrows)* move more slowly. Excitons primarily involve amide I resonant transfer, whereas solitons involve coupling between amide I and lattice vibration. (From *International Review of Cytology,* Vol 106: 183-225, Davydov, Excitons and solitons in molecular systems, Figs. 1 and 2, page 186, 1987, Academic Press, with permission from Elsevier Science.)

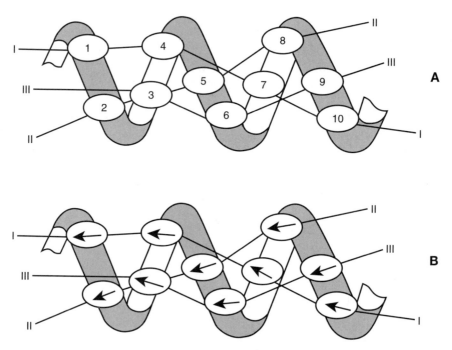

Figure 19-2 **A,** Hydrogen bonds connecting the peptide groups in the α-helix protein molecule. Three chains of hydrogen bonds are formed: I between peptides 1,4,7,10, etc.; II between 2,5,8, etc.; and III between 3,5,8, etc. The forces created by these bonds twist the protein backbone into an α-helical shape. **B,** Peptide groups in the α-helix are electric dipoles, with their moments arranged as shown. There are three separate spines or chains of coupled peptide groups running along the helix. Solitons involve vibrational transfers along these chains. (From *International Review of Cytology,* Vol 106: 183-225, Davydov, Excitons and solitons in molecular systems, Fig. 3, page 187, Fig. 17, page 218, 1987, Academic Press, with permission from Elsevier Science.)

Davydov discusses the various ways energy and information are propagated along a protein. We will not go into detail about these alternatives but simply report Davydov's conclusion that there is only one ideal method for energy and information to propagate along a protein, which is as solitary waves or solitons. These waves travel with constant velocity and do not contain a mixture of frequencies that smear or spread out over time. A series of studies revealed that solitons can transport energy and information along the molecule without loss or distortion. The soliton is self-organizing. Should any component wave have a tendency to move faster than the others, its energy is immediately transferred to slower moving waves, so the assembly retains its coherence.

Early research by Davydov and Kislukha showed that solitons can be initiated by chemical reactions, such as the interaction of a stimulus with a receptor. The ideal place for a chemical reaction to trigger a soliton is at the end of the protein, such as the tip of a microtubule in a sensory cell like that shown in Figure 16-2. The soliton can arise at the very end of a protein molecule, and it can arise in a very short protein (Eilbeck 1979).

Careful study revealed that the conformational waves set up in the protein backbone as the result of bending, rotations, and vibrations around the intrapeptide bonds shown in Figure 19-1 also set up resonance interactions with the amide I vibrations shown in Figure 19-2. These vibrations of the C=O bond are the strongest vibrations and show up in the infrared absorption spectrum in the region 1650 to 1660 cm^{-1}. They last 10^{-13} second. Resonant interactions take place between adjacent amide I groups and are rapidly transferred throughout the chain.

Earlier researchers had considered the possibility that resonant transfer between amide I groups might solve the crisis in bioenergetics, but this did not work out. The amide I vibrations smear out over the entire protein. They are a collective property of the whole protein and therefore are unable to convey energy or information from place to place.

Davydov combined the amide I vibratory interactions with the bending, twisting, and vibration of the intrapeptide bonds to develop a realistic theory of conformational propagation in the peptide backbone. This came about by careful calculations of the interaction of the amide I vibrations with the displacements from the equilibrium positions of the peptide groups as a wave of deformation passes along the protein. The calculations actually characterized two possible types of excitation propagation: *excitons* and *solitons*.

EXCITONS

If the velocity of the conformational wave exceeds that of longitudinal sound, the deformations of the peptide lattice cannot keep up with the intrapeptide amide I interactions. These excitations are called *excitons*. Excitons are produced when a protein absorbs infrared radiation. The vibrations are approximately perpendicular to the length of the polypeptide chain. They are very important for the study of protein molecular structure. Excitons transport only intrapeptide amide I excitation energy. As mentioned earlier, the exciton state is distributed or smeared uniformly along the whole length of the chain; therefore, they are unable to transfer energy and information.

Excitons are unstable. As the vibration smears out over the molecule, energy is lost as phonon radiation because it is coupled between the amide I vibrations and vibrations of the bonds between peptide groups. The exciton degrades into a soliton.

SOLITON PROPERTIES

The soliton occurs because of an interaction between the intrapeptide excitation amide I and lattice deformation. Although the energy of a soliton is less than that of an exciton, the soliton can have greater mass and therefore can transfer great kinetic energy at a small velocity.

The great stability of the soliton arises from the fact that to destroy a soliton, that is, to split it into a free exciton and a deformation that can relax into thermal motion, it is necessary to expend a considerable amount of energy. As a soliton propagates along a protein, say from left to right, all peptide groups to the right are undisplaced from their equilibrium positions, whereas to the left they are displaced equally. In other words, the soliton leaves a wave of rarefaction behind it. In order to destroy a soliton, it is necessary to return all of the peptide groups of the chain to their equilibrium positions. The more closely the vibrations of the amide I groups are coupled to the lattice vibrations, the more stable the soliton. In contrast, large values of coupling shorten the lifetime of excitons because the energy tends to be dissipated as phonons or heat.

Davydov reported that solitons would not move as fast as sound, but more recent research has indicated the possibility of ultrasonic solitons (Christiansen, Eilbeck, Enol'skii & Gaididei 1992). They do not emit phonons. In other words, their energy is not transformed into the energy of thermal motion.

The soliton is particularly suited to energy transfer in soft molecular chains, chains that are stabilized by weak hydrogen bonds, such as the α-helical proteins found in biological systems.

The soliton is very stable in the presence of light. The probability that light will initiate a soliton is small, and, for the same reason, the probability of light being given off by a soliton is small.

Figure 19-3 compares soliton and exciton energies. Excitons slowly become more energetic as their velocity increases. In contrast, as the velocity of solitons approaches the speed of sound, their energetic content rises exponentially.

When two peptide groups are excited at the opposite ends of the polypeptide chain, two solitons move toward each other with a velocity three eighths that of longitudinal sound. When they collide, they pass through each other (Lomdahl, Layne & Bigio 1984). The collision resembles the elastic collision of two particles.

SOLITONS IN REAL α-HELICAL PROTEINS

As shown in Figure 19-2, there are three different chains of peptide groups that interact to transfer energy along a protein. The problem of how the vibrations of the three different chains interact with each other and with the amide I vibrations is intricate, but it has been worked out by Davydov and others. The solution involves coupled differential equations. Solving these equations numerically requires very large supercomputers capable of billions of real number operations per second, and this was first accomplished at the Los Alamos Scientific Laboratory by Fermi, Pasta, and Ulam (1955) and again by Hyman, McLaughlin, and Scott (1979). Subsequent work has permitted highly detailed analysis of progressively more complicated models of exciton and soliton propagation.

Two basic kinds of solitons emerge from the calculations. These are shown in Figure 19-4. The two forms depend on the degree of interactions among the three chains of peptide groups and the amide I vibrations. In the *symmetrical soliton,* the excitations propagate along the three chains in phase and the amide I vibrational energy is shared equally among the three spines. The pitch of the helix decreases and the diameter increases in the region of excitation. In the *asymmetrical soliton,* the amide I vibrations are shared among the three spines. The diameter of the helix increases and a local bend forms at the region of excitation.

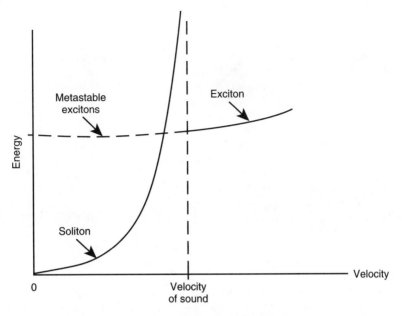

Figure 19-3 Soliton and exciton energies depend on their velocities, particularly in relation to the velocity of sound. As the velocity of the soliton approaches the speed of sound, its energetic content rises exponentially. In contrast, the exciton is unstable and tends to degrade into a soliton. (From *International Review of Cytology*, Vol 106: 183-225, Davydov, Excitons and solitons in molecular systems, Fig. 5, page 198, 1987, Academic Press, with permission from Elsevier Science.)

Symmetrical soliton

Asymmetrical soliton

Figure 19-4 Two types of solitons based on the degree of interactions among the three chains of peptide groups and the amide I vibrations. In the *symmetrical soliton,* the excitations propagate along the three chains in phase and the amide I vibrational energy is shared equally among the three spines. The pitch of the helix decreases and the diameter increases in the region of excitation. In the *asymmetrical soliton,* the amide I vibrations are shared among the three spines. The diameter of the helix increases and a local bend forms at the region of excitation. (From *International Review of Cytology*, Vol 106: 183-225, Davydov, Excitons and solitons in molecular systems, Fig. 6, page 200, 1987, Academic Press, with permission from Elsevier Science.)

REFERENCES

Christiansen, P.L., Eilbeck, J.C., Enol'skii, V.Z.Z., Gaididei, J.B. 1992, 'On ultrasonic Davydov solitons and the Henon-Héiles system', *Physics Letters A*, vol. 166, pp. 129-134.

Davydov, A.S. 1987, 'Excitons and solitons in molecular systems', *International Review of Cytology*, vol. 106, pp. 183-225.

Eilbeck, J.C. has prepared a computer film with the title "Davydov Solitons." The film and the mathematics behind it are described in a paper by Hyman, J.M., McLaughlin, D.W., & Scott, A.C. 1981, 'On Davydov's alpha-helix solitons', *Physica 3D*, vols. 1 & 2, pp. 23-44.

Eilbeck, J.C. 1979, 'Davydov solitons on the alpha-helix', Brief soliton movies prepared by Eilbeck are also available on the web at http://www.ma.hw.ac.uk/solitons/.

Fermi, E., Pasta, J.R. & Ulam, S.M. 1955, Los Alamos Scientific Laboratory Report LA U.S. LA-1940, Los Alamos, NM.

Hyman, J.M., McLaughlin, D.W. & Scott, A.C. 1979, Los Alamos Scientific Laboratory Report LA U.S., Los Alamos, NM.

Korteweg, D.J. & de Vries, G. 1895, 'On the change of form of long waves advancing in a rectangular canal, and on a new type of long stationary waves', *Philosophical Magazine*, vol. 39, pp. 422-443.

Lomdahl, P.S., Layne, S.P. & Bigio, I.J. 1984, *Los Alamos Science*, vol. 10, pp. 2-22.

Oschman, J.L. 1993, 'Sensing solitons in soft tissues', *Guild News, the News Magazine for Members of the Guild for Structural Integration*, vol. 3, pp. 22-25.

Oschman, J.L. & Oschman, N.H. 1995a, 'Somatic recall. Part I. Soft tissue memory', *Massage Therapy Journal, American Massage Therapy Association*, vol. 34 pp. 36-46; Summer issue, pp. 111-116.

Oschman, J.L. & Oschman, N.H. 1995b, 'Somatic recall. Part II. Soft tissue holography', *Massage Therapy Journal, American Massage Therapy Association*, vol. 34, pp. 66-67; Fall issue, pp. 106-116.

Pauling, L. & Corey, R.B., Branson, H.R. 1951, 'The structure of proteins: Two hydrogen-bonded helical configurations of the polypeptide chain', *Proceedings of the National Academy of Sciences USA*, vol. 37, pp. 205-211.

Scott Russell, J. 1845, 'Report on waves', *Report of the XIVth meeting of the British Association for the Advancement of Science*, London, pp. 311-320.

Stern, T.E. & Bala, K. 1999, *Multiwavelength Optical Networks: A Layered Approach*, Addison-Wesley, Reading, MA.

Voss, D. 1995. 'You say you want more bandwidth? Solitons and the erbium gain factor', *Wired*, vol. 6, p. 4.

Wilhelmsson, H. (ed.) 1979, 'Solitons in physics', *Physica Scripta*, vol. 20, pp. 280-562.

20 Solitons and muscle contraction

CHAPTER OUTLINE

Chapter 13 described the "crisis in bioenergetics," which involves the mechanism by which the energy from adenosine triphosphate (ATP) hydrolysis is transformed into the mechanical energy of movement. The discovery, made using electron microscopy and time-resolved x-ray diffraction, of the tilting of the myosin head (see Figures 18-5 and 18-6) has left unresolved the precise molecular mechanism of force transduction.

Davydov developed a hypothesis for muscle contraction that uses all of the modern information on the role of ATP hydrolysis at the myosin head with soliton conduction along the length of the myosin molecule. The model was presented first in 1973 (Davydov 1973) and is elaborated upon in the 1987 review by Davydov (1987). The earlier paper discusses the variety of models of muscle contraction that have been proposed and the difficulties of each of them. The earlier paper precedes the development of Davydov's soliton model and refers to the wave of excitation as an exciton. Subsequently, Davydov recognized that the soliton is more appropriate for explaining muscle contraction.

The hypothesis is simply this: Calcium ions reaching the myosin heads at the ends of the thick filaments initiate the hydrolysis of ATP molecules attached to them. The energy released generates solitons in the long helical sections of the myosin molecules shown in Figure 18-3. These solitons move from the myosin heads to the tails. The motion of the solitons is accompanied by a local bending and thickening as shown in Figure 19-4. The motion of this "swollen" region of the thick filament from its end to the center is what causes the displacement of the thin (actin) filaments attached to the Z lines.

According to Davydov's model, bending of the myosin heads is not due to elongation, tilting, and contraction of the heads themselves, as in the widely accepted model shown in Figure 18-5 and 18-6. Instead, the tilting arises as a consequence of the traveling of the of swollen regions along the myosin molecules, from their heads to the tails. "In the region embraced by an excitation, the pitch of the helix decreases, and the 'contracted' region of the helix moves along the molecule. . . ." In the Davydov model, contraction arises from the properties of the whole myosin molecule, not just its head, which constitutes only 7% of the muscular substrate. Hence the contraction of the muscle arises from the kinetic energy of the solitons propagating along the myosin filaments, which causes the thick filaments to move along the actin filaments.

Davydov's model is shown in Figure 20-1. The motion resembles the movement of a snake (called *concertina progression* by Gans 1970) in a straight narrow channel. It also resembles the movement of a worm in its burrow (see Figure 20-1).

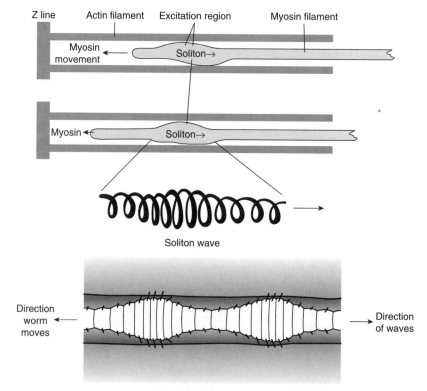

Figure 20-1 The Davydov model of muscle contraction *(top)*. A soliton wave moves to the right from the site of ATP hydrolysis in the myosin head toward the tail. The motion of the solitons is accompanied by a local bending and thickening *(center)*. The motion of this "swollen" region of the thick filament causes the displacement of the thin (actin) filaments attached to the Z lines. "In the region embraced by an excitation, the pitch of the helix decreases, and the 'contracted' region of the helix moves along the molecule. . . ." The motion resembles the movement of a snake, called *concertina progression,* in a straight narrow channel. It also resembles the movement of a worm in its burrow *(bottom)*. Earthworms *(Lumbricus)* move in their burrows (right to left) by extending and retracting the bristles of each segment as muscular waves pass down the body from head to tail (left to right). In all of these movements, the wave passes in the opposite direction to the direction of movement. These are called *retrograde waves.* Other researchers cited in the text refer to this as *reptation.* (Lower illustration is from page 292 in Buchsbaum, R., Buchsbaum, M., Pearse, J., Pearse, V. 1987, *Animals without backbones,* 3rd ed. The University of Chicago Press, Chicago, used by permission of Vicki Buchsbaum Pearse. Middle illustration is from *Journal of Theoretical Biology,* Vol 38: 559-569, Davydov, The theory of contraction of proteins under their excitation, Fig 2, page 568, 1973, Academic Press, with permission from Elsevier Science.)

Recently a noted expert on muscle contraction published a similar model using snakelike motions in actin rather than myosin (Pollack 2001), although he does not cite Davydov's work. The model Pollack developed is termed *reptation* because the actin filament *snakes* its way toward the center of the sarcomere. Pollack cites two other studies that demonstrated reptation in artificial polymers and protein filaments (Kas, Strey & Sackmann; Yanagida et al. 1984).

SOLITONS IN NONMUSCLE CELLS

One of the most important discoveries in cell biology is that nonmuscle cells contain the same contractile proteins as muscles, actin, and myosin. This has opened up a whole new area of research into cell motility. The phenomenon is basic to the defense and repair systems in the body. White blood cells migrate to areas of injury to destroy bacteria, as do various other kinds of cells that repair the damage.

Davydov believes the same mechanism involved in muscle contraction applies to nonmuscle cells. In the presence of ATP and local increases in calcium ion concentration, tensions develop between actin and myosin filaments and cause them to move relative to each other (Figure 20-2). This gives rise to the movements of organelles within cells, as well as cell motility.

EFFECTS OF DRUGS ON SOLITON PROPAGATION

A final aspect of Davydov's theoretical work concerns the effects of drugs such as anesthetics. In spite of decades of careful research, the mechanism by which anesthetics have their effects remains unknown.

From the considerations raised in previous chapters, we suspect that the movement of solitons in connective tissue and in the living matrix as a whole are involved in

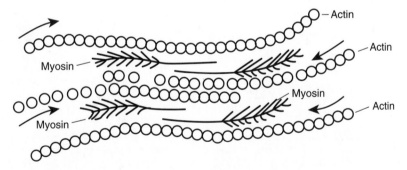

Figure 20-2 Cytoskeletal movement system. The structure of the contractile system in nonmuscle cells is far less organized than in muscle, but the same proteins, actin and myosin, are present. In the presence of ATP and local increases in calcium ions, cellular movements are created by tensions developed between actin and myosin filaments. Davydov suggests the mechanism is the same as in muscles: solitons moving along the myosin cause translation relative to actin filaments anchored in the cell membrane or anchored to organelles. (Modified from *International Review of Cytology*, Vol 106: 183-225, Davydov, Excitons and solitons in molecular systems, Fig 12, page 210, 1987, Academic Press, with permission from Elsevier Science.)

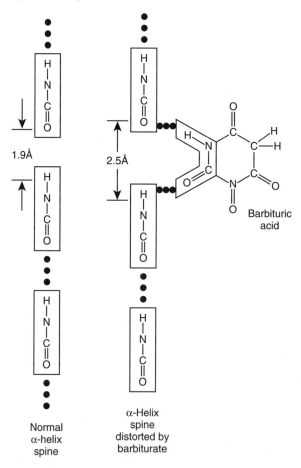

perception. Many bodyworkers comment on changes they detect in the quality of the connective tissue in individuals who are taking various kinds of drugs. Davydov suggests that barbiturates act by altering the relations between the peptide groups in proteins. Specifically, the binding of an anesthetic such as barbituric acid changes the localized structure within the α-helix. In essence, the barbiturate weakens or even breaks the hydrogen bonding in the helix.

Figure 20-3 Binding of an anesthetic such as barbituric acid changes the localized structure within the α-helix. In essence, the barbiturate weakens or even breaks the hydrogen bonding in the helix. The illustration shows that the bond length can increase by about 0.6 to 0.8 Å. The energy of the hydrogen bonding is deceased by 55%. These changes cause the soliton to slow down or stop, disrupting the energy and information transfer in the protein matrix. Numerical calculations indicate the soliton degrades significantly by emitting phonons as it traverses the region where the barbiturate is attached. This will disrupt the energetic and informational activities taking place in mitochondrial membranes, neuronal membranes, and excitation and communication in the connective tissue system. The concentration of barbiturate need only be sufficient to associate with 1% of the proteins to have effects. (Modified from *International Review of Cytology*, Vol 106: 183-225, Davydov, Excitons and solitons in molecular systems, Fig 15, page 215, 1987, Academic Press, with permission from Elsevier Science.)

Figure 20-3 shows that binding to barbituric acid can increase the bond length by about 0.6 to 0.8 Å. The energy of the hydrogen bonding is deceased by 55%. These changes cause the soliton to slow down or stop, disrupting the energy and information transfer in the protein matrix and leading to a disruption in normal consciousness.

Calculations indicate the soliton degrades significantly by emitting phonons as it traverses the region where the barbiturate is attached. This will disrupt energetic and informational transfers taking place in mitochondrial membranes and neuronal membranes, as well as excitation and communication in the connective tissue/cytoskeletal/nuclear matrix systems. The concentration of barbiturate need only be sufficient to associate with 1% of the proteins to have effects.

REFERENCES

Davydov, A.S. 1973, 'The theory of contraction of proteins under their excitation', *Journal of Theoretical Biology*, vol. 38, pp. 559-569.

Davydov, A.S. 1987, 'Excitons and solitons in molecular systems', *International Review of Cytology*, vol. 106, pp. 183-225.

Gans, C. 1970, 'How snakes move', *Scientific American*, vol. 222, pp. 82-96.

Kas, J., Strey, H. & Sackmann, E. 1994, 'Direct imaging of reptation for semi-flexible actin filaments', *Nature*, vol. 368, pp. 226-229.

Pollack, G.H. 2001, *Cells, Gels and the Engines of Life*, Ebner & Sons, Seattle, Washington.

Yanagida, T., Nakase, M., Nishiyama, K. & Oosawa, E. 1984, 'Direct observation of motion of single F-actin filaments in the presence of myosin', *Nature*, vol. 307, pp. 58-60.

Sensing solitons in soft tissues

We learned much by studying the separate parts of life's fabric, but there exist profoundly important properties that cannot be discovered this way. These properties are consequences of relations rather than of parts. The most significant of these relations are transparent, invisible to our senses. But their outcome is our essence. The hands-on therapist is daily and quietly and deeply immersed in every aspect of the system we are describing, and is therefore the cutting-edge explorer of this domain. Oschman (1995)

It has been valuable to talk with hands-on therapists about the various phenomena discussed in this book. Many important observations take place in the quiet of a one-on-one therapeutic session, such as happens in massage, Zero Balancing, Healing Touch, Polarity, Structural Integration, Acupuncture, etc. These observations are a fascinating and virtually untapped resource for exploring unanswered scientific questions.

An energetic approach is leading to a deeper and simpler understanding of all of these observations. Like the way many therapists experience their practices, the biophysicist must alternate between material and energetic realities. The great physicist Niels Bohr developed the principle of complementarity, which states that objects can be viewed as either waves of energy or as particles. In Bohr's time, these two views could not be logically connected, and Bohr said that to understand reality one must be able to look at it in both ways.

Several discussions with teachers of Structural Integration have concerned the nature of the visible or palpable wave or ripple that sometimes propagates through tissues in advance of the place where the hands-on work is being done. These discussions have been conducted in the context of an interesting definition of the word *phonon*.

The *phonon* is the quantum or particle of sound, the smallest unit of energy of a sound vibration. Do not confuse it with the unit of light, the *photon,* although energy can be converted back and forth between phonons and photons in materials, such as connective tissue, which are technically described as dielectric semiconductors. The movement of sound or other vibrations through materials alters their optical properties.

Normally we think of sound as a wavelike disturbance passing through a medium such as a solid, liquid, or gas; however, sometimes it is useful to look at sound as though it were composed of particles. For example, the phonon concept is useful for considering how vibrations move through solids. Phonons can collide with each other (phonon-phonon interactions) and with impurities (such as stored toxin molecules or out-of-alignment collagen fibers, in the case of connective tissues).

In some instances, heat can be transferred by phonons. This is interesting because heat is one of the forms of energy that radiates from the hands of therapists during bodywork. Hence when we think biophysically about the possible effects of touch on connective tissue, the phonon can be a valuable concept.

The phonon can be described as "an electromechanical wave in a piezoelectric medium." This seems relevant, because connective tissue is piezoelectric, that is, it generates an electric field when it is compressed, and vice versa; that is, it compresses in the presence of an electric field.

Pressing on a tissue, as in bodywork, can create an electric field because of the piezoelectric effect. This field can propagate as a wave through the surrounding tissue. When this happens, there also will be a wave of mechanical compression because of the reverse piezoelectric effect. (For example, piezoelectric transducers are used to convert electrical pulses into sound waves that travel through the air or over the ocean.) The piezoelectric compression wave could create the ripple seen in a client's tissue.

This is an interesting idea, as far as it goes; however, there may be more to the story because of the possibility that the ripple effect also may be a soliton.

We have seen that the soliton is a fascinating type of wave. It is a solitary or singular wave that can occur on the ocean or in any other medium.

Figure 21-1 shows the way a soliton (called a Tsunami, or tidal wave) is formed in the ocean environment by a displacement of the sea floor. A displacement of the sea floor, caused by an earthquake or slipping of a fault line, creates a depression on the ocean floor. Because water is not stretchable or compressible, a corresponding depression forms at the water surface. This depression propagates along the surface. Observers at the shore first note the arrival of the Tsunami by a drastic drop in the water level. The depression is followed by large wave that can be very powerful and destructive.

Consider the sources of the soliton in the ocean and in the body. In both cases, the conditions giving rise to the soliton are often related to the gravitational field. It is gravitational stresses that give rise to the slippage of geological strata or planes on the ocean floor, and it is the gravitational pull on the water above the slippage that energizes the formation of the wave. The situation in the body that creates the conditions for the formation of a "tissue soliton" may arise from gravitational stresses that have created densities or distortions in the connective tissue, either because of the way the body has been used in moving in the gravitational field or because of injuries produced by falling. For a detailed discussion of the role of gravity in health and disease, see Chapters 11 and 12 in Oschman (2000).

In contrast to normal waves, solitons do not disperse or dissipate their energy by spreading out. The usual concentric or bull's-eye–shaped wave, such as that produced by dropping a pebble in a pond, loses its energy as it spreads over the water surface.

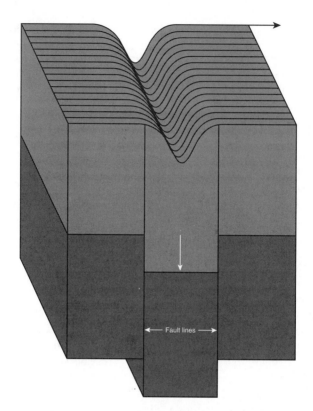

Figure 21-1 The way a soliton or Tsunami or tidal wave is formed in the ocean environment. A displacement of the sea floor, triggered by an earthquake or volcanic eruption can cause slipping of a fault line, creating a depression on the ocean floor. Because water is not stretchable or compressible, a corresponding depression forms at the water surface. This depression propagates along the water surface. Observers at the shore first note the arrival of the Tsunami by a drastic drop in the water level. The drop is followed by large wave that can be very powerful and destructive.

Another extraordinary feature of the soliton is that two solitons can collide, but they emerge from the collision unchanged in shape and speed.

A final feature of the soliton that is of great interest to us is that it possesses magnetic properties. To be specific, quantum physical theory predicts that each soliton is a magnetic monopole, an isolated north or south magnetic pole.

You might ask, "What has this to do with me?" At present we are looking into the soliton and related wave phenomena as a possible explanation for the way bodywork seems to be able to release or unlock repressed or emotionally charged memories. We take a close look at this phenomenon in the next chapter.

There is a theory, popularized by Karl Pribram, that memories are stored in the nervous system as wave interference patterns comparable to those used in holography. Neurosurgeons have had a lot of difficulty understanding why they can remove large amounts of brain tissue, yet their patients can still remember everything. Pribram and others have suggested that memory is due to a "distributed property" of the brain rather than to a localizable set of neural connections. In a hologram, the image is recorded over the entire surface of a piece of film, and the entire image can be reproduced from any tiny piece of the film.

A problem with Pribram's holographic memory model is that forming a hologram requires a laser beam. Laser light is highly coherent or organized. How can the nervous system produce coherent waves?

A most exciting answer to this question has arisen from the research by the physicist Herbert Fröhlich, as described in Chapter 14. According to Fröhlich, molecules such as collagen and the phospholipids of cell membranes and the other crystalline materials in the body (see Figure 9-2) should produce stable large-scale coherent vibrations that will be conducted through the living matrix and radiated into the environment of the organism. Fröhlich and a number of other scientists think that these vibrations communicate regulatory information, that they are responsible for the integration of function taking place at various levels within the organism. Fröhlich predicted that the coherent signals would be in the microwave and visible light portions of the electromagnetic spectrum. Such signals have been detected in living systems and have been shown to regulate cellular activities.

We now are researching a model of memory that is based on Pribram's holographic concept, together with the idea that memory may reside in *all* of the tissues of the body rather than just in the nervous system. We are considering the Fröhlich oscillations set up in the cellular and connective tissue matrices as potential sources of the coherent oscillations needed to create and read holographic memories within the tissues.

A number of cell biologists are studying how non-neural cells can store memories in their cytoskeletons. This could provide a basis for the release of stored memories in various forms of bodywork. This idea relates to manual therapies because the cytoskeleton of every cell in the body contains a number of components that are gels; that is, they are polymers constructed of smaller units called *monomers*. Microfilaments and microtubules are two components of the cytoskeleton that are polymeric.

Dr. Ida P. Rolf considered the immediate effect of her work to involve a reversible, pressure-induced, gel-to-sol transition in the ground substance of connective tissue (Rolf 1977). Her idea has been extended by suggesting that the gel-to-sol transition also may release toxins that have been stored or trapped in the gel phase (Oschman 2000, Fig. 12.4, p. 172). We also summarized the literature supporting her idea that pressure and other sorts of energy fields can trigger the gel-to-sol transition.

We consider the ground substance of cells to be an extension of the ground substance of the connective tissue. Hence manual therapies could erase cellular memories by the same mechanism that it releases toxins that have been trapped in the ground substance. We suggest that when the gel reforms, after a few minutes, it will be rejuvenated in the sense that toxic materials and painful memories or attitudes will be gone.

Memory is a phenomenon that takes place in the domain, or space, we refer to as *consciousness*. This is one of the more controversial unsolved phenomena of biology. Consciousness is one of the things we "do" all of the time, yet its nature is one of the most elusive topics in science and philosophy.

Perhaps memory and consciousness are both wave phenomena. Perhaps they are elusive because they are, in biophysical terms, "virtual images" constructed of interfering wave forms rather than localizable "real images" such as those produced by a camera lens. This could explain why scientists have been unable to locate a specific structure responsible for memory. We shall take up this topic again in a concluding chapter, after a few more of the puzzle pieces have been described.

Finally, we have long been fascinated with the martial arts and the projection of *Qi* or *Chi* as occurs in methods such as Qi Gong. A Qi Gong "master" can project healing energy, *Qi*, a dozen feet or more, where it can be felt by another person. Reports from Japan and China have shown that the projected Qi includes a strong magnetic component that can be detected with a simple magnetometer.

Qi projection could be produced by a soliton wave that passes through the connective tissue and is radiated into space. The magnetic aspect discovered by the Japanese workers is consistent with the quantum physical prediction that each soliton wave will be a magnetic monopole.

There are many good reasons, from the work of Fröhlich and Davydov, for suggesting that connective tissue could generate solitons. We shall see that the film of water that coats the collagen molecules may play a key role in generating solitons. The healing nature of the projected energy would arise because the frequency of the oscillations is the same as that used by the body to communicate regulatory messages, such as the signals needed for the healing of injuries.

A number of energy therapists suspect that the cranial-sacral pulse may be a soliton.

This brief summary of some current thinking about solitons in relation to manual therapies may whet your appetite for details that will follow in the next chapters. A key point is that there is great value and interest in the observations of manual therapists on the nature of living tissue.

For an excellent discussion of the interactions of light and sound, see Chapter 20 in Saleh and Teich (1991). Karl Pribram's work on holographic memory is summarized in various works (Pribram 1966, 1969, 1977). Holography is described in Leith and Upatnieks (1965). An important collection of articles on the biophysics of whole systems has been edited by Fröhlich (1988) and is discussed in detail in Chapter 14 in this book. The magnetic component of Qi projection was described by Seto et al. (1992). Solitons are described by Rebbi (1979) and in a symposium edited by Wilhelmsson (1979). The propagation of solitons through proteins and its implications for energetics are discussed in Hyman, McLaughlin, and Scott (1981). Studies of cellular memory are summarized by Hameroff (1983, 1987). Dr. Rolf's work on Rolfing or Structural Integration is described in her book (Rolf 1977).

REFERENCES

Frohlich, H. (ed) 1988, *Biological Coherence and Response to External Stimuli,* Springer-Verlag, Berlin.

Hameroff, S.R. 1983, 'Coherence in the cytoskeleton: Implications for biological information processing', in *Coherent Excitations in Biological Systems,* eds H. Fröhlich & F. Kremer, Springer-Verlag, New York, pp. 242-265.

Hameroff, S.R. 1987, *Ultimate Computing: Biomolecular Consciousness and Nanotechnology*, Elsevier-North Holland, Amsterdam.

Hyman, J.M., McLaughlin, D.W. & Scott, A.C. 1981, 'On Davydov's alpha-helix soliton', *Physica D*, vol. 30, pp. 23-44.

Leith, E.N. & Upatnieks, J. 1965, 'Photography by laser', *Scientific American*, vol. 212, pp. 24-35.

Oschman, J.L. 2000, *Energy Medicine: The Scientific Basis*, Churchill Livingstone, Edinburgh.

Oschman, J.L., Oschman, N.H. 1995, 'Approaching the toes (theories of everything)', *Guild News, the News Magazine for Members of the Guild for Structural Integration*, vol. 5, pp. 13-16.

Pribram, K. 1966, 'Some dimensions of remembering: Steps toward a neuropsychological model of memory', in *Macromolecules and Behavior*, ed. J. Gato, Appleton, Century, Crofts, New York, pp. 165-187.

Pribram, K. 1969, 'The neurophysiology of remembering', *Scientific American*, vol. 220, p. 75.

Pribram, K. 1977, *Languages of the Brain*, Wadsworth Publishing, Monterey, California.

Rebbi, C. 1979, 'Solitons', *Scientific American*, vol. 240, pp. 92-116.

Rolf, I.P. 1977, *Rolfing: The Integration of Human Structures*, Dennis-Landman, Santa Monica, CA.

Saleh, B.E.A. & Teich, M.C. 1991, *Fundamentals of Photonics*, John Wiley & Sons, New York, pp. 799-831.

Seto, A., Kusaka, C., Nakazato, S., Huang, W., Sato, T., Hisamitsu, T. & Takeshige, C. 1992, 'Detection of extraordinary large bio-magnetic field strength from human hand', *Acupuncture and Electro-Therapeutics Research International Journal*, vol. 17, pp. 75-94.

Wilhelmsson, H (ed.) 1979, 'Solitons in physics', *Physica Scripta*, vol. 20, pp. 290-562.

Soft tissue memory

A *PETITE MADELEINE*

In *Swann's Way,* a taste of a small cake, a *petite Madeleine,* causes Marcel Proust (1924) to be flooded with memories from his past. At first he is baffled, but then he remembers his aunt giving him Madeleines when he was small. Obviously, the association triggered his memory.

Most of us have had similar experiences in which a glimpse of some long-forgotten place or object, or a particular odor, taste, sound, or even a movement, elicits the recall of a scene from our distant past. This chapter concerns a related phenomenon that frequently is experienced by many kinds of therapists.

SOMATIC RECALL

Massage therapists, acupuncturists, Rolfers, and other somatic practitioners frequently report uncanny experiences in which vivid images flood into their consciousness as they are working on some part of a client's body. Sometimes there is a transient sensation that "something has happened" within the body they are touching. An avalanche of detailed sensory material may be triggered. The images may be so striking that the practitioner asks the client about them, only to discover that their client is simultaneously having a similar or identical flashback. Rolfer Randy Mack describes this as "... the recall of deeply repressed, highly charged emotional material with full sensory detail possibly including visual, auditory, tactile, gustatory, and olfactory components."

Practitioners who repeatedly have these somatic recall experiences with their clients begin to suspect that "memories" of traumatic or other events may be stored in or accessed by the soft tissues of the body. Sometimes the flashback is associated with erasure of the memory. When this happens, the emotional charge surrounding the memory may disappear. The client may even forget, by the end of the session, that the recall occurred. In other cases, the recollection begins a therapeutic process that resolves the associated trauma, pain, or psychological attitudes. In other instances, the flashbacks may occur a day or two after a session of massage or other bodywork.

It long has been recognized that our individual memories shape our sense of who we are, as well as what we do and how we do it, on a moment-to-moment basis. Our personal identity, our comprehension of the world around us, our place in that world, what we can and cannot accomplish, and our every act and decision all are referenced to what we have learned and remembered. If these references are to traumatic past experiences and to the resulting pains, secrets, fears, judgments, mistruths, guilts, angers, and narrow attitudes or beliefs, our physical and behavioral flexibilities are limited. Freedom of movement and thought and awareness of what is happening inside and outside of us are compromised. To the extent that our mental lives influence our physical bodies and vice versa, any therapeutic practice that has an effect on traumatic memory can have a profound, dynamic, and multidimensional influence on every attribute of the organism.

Memory and consciousness are among the most fascinating and controversial topics for scientific inquiry and somatic exploration. Reports that touching someone can release memory traces and even communicate them to another person are of great interest. Of course, conventional science labels such experiences as anomalies or hallucinations because they do not fit with our normal theories about how the brain and nervous system work. However, we have spoken to enough practitioners who report similar experiences that we have come to regard somatic recall as a frequently occurring phenomenon. Some massage therapists have these experiences daily or even with every client. Not only is somatic recall widespread, but we think it is an important clue to unsolved mysteries of learning, memory, consciousness, the ways parts of the body communicate with each other, and the effects of touch.

We now are exploring somatic recall in light of recent progress in biophysics and cell biology. New discoveries are pointing to a simple yet scientifically logical explanation for a variety of phenomena related to massage and other kinds of bodywork. The emerging concepts have far-reaching implications for scientific and philosophical inquiries into the nature of consciousness and for a variety of approaches to the body.

THE LIVING MATRIX CONCEPT CLARIFIES THE STORY

The living matrix system described in previous chapters provides a basis for exploring consciousness at the level of the whole organism. Our customary division of the body into organs, tissues, cells, and molecules separates the study of life into subdisciplines. The corresponding division of biomedical inquiry into areas such as physiology, cell biology, pharmacology, genetics, molecular biology, and bioenergetics has slowed our comprehension of the vital universal cross-disciplinary integrating principles that must exist for an organism to maintain unity of function at all levels. Dividing the living system into mind and body domains further confounds and confuses us. I believe that research into mind-body connections is conceptually absurd, simply because a thing cannot be connected to itself. Chronic disorders and diseases such as cancer, acquired immunodeficiency syndrome (AIDS), atherosclerosis, osteoporosis, stroke, and heart disease persist because our understanding of whole systems physiological integration is incomplete. The consuming and expensive search for new and profitable pharmaceuticals has closed off other potentially useful avenues of investigation. As a result, wound healing, regeneration, and recognition of self and nonself continue to be poorly understood phenomena. What they have in common is that they involve communications within the living matrix. Massage therapists and other bodyworkers have been exploring these communications for a long time and have made valuable insights about them. A partial list of pioneers in this area includes F.M. Alexander, Mantak Chia, Moshe Feldenkrais, Dolores Krieger, A. Lowen, F.A. Mesmer, B.J. Palmer, Wilhelm Reich, Ida P. Rolf, Fritz Smith, Andrew T. Still, Randolph Stone, W.G. Sutherland, and Milton Trager. Some scientists who have contributed along these lines include Robert O. Becker, Harold Saxon Burr, Valerie Hunt, Hiroshi Motoyama, Candace Pert, Bruce Pomeranz, Albert Szent-Györgyi, and J.E. Upledger.

CONTINUUM COMMUNICATION

In terms of both manual therapeutics and biomedicine, the most exciting property of the tensegrous living matrix is the ability of the entire network to generate and conduct vibrations. The vibrations occur as mechanical waves or sounds, called *phonons,* electrical signals, magnetic fields, electromagnetic fields, heat, light, and solitons. In contrast to chemical messengers, energy fields propagate extremely rapidly and do not require complex enzymatic systems to break down the messages so the activated process can be turned back off again. For the most part, these forms of energy obey established laws of physics that describe fields from any source. Signals are produced and distributed throughout the body because of properties that are common to all of the components of the living matrix. To summarize:

1. *Semiconduction:* All of the components are semiconductors. This means they can both conduct and process vibrational information, much like an integrated circuit or microprocessor in a computer. They also convert energy from one form to another.

2. *Piezoelectricity:* All of the components are piezoelectric. This means that waves of mechanical vibration moving through the living matrix produce electrical fields and vice versa, that is, waves of electricity moving through the lattice produce mechanical vibrations.

3. *Crystallinity:* Much of the living matrix consists of molecules that are regularly arrayed in crystallike lattices. This includes lipids in cell membranes, collagen molecules of connective tissue, actin and myosin molecules of muscle, and components of the cytoskeleton.

4. *Coherency:* The highly regular structures just mentioned produce giant coherent or laserlike oscillations that move rapidly throughout the living matrix and that also are radiated into the environment. These vibrations are called *Fröhlich oscillations.* They occur at particular frequencies in the microwave and visible light portions of the electromagnetic spectrum. A number of scientists have detected these signals (Popp, Li & Gu 1982; Popp et al. 1981).

5. *Hydration:* Water is a dynamic component of the living matrix. On average, each matrix protein has 15,000 water molecules associated with it. Because many of the proteins are highly ordered, as we have just seen, the associated water molecules also are highly ordered. Water molecules are polarized (dipoles). The living matrix organizes the dipolar water molecules in a way that constrains or restricts their ability to vibrate, rotate, or wiggle about in different spatial planes. Water molecules are only free to vibrate or spin in particular directions.

6. *Continuity:* As we have seen, the properties just listed are not localized but are spread throughout the organism. Although we may distinguish individual organs, tissues, cells, and molecules, the living matrix is a continuous and unbroken whole.

A consequence of continuum communication is that every process taking place anywhere in the organism produces a characteristic pattern of vibrations that travels throughout the living matrix and that undoubtedly distributes regulatory information. In terms of electronics, the signals are FM (frequency modulated) rather than AM (amplitude modulated). The frequency changes every time a cell moves or alters its shape, an organ shifts its functional state, a muscle contracts, a gland secretes, a nerve conducts an impulse, or a cell metastasizes (Pienta & Coffey 1991). Transmission of vibratory signals through the living matrix imparts unity of function to the organism.

According to the continuum communication model, the living matrix creates a veritable symphony of vibratory messages that travel to and fro, alerting each part of the organism about the activities taking place in each other part. What we refer to as *consciousness* is the totality of these vibrations. Disease, disorder, and pain arise within portions of the vibratory continuum where information and energy flows are restricted. Restrictions occur locally because infections, physical injury, and emotional trauma alter cellular and extracellular properties of the fabric.

The living matrix retains a record or memory of the influences that have been exerted upon it. When vibrations pass through tissues, they are altered by the signatures of the stored information. In this way, our consciousness and our choices are influenced by memories stored in soft tissues.

An important property of the living matrix is its ability to regenerate or restore itself. Various kinds of bodywork facilitate these processes.

CELLULAR MEMORY

In the past, memory has been attributed to the nervous system, but biologists are realizing that *all* cells in the body have the capacity to store information in their cytoskeletons (reviewed by Hameroff 1988; Hameroff et al. 1988). The cytoskeleton frequently is referred to as *the nervous system of the cell.* Because the cytoskeleton is continuous with all of the other molecular networks in the body, as we have seen, memories stored within any individual cell are accessed and communicated via the living matrix.

The cytoskeleton is made up of a number of components, each of which can store and process information. Most of the focus has been on microtubules, which are relatively stiff rods. Microtubules are the structures that give each cell its characteristic shape, much like the bones give form to the body as a whole.

Hameroff describes in detail how the microtubules can act like computers. Microtubules are made up of monomeric subunits known as *tubulin*. These subunits are polymerized into microtubules at specific sites known as *microtubule organizing centers*. Microtubules are polymers (*poly* = many) formed when many identical units, called *tubulin monomers* (*mono* = one), join together.

Each tubulin monomer is polarized and has two different ways of fitting into the polymer (Figure 22-1, *A*). Additional proteins, called *microtubule-associated proteins*

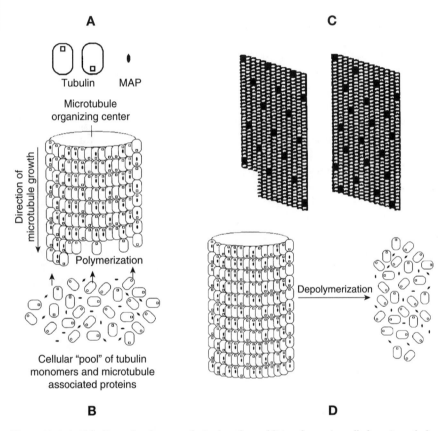

Figure 22-1 A, Tubulin molecules are polarized, and an additional protein, called a *microtubule-associated protein (MAP)*, can attach to the microtubule. **B,** Microtubule assembly occurs by polymerization of tubulin and MAPs from the cellular pool. **C,** Electron microscopic studies of MAP distribution on microtubule surfaces show different patterns in different microtubules, suggesting a form of memory. **D,** Depolymerization of microtubules during cell division or when under hydrostatic pressure can cause loss of memories stored in the tubule structure. (C is taken from Fig. 4, page 246, in Hameroff, S.R. 1988, 'Coherence in the cytoskeleton: Implications for biological information processing.' In Fröhlich, H. (editor): *Biological coherence and response to external stimuli,* © Springer-Verlag, Heidelberg. Used with permission of the publisher and Stuart Hameroff.)

(MAPs), can attach to the microtubule. Information is stored by the orientation of the tubulin monomers and by the position of attachment of the MAPs. The result is a record of the conditions in the cell and in the environment at the time of microtubule assembly (Figure 22-1, *B*).

Remarkable studies have used immunogold tracers that stick to proteins attached to the microtubules. Because of its density, the gold shows up under electron microscopy. This method reveals a variety of different patterns of MAPs attached to microtubules (Burns 1978; Geuens et al. 1986). Figure 22-1, *C,* shows two different patterns of MAPs, representing two different sets of information, attached to microtubules.

Hameroff describes how the patterns of microtubule subunits form "information strings" comparable to those in the word processor being used to write this book. In the computer, the information is stored on a magnetic medium in the form of a series of magnetic particles that can be oriented in either of two polarities, north-south or south-north. The disk drive can read these digital "character strings" and reproduce the sequence of letters and words of the manuscript. Similarly, information is stored as the orientation of tubulin monomers along microtubules. The information is in strings that can move along the microtubules. In nerves, very long microtubules and associated neurofilaments can function as devices that are known in computer terminology as *string processors* (Hameroff 1988).

SHORT-TERM AND LONG-TERM MEMORY

Neurophysiologists distinguish between short-term and long-term memory, and they look for different mechanisms to account for them. The possibility that individual cells can distinguish shorter-term and longer-term memories arises from studies on the turnover of microtubules. Using fluorescent labeling techniques, Schulze and Kirschner (1988) were able to demonstrate two distinct populations of microtubules in cultured cells. One population was highly dynamic; the second was relatively stable. Hence information stored on the more stable microtubules by the mechanisms Hameroff and others described could, in principle, represent longer-term memory, whereas the more dynamic microtubules could represent shorter-term memories (Figure 22-2).

ERASING MEMORIES AND RELEASING TOXINS

I can erase the character strings in my computer disk drive with a magnet that turns all of the magnetic particles to the same orientation. The information stored on a microtubule can be erased by depolymerizing it (making it fall apart) into its monomeric units (see Figure 22-1, *D*).

Depolymerization of microtubules occurs every time a cell divides. In essence, the cytoskeletal framework falls apart temporarily so that the deoxyribonucleic acid (DNA) can replicate and the daughter cells can separate. When this happens, all of the information encoded as the direction of orientation of tubulin monomer units and as patterns of MAPs is lost. Obviously, tissues whose cells divide rapidly, such digestive

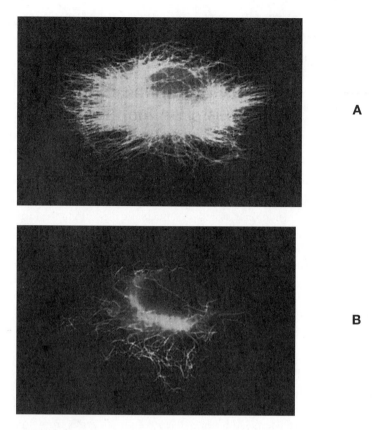

A

B

Figure 22-2 A possible basis for short-term and long-term memory storage within cellular microtubules arises from studies of Schulze and Kirschner (1988) that distinguish two distinct populations of microtubules in cultured cells. One population (A) is highly dynamic; the second (B) is relatively stable. Hence information stored on the more stable microtubules by the mechanisms shown in Figure 22-1 might represent longer-term memories, whereas the more dynamic microtubules could represent shorter-term memories. (Copyright 1989, from *Molecular biology of the cell*, by Alberts B, Bray D, Lewis J, Raff M, Roberts K, Watson JD. Reproduced by permission of Routeledge, Inc., part of The Taylor & Francis Group.)

tract, cornea, and skin, will not be able to retain information for long periods, in contrast to tissues that have a low rate of cellular turnover. In this way we can see how time heals as traumatic memories stored in tissues gradually are weakened or lost as cell and tissue constituents are replaced.

For massage and other kinds of bodywork, the important property of polymers is that pressure, temperature change, magnetic fields, and other forms of energy can cause them to depolymerize or fall apart (Tanaka 1981). This is the gel-to-sol transition that Dr. Ida P. Rolf (1977) used to account for the immediate effects of Rolfing on body structure. Rolf's focus was on the connective tissue ground substance (including the proteoglycan molecules shown in the first illustration) that lies between the collagen fibrils. She thought the ground substance depolymerizes from the pressure applied by the Rolfer or Structural Integration practitioner, and that this allows tissues to lengthen and soften. When the practitioner stops pressing on the tissue, the gel quickly repolymerizes. Similar changes can take place during massage and other hands-on methods.

One can ask what distinguishes the effects of Structural Integration and other forms of systematic bodywork from the ordinary pressures developed during everyday activities, such as moving in the gravity field, wearing clothing, contacting furniture, random scratching and rubbing, and swimming underwater. In general, our individual body structure and the structural problems that bring us to a massage therapist or other health care professional (pains, compensations, imbalances, loss of flexibility, inefficient movements) arise from the ways we use or misuse our bodies in our everyday activities. Cumulative structural problems also can arise from long-standing emotional attitudes. For example, a man has stiff shoulders that are as hard as concrete. His tissue has responded to years of fear of being suddenly struck from behind, something his brother often did to him many years before. A skilled therapist reads these structural/emotional patterns and has methods of resolving or freeing us from them. A number of sessions may be required, but when the job is done the tissue and the mental attitude toward life can be completely rearranged. Repeated gel-sol-gel transitions in cells and tissues allow the structure to reorganize, soften, lengthen, and become more flexible.

Repeated cycles reorganization, triggered by the therapist's precise, systematic, and intentional application of pressure and other forms of energy to tissues, has structural and emotional consequences and can bring about the release of toxic materials that have been stored within the tissues for many years. When we use the term *toxins* we are referring to foreign substances that have become trapped in the connective tissue meshwork. This meshwork traps substances because it has many electrically charged components that toxins can stick to, and because it has many tiny pockets in which toxic molecules become lodged. Therapists with a keen sense of smell often detect odors of alcohol when they work on tissue that has been repeatedly sterilized before insulin injections, or of ether when working on the tissues of someone who has been anesthetized, or of insecticides when working on someone who has been repeatedly sprayed. This process in connective tissues is called *storage excretion,* which involves the trapping and storage of toxins to prevent them from entering the blood and being carried throughout the body.

We have described how temporary gel-to-sol changes can release toxic material that has been stored in the cells and tissues, sometimes for many years. We can suggest that one of the effects of Rolfing, massage, and other methods is to reversibly cause gel-to-sol transformations in both in the connective tissue ground substance and in cytoskeletons. The result is release of both stored memories and toxic substances. The latter are released into the environment or are carried away by the circulatory system and broken down or excreted.

SOFT TISSUE MEMORY

Microtubules are not the only components of soft tissues that are capable of storing information. A highly respected physiologist has described how records of the ways the body has been used (or misused) are incorporated into the structure of connective tissue. In his well-known book, *The Life of Mammals,* Young (1975) provides an eloquent account of the plasticity of connective tissue and its ability to store information.

Young states that the structure of any tissue depends on how it developed and the forces exerted on it by other tissues and the environment. Collagen is deposited along the lines of tension in connective tissues, such as fascia, tendons, bones, ligaments, and cartilage.

Paul Weiss (1961) studied tissue cultures and healing wounds and documented the phenomenon Young described. Wound repair begins with the formation of a clot containing fibrin filaments. At first the fibers are oriented randomly. As the clot dissolves, fibers that are not under tension are dissolved first, leaving behind a web of oriented fibrin fibers. Fibroblast cells migrate into this web, become oriented along the fibers, and deposit collagen, primarily along tension lines. Any collagen fibers that are not oriented along tension lines are removed by a process similar to the readjustment that took place in the clot. The result is a tissue composed of fibers oriented in the direction that is appropriate to the tensional forces produced by normal movements (Figure 22-3).

Therapists from many disciplines know that it is beneficial to resume normal use of the body as soon as possible after an injury. Normal motion helps guide appropriate deposition of collagen fibrils. In immobilized tissues, randomly oriented fibers persist and disused muscles begin to stick to each other, particularly where there has been damage or scarring. Cyriax and Russell (1977) refer to this as *formation of adhesions,* and Ida Rolf calls it *gluing.* Both terms describe a random web of connections that form between the myofascial layers of adjacent muscles. This webwork compromises the thin layer of lubricating fluid that normally allows adjacent muscles to slide over each other. When a muscle contracts, it tends to drag adjacent muscles along with it, reducing muscular efficiency and precision of motor control. These effects are particularly important for athletes and other performers, who strive to achieve optimum control and efficiency of motion. As in Weiss's blood clots, normal tensions are needed to facilitate resorption of unnecessary collagen fibers after an injury. In Rolfing and in the Cyriax method, deep cross-fiber friction breaks the adhesions and restores mobility. Movement therapies accomplish a similar result by bringing movement and awareness to those parts of the body that have been immobilized (or held still relative to surrounding parts) for whatever reason.

From Young's work we can see these as examples of the way the organism makes predictions, or forecasts, about the future that will help survival. Genetic information programs the fibroblasts to deposit collagen in the direction of tensions, and forces from the environment generate those tensions. Disuse or injury promotes a more random deposition of fibers, and this causes adjacent layers to adhere or become glued to each other. Of course, this gluing has a biological purpose. As muscles atrophy from lack of use, they tend to stick to each other, forming a built-in crutch that stabilizes and supports the body.

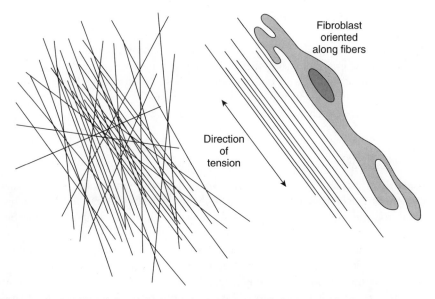

Direction
of
tension

Fibroblast
oriented
along fibers

Figure 22-3 Weiss (1961) found that wound repair begins with formation of a clot containing randomly oriented fibers. As the clot dissolves, fibers not under tension dissolve first, leaving only fibers oriented along the lines of tension. Fibroblast cells migrate into the web and become oriented along the tension fibers. Fibroblasts lay down collagen along the lines of tension. (From *Energy Medicine: the scientific basis,* Oschman, Figure 11.8, page 158, Churchill Livingstone, Edinburgh, 2000, with permission from Elsevier Science.)

Connective tissue structure is a record or memory of the forces imposed on the organism. This historical record has two components. The genetic part recapitulates the story of how our ancestors successfully adapted to the gravitational field of the earth. The acquired component is a record of the choices, habits, and traumas we have experienced during our individual lifetime. The collagen fibers orient in a way that can best support future stresses, assuming the organism continues the same patterns of movement or disuse.

It is widely thought that the phenomena Young described are not confined to healing wounds (reviewed by Bassett 1968). Readjustment of collagen deposition takes place in all portions of the living matrix at all times. This readjustment is the primary method by which body structure adapts to the loads imposed on it and the ways the body is used (Oschman 1989a, 1989b, 1990). Young stated that memories are stored not only in the collagen network but also in the elastin fibers and even in the various cells found throughout the connective tissue (histocytes, fibroblasts, osteoblasts, plasma cells, mast cells, fat cells).

Young's concept of memory in connective tissues and cells provides a physiological basis for the way the stresses of life, injuries, diseases, muscular holding patterns, emotional attitudes, and repeated unbalanced movements can influence the form of the body. It also explains some of the dramatic effects of various movement therapies. One has the impression that every movement of the body is recorded in the living matrix. Repeated or habitual movements result in a particular connective tissue and cellular architecture. Any change in those habits, no matter how slight, will forever alter that architecture.

Can "memories" encoded in connective tissue and cytoskeletal structures lead to a conscious mental image of past events? How might such information be released during massage or other kinds of bodywork? How is such information communicated from the tissue being worked upon to the consciousness of both the client and the practitioner? The next chapter will begin to answer these questions.

CONCLUSIONS

Massage therapists and other bodyworkers have daily and remarkable experiences of physiological integration in action. Intuition and sensitivity have led to practical methods for interacting with fundamental and evolutionarily ancient communication systems in the body. These communication systems integrate and unify structure and function. The integrity of these systems is profoundly important in the healing of injuries of all kinds. From the information presented so far, we can see how massage and other methods can simultaneously open lines of communication, clear the body of toxic materials that have been stored for a long time, help resolve memories of emotional and physical traumas, restore flexibility, and reduce pain.

Historically, physiological integration has not been a topic of great interest for biomedical research, which focuses on parts rather than wholes. Recent work of biophysicists around the world now is providing a context in which the experiences of various therapists can be validated scientifically, and the experiences of the practitioners can provide important clues for researchers as well.

The realization that the cytoplasmic matrix is but an extension of the connective tissue, and vice versa, opens up a whole new dimension for research. It also resolves a long-standing confusion about the fundamental unit of life. The dilemma began in 1839, when Schwann declared that the extracellular matrix is the source of all life and that cells are created within it "according to definite laws." In 1859, Virchow disagreed. The extracellular matrix depends on the cells, which truly are the elementary units, the atoms of life. This idea was shattered by the discovery that fermentation could take place in a cell-free extract composed only of molecules and enzymes. There arose a molecular prejudice: living matter, being built of molecules, must have at its basis a set of molecular reactions. Others, such as Albert Szent-Györgyi, found this point of view inadequate and looked to electrons, protons, and other subatomic particles as building blocks, units of energy and information, and components of consciousness. Our inability to cure major diseases stems from our failure to include such phenomena in our thinking about structure and function.

Biophysics now is progressing rapidly because of a whole systems perspective. The search for fundamental units is replaced by study of the web of relations among the various parts of the whole. Inquiries at all levels are equally relevant and important.

The world thus appears as a complicated tissue of events, in which connections of different kinds alternate or overlap or combine and thereby determine the texture of the whole. HEISENBERG (1958)

The realization that the cytoplasmic matrix is but an extension of the connective tissue, and vice versa, opens up a whole new dimension for research. The continuous living matrix, extending throughout the organism, is the context for the web of relations that is the subject of all research. The living matrix has no fundamental unit, no central aspect, no part that is primary or most basic. The integrity of the network depends on the activity of all components, and all components are governed by relations with the whole.

Subatomic particles, and all matter made therefrom, including our cells, tissues, and bodies, are in fact patterns of activity rather than things. CAPRA (1982)

The biophysical properties of the living matrix can explain a variety of phenomena that have been elusive in the past: learning, memory, consciousness, and unity of structure and function. Little of the important biophysical research related to complementary medicine is being performed in the United States. For example, four important books (edited by Fröhlich 1988; Popp, Li & Gu 1992; Endler & Schulte 1994; Ho, Popp & Warnke 1994) contain contributions of 94 scientists from around the world, but only 8 are from the United States (they are J. Schulte, East Lansing, MI; J.K. Pollock & D.G. Pohl, Milledgeville, GA; W.R. Adey, Loma Linda, CA; S.R. Hameroff, Tucson, AZ; R.P. Liburdy, Berkeley, CA; T.Y. Tsong, St. Paul, MN; and T.M. Wu, Binghamton, NY). An international symposium edited by Allen, Cleary, and Sowers (1994) contains contributions from well over a hundred authors, and again only a handful are from the United States. An accessible account of some of this literature can be found in *The Rainbow and the Worm* by Mae-Wan Ho (1993).

Although the concepts presented here are not yet a part of normal biomedicine, they have a sound scientific foundation and they go a long way toward explaining some of the phenomena that arise in complementary medicine. We believe the bodywork practice is one of the best "laboratories" for testing these concepts. We look forward to hearing from those who find these ideas useful in their own process of refining and expanding techniques and understandings. Research is fun when it leads to new questions and opens us to new possibilities.

REFERENCES

Allen, M.J., Cleary, S.F. & Sowers, A.E. 1994, *Charge and Field Effects in Biosystems-4,* World Scientific, River Edge, New Jersey.

Bassett, C.A.L. 1968, 'Biologic significance of piezoelectricity', *Calcified Tissue Research,* vol. 1, pp. 252-272.

Burns, R.B. 1978, 'Spatial organization of the microtubule associated proteins of reassembled brain microtubules', *Journal of Ultrastructure Research,* vol. 65, pp. 73-82.

Capra, F. 1982, *The Turning Point,* Simon and Schuster, New York.

Cyriax, J. & Russell, G. 1977, *Textbook of Orthopaedic Medicine,* vol. 2, 9th ed, Bailliere Tindall, London.

Endler, P.C. & Schulte, J. (eds) 1994, *Ultra High Dilution Physiology and Physics,* Kluwer Academic, Dordrecht, The Netherlands.

Fröhlich, H. (ed.) 1988, *Biological Coherence and Response to External Stimuli,* Springer-Verlag, Berlin.

Geuens, G., Gundersen, G.G., Nuydens, R., Cornelissen, F., Buklinski, V.C. & DeBrabander, M. 1986, 'Ultrastructural colocalization of tyrosinated and nontyrosinated alpha tubulin in interphase and mitotic cells', *Journal of Cell Biology*, vol. 103, pp. 1883-1893.

Hameroff, S.R. 1988, 'Coherence in the cytoskeleton: Implications for biological information processing', in *Biological Coherence and Response to External Stimuli*, ed. Fröhlich H, Springer-Verlag, Berlin, pp. 242-263.

Hameroff, S., Rasmussen, S. & Mansson, B. 1988, 'Molecular automata in microtubules: Basic computational logic of the living state?', in *Artificial Life, SFI Studies in the Sciences of Complexity*, vol. 6, ed. C. Langton, Addison-Wesley, Redwood City, California, pp. 521-553.

Heisenberg, W. 1958, *Physics and Philosophy*, Harper Torchbooks, New York.

Ho, M.-W., Popp, F.-A. & Warnke, U. 1994, *Bioelectrodynamics and Biocommunication*, World Scientific, River Edge, New Jersey.

Ho, M.-W. 1993, *The Rainbow and the Worm*, World Scientific, River Edge, New Jersey.

Oschman, J.L. 1989a, 'How does the body maintain its shape? Part 1, Metabolic pathways', *Rolf Lines*, vol. 17, pp. 27- 29.

Oschman, J.L. 1989b, 'How does the body maintain its shape? Part 2, Neural and biomechanical pathways', *Rolf Lines*, vol. 17, pp. 30-32.

Oschman, J.L. 1990, 'How does the body maintain its shape? Part 3, Conclusions', *Rolf Lines*, vol. 18, pp. 24-25.

Oschman, J.L. 1994, 'Sensing solitons in soft tissues', *Guild News, Guild for Structural Integration, Boulder, Colorado*, vol. 3, pp. 22-25.

Pienta, K.J. & Coffey, D.S. 1991, 'Cellular harmonic information transfer through a tissue tensegrity-matrix system', *Medical Hypotheses*, vol. 34, pp. 88-95.

Popp, F.A., Li, H.H. & Gu, Q. 1992, *Recent Advances in Biophoton Research and Its Applications*, World Scientific, Singapore.

Popp, F.A., Ruth, B., Bahr, W., Böhm, J., Grass, P., Grolig, G., Rattemeyer, M., Schmidt, H.G. & Wulle, P. 1981, 'Emission of visible and ultraviolet radiation by active biological systems', *Collective Phenomena*, vol. 3, pp. 187-214.

Proust, M. 1924, *Remembrance of Things Past*, trans. C.K. Scott Moncrieff, Random House, New York.

Rolf, I.P. 1977, *Rolfing: The Integration of Human Structures*, Dennis-Landman, Santa Monica, California.

Schulze, E. & Kirschner, M. 1988, 'New features of microtubule behavior observed in vivo', *Nature*, vol. 334, pp. 356-359.

Tanaka, T. 1981, 'Gels', *Scientific American*, vol. 244, pp. 124-138.

Weiss, P. 1961, 'The biological foundation of wound repair', *Harvey Lectures*, vol. 55, pp. 13-42.

Young, J.Z. 1975, *The Life of Mammals: Their Anatomy and Physiology*, 2nd ed, Clarendon Press, Oxford.

23 Soft tissue holography

In Chapter 22, we defined *somatic recall* as the release during massage and other kinds of bodywork of repressed and often highly emotional memories. Often such flashbacks can be beneficial, leading to resolution of old trauma, pain, or psychological attitudes. Sometimes therapist and client simultaneously detect an identical avalanche of sensory information. We described some ways that soft tissues can store information and how touching certain parts of the body could trigger and then erase memories at the same time that toxic materials are being released, physiological communication channels are opening up, and both physical and emotional flexibility are being restored.

As a phenomenon, somatic recall seems a bit too peculiar for scientific exploration. Most scientists consider instances of somatic recall to be hallucinations or delusions because they do not seem to fit with normal theories about how the brain and nervous system work. This is frustrating for the therapist who has such "hallucinations" on a daily basis and who would like some scientific validation or explanation for a phenomenon that seems both important and therapeutic. We take the view that the phenomenon is not only valid and therapeutic, but that it is an important clue that could help us answer unsolved questions about the mechanisms of learning, memory, consciousness, and whole systems communication.

In the first part of this book, we described a new way of looking at living tissue as an interconnected molecular continuum, which we refer to as the *living matrix*. This way of looking at the body is the result of an important discovery: the matrix inside cells, known as the *cytoskeleton,* is directly connected to the matrix outside of cells, classically known as *connective tissue*. The living matrix gives the body its overall shape and features; defines the form of each organ, tissue, and cell; and extends into every nook and cranny of the organism. The nucleus and deoxyribonucleic acid (DNA) are a part of the living matrix.

The most exciting property of the living matrix is the ability of the entire network to generate and conduct vibrations. Modern biophysical research is revealing a wide range of properties that enable the body to use sound, light, electricity, magnetic fields, heat, elasticity, and other forms of vibrations as signals for integrating and coordinating diverse physiological activities, including those involved in tissue repair.

According to the continuum communication model, every event in the organism produces vibrations that travel throughout the living matrix. In this way, every part is informed of what all other parts are doing. Hands-on and energy techniques are effective because practitioners have used their intuition and sensitivity to develop methods of interacting with fundamental and evolutionarily ancient communication systems that have not, until recently, had any logical scientific basis.

The cytoskeleton is being referred to as the *nervous system of the cell*. Biologists now are describing ways that specific components of the living matrix can store, process, and erase information. We now continue to develop a theory of how various therapeutic approaches may release memories stored in soft tissues and how these memories reach the consciousness of both the client and the practitioner.

Before doing this, however, we need to summarize the reason neurophysiologist have not looked beyond the brain in their search for the location of memory.

THE BRAIN AS THE SEAT OF MEMORY?

Modern biomedical research focuses on the brain as the location of memory even though there are many signs that this is only part of the story. The reason for the bias is partly historical. It arose from early brain research, some of which was done in the 1920s by the famous Canadian pioneer of neurosurgery Wilder Penfield. Penfield discovered that electrical stimulation of particular areas on the brain surface caused patients to reexperience "memories" from the past (Penfield 1975). These recollections contained vivid details of long-forgotten events that manifested as flashbacks resembling moving pictures.

After years of research along these lines, Penfield concluded that electrical brain stimulation could activate sequential records of consciousness laid down during a person's earlier experience. The detail contained in these recalls was so vivid that Penfield concluded that every experience we have is recorded in the brain.

The vividness of memory recall is familiar to massage therapists and to practitioners of various other somatic methods, including hypnotic regression, rebirthing, acupuncture, and even music and movement therapies. During sessions using these and other methods, clients often relive early traumatic experiences. In some cases, experiences that took place at birth, or even in utero, can be recalled in detail and with observable therapeutic benefit.

Penfield's discovery that electrical brain stimulation elicits specific recollections led to an obvious, but incorrect, conclusion. Memory traces, which are called *engrams,* seemed to be stored as patterns of neural discharge in specific areas of the brain. This idea was supported by research showing that surgical lesions in certain areas of the cortex can seriously disrupt learning.

Modern researchers have repeated Penfield's studies and questioned the original interpretations. "Memories" elicited by electrical stimulation of the brain have a dreamlike quality and may not be memories at all. Sometimes stimulation at different sites produces the same recollection, and at other times repeated stimulation at one site evokes different recollections. Even removal of major parts of the temporal lobe, the location of the stimulation points, did not destroy memories of events that had been elicited by electrical stimulation of the lobe before it was removed.

The brain is part of an intricate system, and the effects of stimulating, damaging, or removing certain parts does not prove that those parts are the locations of memories. Because of the interconnectedness of the nervous system and living matrix, one cannot be certain that a particular evoked experience is stored near a site of electrical stimulation or far away from it. Moreover, each region of the cortex refers to a particular part of the body. The brain and distant tissues are connected by motor and sensory nerves and by other communicating channels within the living matrix. Stimulation of a spot on the cortex may activate processes that take place in an intricate system that includes cells and tissues that are very far from the site of stimulation.

The logical problem of confining the search for memory and consciousness to the brain has exacerbated an already difficult problem. Study of these phenomena is conducted by narrow disciplines, each with methods to study only a small part of the whole problem.

The brain's monopoly on memory has been eroding for many decades. Studies done as early as 1940 demonstrated that certain simple reflexes can be conditioned or learned by spinal cord neurons that have been surgically disconnected from the brain (Shurrager & Culler 1940). This finding led to the conclusion that memory may be found in all parts of the nervous system. We now see that this concept, too, may be limited because of cytoskeletal memory in non-neural cells and because there are other forms of information storage in soft tissues (e.g., as the orientation of connective tissue fibers described in Figure 22-3).

From our point of view, the most significant lines of inquiry arose from studies of neurophysiologists who continued Penfield's search for the location of the engram. Of these, one of the best known was Karl Lashley, the distinguished Harvard psychologist who spent virtually his entire scientific career, 30 years, in an unsuccessful search for the engram (Lashley 1950).

Lashley's basic approach was to train rats to perform tasks such as running in a maze to find food. He then would surgically damage or remove specific parts of the rats' brains, or cut the connections between them, and test again. His goal was to identify the part of the brain where the maze-running engram was stored. Even removal of large amounts of brain tissue, which impaired the rats' motor skills, failed to erase memories essential to running through the maze. Lashley concluded that all parts of the functional area where memory is stored are equipotential.

Karl Pribram was a student of Lashley, and he wanted to continue the search for the engram. After reviewing all of Lashley's work, Pribram concurred that memory somehow must be distributed throughout the brain as a whole rather than localized at specific sites. This view was supported by the repeated observation of neurosurgeons that removal of large portions of the brain for medical reasons can dim a person's memory but never seems to cause a selective loss of particular memories. The engram is so elusive that some neurophysiologist suspect that it does not exist.

Pribram's problem was that there was no concept of memory that was consistent with all of the evidence. This fact had a deep impact on the field of experimental psychology, which had great difficulty advancing without a solid understanding of the mechanisms of processes so basic as learning and memory.

HOLOGRAPHY

All of this changed dramatically with the invention of the hologram. Holography was first postulated in 1947 by Dennis Gabor in London, but it did not blossom into a radically new branch of optics for some 15 years (Gabor 1972). In 1964, Leith and Upatnieks (1964) introduced modern holography.

Holography is technically defined as "photography by wave-front reconstruction." To understand this concept, consider a beam of light shining on an object consisting of a single tiny point (Figure 23-1, *A*). Note how the light will be reflected from the point. The light waves bounce back toward the source in a series of expanding concentric spherical shells called *wavefronts*. They are three-dimensional versions of the circular waves formed on the surface of a pond when a pebble is dropped into the water.

A complex object can be regarded as a collection of points, and light reflected from its surface will produce a reflection composed of an intricate set of spherical wavefronts (Figure 23-1, *B*).

In holography, this intricate pattern, which contains precise information on the shape of the object, is recorded directly on a photographic film. The recording, however, is not a record of an image of the object. In contrast to conventional photography, holography does not use a lens to focus the image on the film; instead, wavefronts reflected from every part of the object flood over the whole film. Each point on the object reflects light onto the whole area of the film, and each point on the film receives light that has been reflected from every point on the object.

The wavefronts reflected from the object create a set of waves known as the *object beam*. This light interacts with a reference beam that consists of light from the same source that has been simply reflected from a mirror. The final image is an interference pattern, created from the interaction of the object and reference beams (Figure 23-1, *C*).

Interference patterns

Let us take a closer look at interference. Most of us have thrown a pebble into a pond and watched in fascination as the ripples spread over the surface. Consider a pond with a smooth surface, with no wind or other disturbance. When we throw a pebble into the pond, the ripples spread concentrically outward. In the language of physics, we have created a series of ripples known as a *wave train*. The leading edge of the wave is called the *advancing wavefront*. It is the place where the moving disturbance produced by your pebble interacts with the placid surface of the pond (Figure 23-2, *A*).

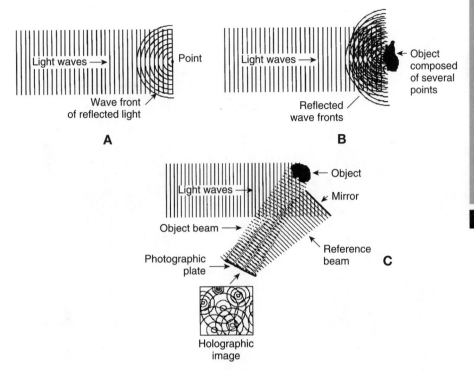

Figure 23-1 Holography is defined as *photography by wave-front reconstruction.* **A,** A beam of light shining on an object consisting of a single tiny point. The light is reflected back toward the source as a series of expanding concentric spherical shells, called *wave fronts.* **B,** A complex object can be regarded as a collection of points, and light reflected from its surface will produce a reflection composed of an intricate set of spherical wave fronts. **C,** In holography, the intricate pattern of reflected wave fronts, which contains precise information on the shape of the object, is recorded directly on a photographic film. The recording, however, is not a record of an image of the object. In contrast to conventional photography, holography uses no lens to focus the image on the film. Instead, wave fronts reflected from every part of the object flood over the whole film. Each point on the object reflects light onto the whole area of the film, and each point on the film receives light that has been reflected from every point on the object. The wave fronts reflected from the object create a set of waves called the *object beam,* and this light interacts with a reference beam consisting of light from the same source that has been simply reflected from a mirror. The final image recorded on the film is an interference pattern, created by the interaction of the object and reference beams.

If there is no wind and no other activity is taking place on the pond, we can watch the waves spread over the entire pond surface, even to the farthest points. The spreading wave contains information. When the wave has spread out over the entire pond, information about the splash you created is distributed throughout the surface of the pond. (Recall that Lashley and Pribram thought memory might be distributed throughout the brain, but they could not think of a way this could be accomplished.) An observer at any point on the pond could study the size and direction of the incoming waves and estimate where your pebble had splashed into the water.

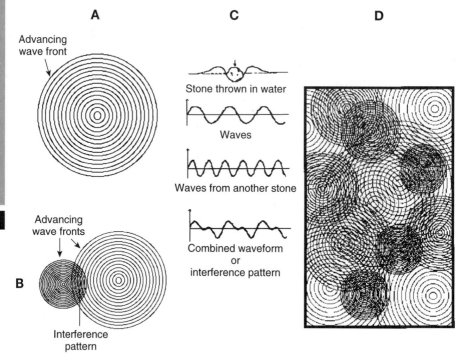

A

Advancing
wave front

B

Advancing
wave fronts

Interference
pattern

C

Stone thrown in water

Waves

Waves from another stone

Combined waveform
or
interference pattern

D

Figure 23-2 Interference patterns. **A,** When we throw a pebble into a pond, ripples spread concentrically out over the surface. Physicists refer to this as a wave train, and the leading edge of the wave is called the *advancing wave front*. When the wave has spread out over the entire surface of the pond, information about this splash is distributed throughout the surface of the pond. **B,** An interference pattern is produced when another pebble is thrown into the pond so that there are now two wave fronts interacting with each other. **C,** A closer look at the way the wave forms interact with each other. Some of the interacting waves add to each other to produce larger peaks, while other waves subtract from each other to produce smaller waves. The result of the interaction is a third combined wave form, which is called an *interference pattern*. **D,** Consider the situation if we throw a handful of pebbles onto the pond surface, creating a series of concentric waves. If we took a snapshot of the pond surface at any instant, we would see a set of advancing wave fronts and interference patterns, much like we observe when it is raining on a puddle of water. The snapshot of the pond surface resembles a holographic image in that it consists of a set of interference patterns produced by the combining of the various wave fronts.

Note that the distant observer studying the incoming waves does not see a real image of your pebble hitting the water; instead they see a wave pattern from which they can *infer* when and where your stone hit the water.

Our observer would have a more difficult problem if there were a breeze, boat, duck, or another pebble thrower creating waves that interact with those from your pebble. Interference patterns form where wavefronts interact (Figure 23-2, *B*).

The interference pattern is produced because some of the interacting waves add to each other to produce larger peaks, whereas other waves subtract from each other to produce smaller waves. Figure 23-2, *C*, shows how two different waves interact to produce a third waveform, the *interference pattern*. This is an intricate array of crests and troughs caused by the collision of the two wavefronts. It is an irregular wave form, but it still contains information from the events that produced the original waves.

Let us ask our distant observer to analyze the interference patterns and reconstruct the events that caused it. The problem is not as complicated as you might think. The Frenchman Jean B.J. Fourier showed that a complex wave can be mathematically separated into its parts. Fourier analysis is a form of calculus that can be applied to any mixture of waves, whether ripples on the surface of water, sounds emanating from an orchestra, or radio signals coming from deep space. The most intricate wave pattern can be separated into the simple waves that created it, and the simple wave forms can be converted mathematically back into the original pattern. The equations are known as *Fourier transforms*. We soon shall see that Fourier transforms also may be used by the brain to process visual and other kinds of sensory information.

Now consider the situation created by many disturbances on a pond surface. To research this, we observed a nearby pond. The wind was blowing gently, creating a wave train moving from right to left. A small twig fell into the water, producing a concentric wave pattern. A similar pattern appeared when the head of a turtle popped above the surface. A frog made three hops in the shallow water near the shore, creating three more concentric waves. Small bugs, each about one-quarter inch long, glided over the surface. (Remarkably, these bugs left no wake except when they changed direction. It looked like they were skimming along, in a state of levitation, just above the water surface, and only made contact when they changed direction by pushing off with one leg.) A flying pair of mating dragonflies dipped down and the tail of the male briefly touched the surface, creating a tiny bull's-eye wave pattern on the surface.

At any instant we could take a photograph looking down on the pond surface and record the advancing wavefronts and interference patterns. The photo might look something like Figure 23-2, *D*.

The picture of waves on the surface of a pond resembles a holographic image. Information about the object is optically encoded as a pattern of concentric rings.

The pattern of light waves reflected from a complex surface is intricate, as we have seen. If white light, which is composed of many colors, is used, each color will produce a separate pattern of fringes. There will be an averaging out or overlap of the information, and the image will be fuzzy. The quality of the image is greatly improved if the source of illumination is monochromatic, that is, of a single color. Even better images are obtained if the light is coherent.

Coherence

What is meant by coherence? Coherent radiations are very different from the random signals produced by random vibrations. This is an important point in relation to various therapeutic approaches and therefore is worthy of an explanation. As an example, compare the light produced by a light bulb with that from a laser. Laser light is monochromatic or of one color. All of the vibrations of the atoms in the laser are coupled with each other, and the waves spread uniformly and concentrically outward from the source. In contrast, the light bulb is simply a hot body producing a jumble of light of many colors. The vibrations of the atoms in the bulb's filament are random, and the waves spread outward chaotically from the source.

For holography, it is essential that the light be both spatially and temporally coherent. Although there are other ways of producing coherent illumination, the laser is far superior to any other and is responsible for the remarkable holograms that are being made today.

When illuminated with coherent light from a laser, the hologram reconstructs an image of the object that contains all of the information in a normal photograph plus additional information that normal photography is unable to capture. The holographic image looks exactly like the original object, but in contrast to a normal photograph, the object appears suspended in space in three-dimensional form, complete with parallax. (This is the way the parts of an object become displaced relative to each other when they are observed from different angles.) The image has depth and cannot be distinguished from the original object. You can view a holographic projection from different angles and you will see its front and sides, just like a real object; however, when you attempt to touch the object, you find there is nothing there. In terms of optics, the holographic projection is a virtual image rather than a real image.

The other remarkable aspect of the hologram is that each part of the image, regardless of how small it is, can reproduce the entire image. A corner cut from a normal photographic image contains only a portion of the original scene, but a fragment of a holographic film contains the entire image. This is comparable to placing observers at various places around our pond and having them watch the interference patterns produced by disturbances of the water's surface. The story of these disturbances is present everywhere on the surface.

As the pieces of a hologram are made smaller, there is a loss of detail and image intensity. The entire image is still available, but it must be reconstructed from a smaller number of interference fringes.

Holographic memory

Leith and Upatnieks published a technical article on holography in 1964 (Leith & Upatnieks 1964) and a popular article in *Scientific American* in 1965 (Leith & Upatnieks 1965). Pribram saw the second article and immediately recognized that holography provided a single conceptual framework that could account for many of the remarkable aspects of memory (Pribram 1969).

Pribram's subsequent work is responsible for the widespread application of the holographic model to brain function, but he was not the first to suggest it. In 1965, immediately after the first technical paper by Leith and Upatnieks, two other scientists, Julesz and Pennington (1965), made the explicit suggestion that memory is stored in the brain as interference patterns comparable to those used in holography.

Now we can see why Pribram was so excited by holography. He had concluded that memory is a distributed property of the nervous system. He and others had noted that removing large parts of the brain only dims memory but does not erase it, and these properties are the hallmarks of holography.

Holography is a highly sophisticated way to store information, and Pribram saw immediately that the brain could exploit holographic principles. Perhaps physical structures responsible for memory were elusive for the same reason that the patterns on a holographic plate are unintelligible and bear no relation to the images they encode. Perhaps brain structures and patterns of nerve impulses contain no first-order information (like a regular photograph) about memory and learning. Perhaps memory is to be found not in the patterns of neural activity but in their Fourier transforms (Pribram 1977).

Over the years, a number of different lines of evidence have been developed to support the holographic model of memory and to show that some of the remarkable aspects of memory are virtually impossible to explain by any other concept. For the recent neurophysiological perspective on holographic memory, see L.R. Squire's book, *Memory and Brain*. A very readable description of this story can be found in Michael Talbot's book, *The Holographic Universe*. The following summarizes some of the evidence.

1. *Vision appears to be holographic.* Pribram did a series of studies in which he measured the electrical activity in the brains of monkeys as they performed visual tasks. He found that there was no one-to-one correspondence between visual images focused on the retina and neural impulses in the brain. Like memory, vision appears to be a distributed property of the visual cortex. Subsequent work in a number of laboratories confirmed that cells in the visual cortex respond to the Fourier transforms of visual patterns (DeValois & DeValois 1980; DeValois et al. 1979). Pribram looked at the older literature and found that other senses, including hearing, smell, touch, and taste, seem to use frequency transforms of the Fourier type.

2. *The vast capacity of memory.* We mentioned earlier Penfield's conclusion that we retain a memory of every event that happens to us during our lifetime. The Hungarian physicist and mathematician John von Neumann calculated that during an average human lifetime, 280,000,000,000,000,000,000 bits of information are stored in the brain. In holography, many images can be superimposed on a single piece of film. This is done by using a different laser frequency for each image or by changing the angle of the laser illumination. Each image can be recovered completely, and separate from the others, by using the original frequency or angle of illumination. It has been calculated that a single square inch of holographic film can store the information contained in 50 copies of the Bible (Collier, Burckhardt & Lin 1971).

3. *The ability to forget.* If many memories are encoded in the same tissue, by using different laser frequencies or angles, as mentioned earlier, searching for a particular bit of memory could involve scanning through a series of angles or wavelengths until the desired memory trace is located. Forgetting could be described as an inability to find the appropriate scan angle or frequency to locate the image or memory that is being searched for.

4. *Ability to recognize a familiar face.* How are we able to recognize a familiar face in a huge crowd of people? In 1970, the physicist Pieter van Heerden (1970)

proposed that this feat is explainable if the brain is capable of recognition holography, in which the optics enable two images to be compared in such a way that the degree of similarity is registered as the brightness of a spot of light. A related method, known as *interference holography,* enables one to recognize an image such as a familiar face and, at the same, to highlight those features that have changed since the image was first recorded. Instruments have been constructed that use this principle to detect minute changes or stresses in manufactured objects.

5. *Photographic memory.* People with photographic memories, also called *eidetic memories,* are capable of extremely vivid and detailed visual recall. In 1972, Harvard vision researchers Pollen and Tractenberg (1972) suggested that such individuals have access to larger than normal areas of their memory holograms. This enables them to recall information with high resolution of image detail.

6. *Scaling of body movements.* Trace your signature in the air with your left elbow. You will find that it is easy to do this, even if you never tried it before. This feat is not explainable by a hard-wired nervous system that is only capable of performing a task after repeated practice. You would have no difficulty writing your signature on a black board in letters 3 feet high, even though it requires use of a completely different set of muscles than you usually use in signing your name. Pribram suggested that this ability to transpose a set of movements from one scale to another or from one part of the body to another can be accounted for by a holographic nervous system that converts memories of learned abilities into a language of interfering wave forms. This same phenomenon could account for your ability to recognize a familiar face while viewing it from any distance or any angle. The idea is that the brain contains a three-dimensional holographic recording of your signature or of your friend's features, and this recording can be can be scaled to any size, or rotated, so it can be recreated at any distance or from any perspective.

7. *Movements as waveforms.* Pribram also cited work done by the Russian scientist Nikolai Bernstein showing that movements, as in dancing, may be encoded as Fourier transforms. During the 1930s, Bernstein painted white dots on the black leotards of dancers. The dots were placed over the joints. When the dancers performed against a black background, moving pictures revealed their motions as a series of dots that formed wave patterns that could be analyzed by Fourier calculus. To Pribram, this indicated that the brain stores movement patterns as wave patterns, a mechanism that could explain our ability to rapidly learn complex physical tasks.

All of these ideas seem to fit together, and neurophysiologists are beginning to accept that at least some aspects of memory are stored holographically.

Among those who were skeptical of holographic memory was Paul Pietsch of Indiana University. Pietsch was doubtful of Pribram's theories and set out to disprove them. To do this, Pietsch did a series of experiments with salamanders. The brain of a salamander can be removed without killing it. Without a brain, the salamander is unable to do much, but normal behavior can be restored by returning the brain to the salamander's head.

Pietsch did a series of experiments based on the idea that if the holographic model is correct, it would not matter how the portion of the brain that controls feeding behavior is positioned in the head. He figured he would flip-flop the left and right hemispheres of the brain, and this would disrupt feeding behavior. Pribram's theory then would be out the window. To his great surprise, the experiment caused no change in the salamander's feeding behavior once the animal had recuperated from the operation.

Pietsch then did a series of some 700 operations on salamanders, turning brains upside down, slicing, flipping, shuffling, subtracting, and even mincing brains and reinserting them into the animals. In all cases, the animals' behavior returned to normal after they recovered from the "postoperative stupor." All of his research, which Pietsch details in his entertaining book, *Shufflebrain*, led him, reluctantly, to become an ardent believer in the holographic model (Pietsch 1981).

Soft tissue holography

Now we can return to Young's model of connective tissue memory and Hameroff's model of cytoskeletal memory, described in the previous chapter. In Young's model, the stresses of the environment select the sites where collagen is deposited. Information is stored in the form of oriented collagen fibers. What is stored is a set of structures that reflect the situations, postures, movements, stresses, and strains that have been experienced by the organism. Hameroff and other cell biologists have extended soft tissue memory processes into the cytoskeletal level. Can these ideas be integrated with Pribram's holographic model?

Such an integration can be achieved by recalling that all of the molecules of the living matrix create large-scale coherent or laserlike vibrations; therefore, the orientation and other properties of each fiber in the connective tissue and cytoskeleton, and the forces imposed upon it, will be translated into specific waveforms that will travel through the living matrix and be broadcast into the environment. Because every molecule in the body can act as both a source and a conductor of information that can spread throughout the whole, the entire body can be viewed as a dynamically interacting, three-dimensional, communicating, coherent hologram. The same mechanism that unifies the structure of the body may simultaneously provide for the storage of information or memory. Consciousness and the processes we refer to as *mind* may arise simultaneously as consequences of this dynamic system. Remarkable as this may seem, Mae-Wan Ho has gathered together several lines of research that support the idea. We shall look at this more closely in the concluding chapter of the book.

Implications for therapies

The living matrix is a continuous physical, energetic, and informational network that distributes regulatory signals throughout the body. Every physiological event and every process creates a variety of vibrations that travel through the Tensegrity matrix, much like ripples spreading over the surface of a pond. Some of the coherent vibrations radiate into the space around the body in various ways.

When a therapist approaches a client, vibrations are exchanged back and forth between their respective living matrices, even before there is direct physical contact. This interaction is a natural and inescapable consequence of the fundamental design of the living matrix as a communicating system and is explainable by the laws of physics. Biomagnetic and other field interactions take place all of the time when two or more people are in proximity, even if they are not touching. The fields are the result of electric, magnetic, thermal, photonic, microwave, and other kinds of energy. Such interactions have been documented by McCraty, Atkinson, Tomasino, and Tiller (1998).

Empathy, which is the ability of one person to "tune in" to the physiological or emotional state of another, and other so-called psychic abilities occur because the living fabric radiates energy fields into the environment. These fields are a rich source of information about the history and the present status of the living matrix. The communicating continuum of one person is a receiving antenna, detecting the state of the tissues in another. The transmitters and detectors involved in these exchanges are the molecular components of the living matrix.

Therapists project specific vibrations directly into places in the client's body where energy and information flows are distorted or deficient. The result is the restoration, revitalization, opening up, organizing, balancing, energizing, and tuning of resonant vibratory circuits.

Learning involves various kinds of changes in the living matrix, some of which we have described: tensional patterns in the connective tissues and structural patterns in microtubules and other components of the cytoskeleton. Other kinds of molecular information storage are being investigated in laboratories around the world.

"Remembering" involves manipulating coherent wavefronts to "read" information holographically encoded in cell and tissue structures. "Consciousness" at any instant is the totality of the coherent signaling within the living matrix, including wavefronts reflected from specific information-containing structures. Our behavior, our consciousness, and our experience of the world are shaped, on a moment-to-moment basis, by choices that are referenced to information contained in the reflected wavefronts.

Coherent signals from the hands of a therapist influence wavefronts flowing throughout the molecular fabric of their client's body. When emotionally "charged" regions are contacted, there may be a sudden recall of stored memories. The memory trace is released as an energetic pulse and interacts with other wavefronts present in the body. The memory is erased when various polymers, such as ground substance and microtubules, depolymerize or fall apart.

In Chapter 21, we suggested that some of the more powerful energetic phenomena taking place during various approaches to the body may involve solitons. Sometimes therapists notice a visible or palpable wave or ripple propagating through their client's tissues in advance of the place where they are working. This may be a powerful, self-regenerating, coherent soliton wave.

Soliton waves traveling through the living matrix may restore communication channels. When this happens, the soliton can penetrate into areas of the living matrix that have been closed off or protected for a long time as the result of some trauma or attitude. There may be a sudden recall of some long-forgotten or deeply repressed traumatic event or decision about how the world works. This can reach consciousness in the form of a reflected wave that travels through the living matrix. The reflected wave can contain a detailed virtual holographic image. An entire array of sensations can be conveyed virtually instantaneously. The wavefront does not stop at the surface of the client's skin, but it is radiated into the environment and communicated to the practitioner, who decodes the wavefront into an identical image. The ability of the therapist to detect and decode the avalanche of sensory information probably depends upon the degree of coherence of his or her own structure.

The soliton wave may be energetic enough to bring about depolymerization of information-rich molecules and erasure of the memory. In other cases, the soliton may stimulate metabolic activity in a tissue that has long been dormant. Somatic recall will be delayed for a day or so, until the cells begin to divide, their cytoskeletons depolymerize, and tensional patterns begin to be reorganized in the soft tissues.

The phenomena we are describing are nonlinear in nature. When a system behaves in a linear fashion, a larger input will produce a larger output. Living systems can operate in the reverse of this. A nonlinear system can undergo a huge change in response to a tiny input. Bodyworkers and homeopathic physicians frequently refer to this as "small is powerful" or "less is more." The living matrix can be delicately poised to accept and use small amounts of coherent energy.

CONCLUSIONS: A BIOPHYSICS OF ENERGY THERAPIES

There are many reasons for studying the nature of life, and as many approaches. Progress along all lines of inquiry occurs in fertile spurts, punctuated by times of relative stagnation. Progress slows for a variety of reasons that can be attributed to "human nature":

- Tendency to ignore anomalies
- Tendency to ask the wrong questions
- Tendency to look for answers in the wrong places
- Tendency to create disciplinary and political boundaries
- Tendency to conceal information by creating incomprehensible vocabularies

Biophysics, the physics of biological processes, is by definition an interdisciplinary line of inquiry. Biophysicists combine information from the two great sciences upon which our medicine is founded.

The history of physical, biological, and medical research shows that clever and dedicated investigators have been motivated by similar goals that eventually proved to be erroneous. In physics it was the search for the fundamental building block of matter; in biology it was locating the fundamental particle of life; in medicine it was finding the fundamental cause and cure for every disorder. Expenditure of huge amounts of money and effort yielded useful information, but the original goals remained elusive. As observational methods became more refined, increasingly smaller fundamental units of matter, life, and disease reveal themselves. The focus shifts from one component to another.

The study of memory dramatizes the points just made. It was obvious from the beginning that memory would be found somewhere in the body, and the brain was the best place to look. Logic also dictated that we define a fundamental particle of memory, the engram. Many years of research have failed to locate a single place where memory happens, and the engram is as elusive as the physicist's fundamental building block of matter; however, the search continues, and it remains focused on the brain.

In the meantime, therapists of many traditions have remarkable experiences, of which somatic recall is an example. To us, this is a huge clue in the search for memory and suggests that researchers may benefit by looking beyond the nervous system. The biophysical properties of the living matrix can explain a variety of phenomena that have been elusive in the past, such as learning, memory, consciousness, unity of structure and function, and intuition.

Although the concepts presented here are not yet a part of normal biomedicine, they have a sound scientific foundation and they go a long way toward explaining some of the phenomena that arise in complementary medicine. We believe the therapeutic practice is one of the best "laboratories" for testing these concepts. Before this can happen, biophysical language must be translated into understandable terms, so therapists can explore the new data and concepts. This book aims to do this. We look forward to hearing from those who find these ideas useful for the process of refining and expanding techniques and understanding. Research is fun when it leads to new questions and opens us to new possibilities. Research is exciting when it leads to new levels of athletic or artistic performance.

REFERENCES

Collier, J., Burckhardt, C.B. & Lin LH, 1971. *Optical Holography*, Academic Press, New York.

DeValois, K.K., DeValois, R.L. & Yund, W.W. 1979, 'Responses of striate cortex cells to grating and checkerboard patterns', *Journal of Physiology*, vol. 291, pp. 483-505.

DeValois, R.L. & DeValois, K.K. 1980, 'Spatial vision', *Annual Review of Psychology*, vol. 31, pp. 309-341.

Gabor, D. 1972, 'Holography, 1948-1971', *Science*, vol. 177, pp. 299-313.

Julesz, B. & Pennington, K.S. 1965, 'Equidistributional information mapping: An analogy to holograms and memory', *Journal of the Optical Society of America*, vol. 55, p. 604.

Lashley, K. 1950, 'In search of the engram', in Physiological Mechanisms in Animal Behavior, *Symposia of the Society for Experimental Biology*, vol. 4, pp. 454-482.

Leith, E.N. & Upatnieks, J. 1964, 'Wave front reconstruction with diffused illumination and three-dimensional objects', *Journal of the Optical Society of America*, vol. 54, pp. 1295-1301.

Leith, E.N. & Upatnieks, J. 1965, 'Photography by laser', *Scientific American*, vol. 212, p. 24.

McCraty, R., Atkinson, M., Tomasino, D. & Tiller, W.A. 1998, 'The electricity of touch: Detection and measurement of cardiac energy exchange between people', in *Brain and Values: Is a Biological Science of Values Possible*, ed. K.H. Pribram, Lawrence Erlbaum Associates, Mahwah, New Jersey, pp. 359-379. Also available from the website of the Institute for Heart Math, Boulder Creek, Colorado.

Penfield, W. 1975, *The Mystery of the Mind: A Critical Study of Consciousness and The Human Brain*, Princeton University Press, Princeton, New Jersey.

Pietsch, P. 1981. *Shufflebrain. The Quest for the Holographic Mind*, Houghton Mifflin, Boston.

Pollen, D.A. & Tractenberg, M.C. 1972, 'Alpha rhythm and eye movements in eidetic imagery', *Nature,* vol. 237, p. 109.

Pribram, K. 1969, 'The neurophysiology of remembering', *Scientific American,* vol. 220, p. 75.

Pribram, K. 1977, *Languages of the Brain,* Wadsworth Publishing, Monterey, California.

Shurrager, P.S. & Culler, E. 1940, 'Conditioning in the spinal dog', *Journal of Experimental Psychology,* vol. 26, pp. 133-159.

Squire, L.R. 1993, *Memory and Brain,* Oxford University Press, London.

Talbot, M. 1992, *The Holograph Universe,* Harper Perennial, New York.

van Heerden, P. 1970, 'Models for the brain', *Nature,* vol. 227, pp. 410-411.

24 A continuum pathway for sensation and movement

The extracellular, intracellular, and nuclear matrices together constitute a noiseless excitable electronic continuum for rapid intercommunication and energy flow permeating the entire organism, enabling it to function as a coherent and sentient whole. Ho (1997)

The pieces of the puzzle summarized in the Preface are coming together. Sherlock is pleased. Of course, he would say that all of this is "Elementary, my dear Watson" and would wonder why we didn't put it all together long ago.

We now have a logical, testable, and refutable scientific basis for some remarkable observations on the nature of sensation and movement. The observations were made decades ago and are described in Valerie Hunt's book, *Infinite Mind*. They are also described in Albert Szent-Györgyi's writings, in the literature of the martial arts, in the experiences of athletes, and in Condon's work on speech.

At about the same time all these scientists were making their observations, others were beginning to assemble a new science that can explain them. We now have "connected the dots" to show how all of this information fits together.

Figure 7-2 showed the conventional pathway by which environmental stimuli are thought to lead to muscle contractions and actions. The same illustration also shows another, more rapid pathway that involves semiconduction via the living matrix. We proposed that Valerie Hunt's observations on the seemingly effortless movements of dancers, in the absence of electromyographic signals, document another sensory/movement system that is available under certain conditions. We are not suggesting that the conventional pathways are wrong, but that they can be supplemented or replaced by an alternative parallel system under appropriate conditions.

We referred to this alternative pathway as the *continuum pathway*. The word *continuum* is appropriate on the basis of the definition given at the beginning of Chapter 8. Here *continuum* refers to a continuous pathway for sensory energy/information that includes but is not limited to nerves. It is a pathway that is more ancient, in terms of evolution, than the nervous system. From study of "primitive" organisms, the Russian scientist Y.A. Vinnikov refers to this as the *locomotor-sensory system*. It is a pathway through the living matrix that encompasses neurons but is not limited to them. Figure 24-1 summarizes the differences between neurological consciousness and the *continuum*

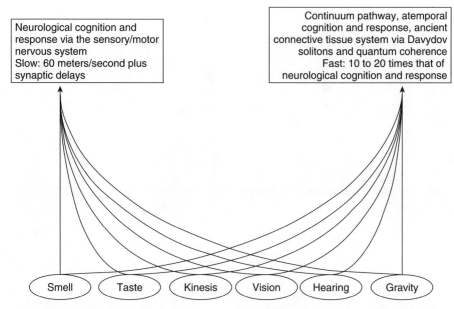

| Neurological cognition and response via the sensory/motor nervous system Slow: 60 meters/second plus synaptic delays | | Continuum pathway, atemporal cognition and response, ancient connective tissue system via Davydov solitons and quantum coherence Fast: 10 to 20 times that of neurological cognition and response |

Smell　　Taste　　Kinesis　　Vision　　Hearing　　Gravity

Figure 24-1 Comparison of neurological consciousness and the consciousness of the living matrix via the continuum pathway.

consciousness that we suggest arises as a result of the *continuum pathway*. Figure 24-2 details soliton transfer through the living matrix from a sensory receptor to a muscle.

The fact that the "baseline" in Dr. Hunt's electromyographic recordings vanished during "trance dancing" may indicate that there is a "switch" somewhere in the circuitry of the body that is able to inactivate normal neuromuscular programs so that they do not conflict with the *continuum pathway*. This is a fascinating possibility for further research.

Many questions arise from the information gathered here. One would like to know the source of the seemingly effortless movement that can arise under certain conditions or states of consciousness. In thinking about this, several possibilities come to mind. One is that amplification, or gain, can take place in what we have referred to as the *continuum pathway*. In this connection, the electric fields arising in the living matrix as a result of the piezoelectric effect could energize amplification in the α-helical protein soliton network, much like the amplifiers that are used in fiberoptic networks of the telecommunications industry. The excitability in the living matrix may be somewhat analogous to the excitability in a grass fire, as described by Winfree (1984). Each part of the medium has an adequate supply of energy to enable an arriving spark to trigger a large flame. In the case of the grass fire, the energy supply is in the form of combustible material. In the living matrix, the energy supply is everywhere present due to the thermal motions of the molecules and the energy stored in the organized layers of water. Another possibility is that the crystalline matrix of the

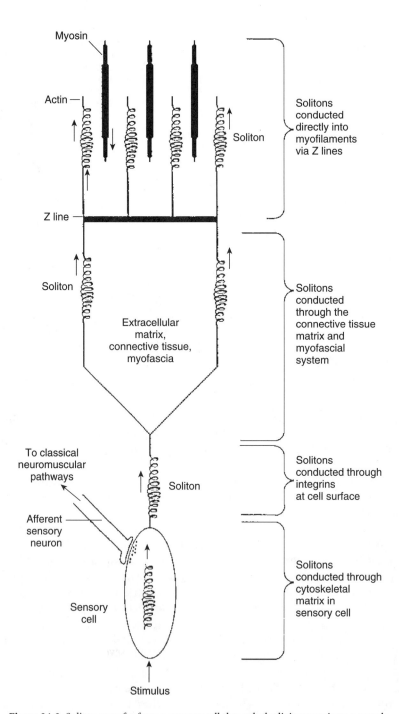

Figure 24-2 Soliton transfer from a sensory cell through the living matrix to a muscle.

connective tissue, cytoskeletons, genetic domains, and muscles can extract energy from the "plenum" of the zero point or quantum vacuum.

There obviously is much more to be learned about sensory motor systems in the body. We have seen that conventional neuroscience regards receptors as the "pass filters" of the body, admitting to the nervous system some kinds of information about the external world and rejecting other kinds. A profound philosophical and biological question raised by the *continuum pathway* is whether *all* of the information about the external world is available to the organism, with the receptors being pass filters for *neurological consciousness*, but not limiting for what we might call *continuum* or *connective tissue consciousness*.

By using fast soliton transfer, which can approach or exceed the speed of sound, the *continuum pathway* would lead to motor responses before perception by the normal neuronal cognitive processes. Estimates vary for the velocity of propagation of both nerve impulses and solitons, but it appears that a signal carried by a soliton could propagate from 10 to 20 times faster than a nerve impulse, not counting synaptic delays. Hence a primordial *continuum consciousness* would be both fully cognitive of the environment in totality and more rapidly responsive to it.

REFERENCES

Ho, M.W. 1997, 'Quantum coherence and conscious experience', *Kybernetes,* vol. 26, pp. 265-276.
Winfree, A.T. 1984, 'Wavefront geometry in excitable media', *Physica,* vol. 12D, pp. 321-332.

25 Quantum coherence in the living matrix

What lies behind us and what lies before us are tiny matters compared to what lies within us.   Ralph Waldo Emerson

This is a concluding chapter in which the pieces of the puzzle presented in the Preface will be connected with quantum theory as best we can at the present time. We also will attempt to look ahead to some implications. In an Afterword we will look beyond the topics we have covered in the book to examine some broader implications for other areas of study.

AN EXCITABLE MEDIUM

Recall that the living matrix is an excitable medium (Chapters 7 and 9) and that Winfree (1984) began to teach us of a temporal anatomy (Chapter 9). Every organ, tissue, cell, molecule, atom, and even "empty" space has a rhythm of its own. The living matrix is a medium that can propagate rhythmic signals of many kinds, small or large, fast or slow, to every portion of the body. In this manner, every part can inform every other part of its activities in a variety of time frames. The temporal anatomy includes rhythms with frequencies measured in decades, years, hours, minutes, seconds, and the smallest measurable fractions of a second. Some communications may be at the speed of light; some may be instantaneous.

In times of emergency or peak performance or in a therapeutic encounter, this entire system can be called upon to do special things. The concept of a systemic living matrix provides us with a language to begin to explore life in new ways and to look for new possibilities. This language involves concepts of cooperation and synergy and the emergence of phenomena that we might not expect from what we have learned by studying what the parts of the system can do. It is a language of whole systems. The emergent phenomena are vital, and they can be breath taking and life saving. Their emergence can help us live happier and more successful lives.

Biological signals can originate anywhere within the organism or outside of it. Because the living matrix is excitable and can store energy, it can be poised to generate a large signal from a tiny input. Signals can propagate in many different forms, some of which have been touched upon in earlier chapters. One form, called the *soliton,* can propagate indefinitely without losing energy. In any case, the undistorted and unfettered propagation of energy and information is a sign of health and is essential for extending human performance beyond everyday limits.

Certain kinds of solitons can collide and emerge from the collision unchanged in shape and speed. Others will annihilate each other under the same conditions. This annihilation does not make the energy of the colliding waves disappear, because the law of conservation of energy only permits energy to be converted from one form to another. The other form of energy produced by colliding waves may be a scalar wave (Oschman 2000, pp. 203-208).

The entire living matrix, including the cytoskeletons of all cells and the fibers in the connective tissues, forms a *continuous excitable* medium. When we think of excitable media, we usually think of the neuronal membrane; however, as Hameroff pointed out (Chapter 15), this is a "skin-deep portrayal" that leaves out what takes place *within* neurons. The same is true for the rest of the cells and tissues in the body. In fact, the distinctions of different levels and different systems within the body are intellectual divisions we have created for our own convenience. The sentient being has no real experience of such systems or levels; it is a continuous whole.

Excitable media oscillate between different states. The nerve membrane has a *resting state* with a large membrane potential. A stimulus electrically depolarizes the membrane, resulting in an action potential that propagates nondecrementally along the cell surface as a wave. In the wake of the wave, the membrane potential is returned to the resting level in a fraction of a second.

The living matrix also is excitable and capable of propagating waves of energy and information, and it oscillates between two states. Our understanding of this oscillation is in an early stage. Were it not for the work of a few brilliant quantum theorists and biophysicists, we would know nothing about this subject. Key studies have been performed by groups in Italy, England, and Germany. What is emerging is a description of phenomena that are unique to life.

Although there some counterparts in the nonliving world, life has developed some unique properties that we now can talk about with more precision than in the past. Water is intimately involved (Ho & Knight 1998). Because every property of water is anomalous and because water is the most mysterious substance we know of, the mystery of life and the mystery of water converge into a single mystery.

> *If there is magic on this planet, it is contained in water . . . its substance reaches everywhere; it touches the past and prepares the future.* LOREN EISELEY

Quantum coherence

Now we look at the resting and excited states involved in the propagation of energy within the living matrix. The discussion will challenge the physicist because the microscopic temporal anatomy we are dealing with in living matter is more intricate than the situations physicists normally look at. This complexity should not deter us, however, because of the fundamental and vital lessons we will learn if we can decipher what is taking place as energy migrates through the living matrix. Fortunately some major authorities in the field of quantum electrodynamics have focused on this important problem, and their conclusions are profound.

> *Quantum electrodynamics is a cornerstone of modern chemistry and condensed-matter physics and thus is the ultimate foundation for current theories of almost all phenomena perceived by the senses, as well as many biological processes and perhaps even consciousness itself.* PREPARATA (1995)

Application of quantum field theory and quantum electrodynamics to living systems leads to important principles of dynamical ordering. Genes give rise to parts such as proteins, but larger structures such as organelles, cells, tissues, and organs are thought to arise spontaneously by self-assembly of the constituents. These concepts mask a variety of fundamental questions on the origin of structures. Important collective properties, which are the essence of life, are not explained. The quantum theory discussed by Preparata, del Guidice, and others explains how microscopic order gives rise to macroscopic order.

Consider the Davydov soliton, an electrosoliton (Eremko & Brizhik 2000), or any other form of energy that can propagate along a protein molecule such as a collagen triple helix (Figure 25-1, *A*). Many of the molecules in the body are arrayed close together in lattices (Figure 25-1, *B*), and the path the energy follows more often than not is helical (Figure 25-1, *C*).

Water is intimately associated with all protein molecules. This "living" water has unique properties as a consequence of its association with the molecules in the living matrix. The study of the water aligned with proteins is a topic that could engage a team of scientists, using the most sophisticated equipment available, for a very long time before even a superficial knowledge of the subject would emerge. Such a study would be extremely worthwhile, however, because it would disclose many keys to understanding health and disease and human functioning.

Fortunately a series of studies using the latest theories in quantum electrodynamics has given us an indication of the nature of the protein-water interactions that come under the heading of "quantum coherence." The major researchers involved in this work include Ho, del Guidice, Doglia, Milani, Vitiello, Fröhlich, Preparata, Smith, Popp, and Warnke.

Water has a specific and dynamic structure by virtue of its relationship with proteins, ions, and the other molecules that make up the living matrix and the fluids within it.

We usually do not think that the water in the glass of water we drink could possibly have any structure to it, but it does. A glass of water at room temperature contains

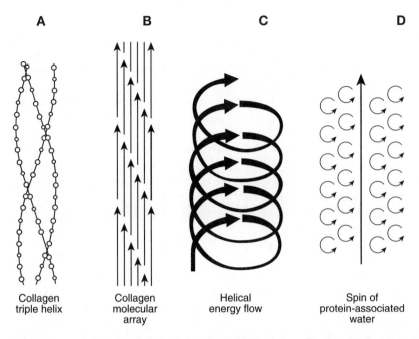

A	B	C	D
Collagen triple helix	Collagen molecular array	Helical energy flow	Spin of protein-associated water

Figure 25-1 **A,** Collagen triple helix. **B,** In connective tissue, the collagen molecules are arrayed close together in crystallike lattices. **C,** At a microscopic level, the energy follows a helical path through tissues. **D,** As energy spirals through the protein, the orientation of the surrounding water molecules will be affected. There will be a buildup of order in the spins of the water molecules. The water spins become organized, coherent, and aligned (see Figure 25-2). (**A** is modified from Figure 2, page 351, in *Biophysical Science—A Study Program*, edited by JL Oncley and others and published in 1959 by John Wiley & Sons, Inc., New York. Reproduced by kind permission of Alexander Rich.)

clusters of water molecules of various sizes and shapes. In the absence of impurities, these icelike water structures are ephemeral. They flicker into existence and last but a fraction of a second before thermal agitation causes them to disperse, only to be replaced by other temporary structures. Add a bit of protein, some other molecule, or salt to the water, however, and water structures become stabilized around the dissolved substance.

Other molecules, called *polyelectrolytes,* can be very long and tangled so that a single molecule can organize a vast amount of water to form a gel. Gelatin is one such material. Connective tissue has a gel phase composed of polyelectrolytes. Hence the "empty" spaces one sees in drawings and in electron micrographs, the apparent gaps between collagen molecules and between collagen fibrils, are composed of a polyelectrolyte gel that structures or organizes large volumes of water into intricate sheets or layers.

Proteins in the body and the water around them are essentially inseparable. The water has a structure that mirrors the protein structure, and the protein structure is dependent on the water to shape it and hold it together. If anything happens in or on the protein, things will happen to the surrounding water structure. If anything happens in the water system, it will alter the architecture of the protein.

We saw in Figure 14-2, *A*, that the water molecule is electrically polarized because the oxygen end is strongly electronegative, and the water molecule can oscillate or vibrate in various planes (see Figure 14-2, *B*). It also can spin or tumble in different planes (see Figure 14-2, *C*). Now we take a closer look at these phenomena as they are described in relation to quantum coherence.

Energy can flow along a protein. The flow can be described in terms of vibrations or waves. As a portion of a protein becomes energized or excited, the resulting field in the surrounding space will affect the orientation of nearby water molecules. There will be a buildup of order in the spins of the water molecules. The water spins become organized, coherent, and aligned (Figure 25-1, *D*). This order in water structure coincides with the propagation of energy along the protein. In living systems, as in oceanic tidal waves, the soliton is a coherent *spin wave* with interesting magnetic properties. When the wave of energy has passed, the organized water structure tends to decompose or randomize.

Figure 25-2, *A*, conceptualizes how energy spirals through a helical protein such as collagen or a helical nucleic acid such as deoxyribonucleic acid (DNA). Figure 25-2, *B*, shows the corresponding order that can arise in the spins of the water molecules adjacent to the helical protein. These illustrations must be recognized as preliminary efforts to conceptualize, in two dimensions, phenomena that are quite intricate and three-dimensional.

In the wake of the soliton is a collapse back to a less ordered or chaotic state. When the collapse takes place, a Fröhlich wave is emitted. The proteins forming the living matrix oscillate between a highly excited and strongly polarized state and a weakly polar disorganized state. Again, Figure 25-2, *C*, shows a preliminary effort to conceptualize this oscillation, recognizing the limits of a two-dimensional drawing.

Preparata (1995) refers to these two states as a *fully random perturbative ground state* (left side of Figure 25-2, *C*) and a *coherent ground state* (right side of Figure 25-2, *C*).

The oscillation between the two states was predicted earlier by del Guidice et al. (1985). My impression is that this oscillation process is a unique and characteristic property of living matter and that its further elucidation may become a cornerstone for a new quantum-based theory of life.

Figure 25-2, *C*, may seem a bit obscure to the reader, but it may be the most significant illustration in this book. It unites quantum theory with all of the pieces of the puzzle that were presented in the Preface, and it has many implications for future study. We can touch upon a few of these now.

What is emitted?

The coherent Fröhlich emission has two components, an electromagnetic field with a frequency at or near that of visible light, and a massless particle called a *Goldstone boson*. This is a disturbance in the quantum vacuum that can be propagated throughout space virtually instantaneously. According to del Guidice and his colleagues in Milan, what we see as structure is, in fact, a *consequence* of coherent focusing of polarized waves of energy. What we observe as ordered liquid crystal networks arise, in part,

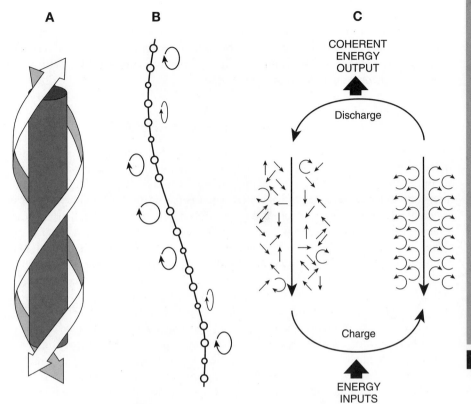

Figure 25-2 A, Three-dimensional conceptualization of the way energy spirals through a helical protein such as collagen or a helical nucleic acid such as deoxyribonucleic acid. **B,** Spin orientations that can arise in the water molecules adjacent to the helical protein. **C,** As waves of energy in the form of solitons or electrosolitons flow through tissues, the proteins forming the living matrix and their associated water molecules oscillate between a relatively disorganized and weakly polarized ground state *(left)* and a highly excited and strongly polarized state *(right).* The collapse of order gives rise to the emission of a Fröhlich wave *(top).* My impression is that this oscillation process is a unique and characteristic property of living matter. (**A** is a drawing by J.C. Collins and is used with his kind permission).

because of the alignment of *rotating* components such as water (del Guidice, Doglia & Milani 1982, 1983, 1988; del Guidice et al. 1985, 1986a, 1986b, 1988, 1989, 1991).

Fröhlich has described how light emitted by this process can be used to communicate between different cells and tissues, regulating cell division and a host of other vital processes. Coherent Fröhlich interactions are thought to regulate the orderly and efficient movements and actions of enzymes throughout the body (see the section on "The genetic code as language" in Chapter 14). Fröhlich developed a theory of cancer based on this concept (Fröhlich 1978). His work provides a basis for endogenous light playing a key role in a host of regulatory processes.

There are many advantages to the use of coherent light signals for biological communications and regulations. This has become a topic of research (Ho, Popp & Warnke

1994). Cyril Smith (1988), for example, discusses the ways coherence can protect electromagnetic communications within the body from interference from natural and artificial fields in the environment. These concepts are extremely important in view of the extent to which technological innovations are bathing us in electromagnetic radiations never before encountered by living systems.

STRUCTURE AND CONSCIOUSNESS ARE RELATED

Now we move on to one of the most significant aspects of our inquiry. The significance arises from several sources. One is the simple fact that virtually all schools of hands-on and energetic bodywork recognize the profound connection among human structure, consciousness, and emotional states, whereas few in the scientific community have any inkling that such a profound connection might exist. Second, there are the brilliant insights from physiology and biophysics articulated by Mae-Wan Ho and the earlier researchers she acknowledges in her various articles. Her book entitled *The Rainbow and the Worm* (Ho 1993) and her various essays (Ho 1996a, 1996b, 1997, 1999) eloquently describe the basis for a single energetic phenomenon simultaneously underlying living organization and conscious experience. The impasses in the research on structure formation and consciousness have arisen, she says, because we have failed to see that they are the same problem. For Ho, the systemically interconnected living matrix has provided a *launching pad* for the development of a unitary theory of living structural and functional organization, as well as conscious experience. Quantum coherence as described earlier provides a basis for this unitary theory. This book has added to this emerging paradigm some applications involving sensation and movement.

The *impasses* that have slowed scientific progress in the past are being resolved, paving the way for the emergence of a new biomedicine. The implications are staggering. We are witnessing a major step in the evolution of our species as we become cognizant of the organizing and energizing principles of life and of our cognition of our aliveness. This emerging understanding does not reject earlier accomplishments in science; it acknowledges and builds upon them.

Think about the best way to regulate living processes

The logic we have followed so far leads us to a creative possibility for the study of regulatory pathways. Imagine what would be the best way to regulate a biological process. Ask some specific questions about the process. How many subprocesses are involved to add up to a successful whole? Which of these contributing processes should be done rapidly, and which should be done slowly? For those processes that should be done as fast as possible, how fast should they be accomplished? What are the optimal time courses for the slower processes?

Common sense suggests that some processes must take place extremely rapidly, whereas others must take place slowly. We had a good example of this in the Prologue, in the discussion of the regulation of circulation to the muscles. We found that it really was clever of nature to use nitric oxide, a highly diffusible but short-lived gas, to regulate the blood supply to muscles. The reason for this is that regulation involves two equally important aspects, turning a process on and turning it off. It would be a disaster to activate a process with no means of turning it off when it is no longer needed. Nature uses various tricks to avoid this potential difficulty.

An athletic perspective. Look at this from the perspective of an athlete. Consider two extremes. Some process vital to optimum performance must be regulated slowly to sustain the athlete throughout the time course of the event. Other processes must be regulated extremely rapidly to achieve the winning edge. The totality of a successful competitor's advantage may arise from appropriate rhythmic activity at a variety of relevant time scales.

As an example of a slow regulation, the structure of the musculoskeletal system is regulated according to the way the body is used. Practice a particular activity, and the muscles, bones, tendons, circulatory system, and nervous system gradually become structurally and functionally optimized for the kind of activity being practiced; but this is a slow regulation. A single practice session is not sufficient to permanently set the process in motion, nor would that be desirable. It is regularly repeated practice that leads to a body that is optimized for a particular activity, and taking a day off from practice should not cause the structure to immediately revert to the way it was before the training program was begun.

In essence, the body takes notice of what you are doing in a particular practice session and makes a prediction that you may engage in similar activities in the future. Your body is forming a *memory* of the movements you made by modifying your body structure accordingly. This memory is not stored in some part of the brain; it is recorded throughout the body in the body's architecture. The adaptive process has been described in a scholarly way by Young (1975).

On the opposite end of the scale are regulations that demand extremely rapid adjustments. During an athletic competition or an artistic performance such as a dance routine, adjustments often must be made in a tiny fraction of a second. The ability to make an adjustment a thousandth or a ten-thousandth of a second faster than a competitor can make all the difference. These are regulations that demand the fastest kinds of communication systems that living nature can provide. For these fast reactions to take place, the matrix through which signals travel must be open and continuous.

Ho (1996b) presents a scheme for the many-fold cycles of life coupled to energy flows (Figure 25-3). Each of these cycles conveys both energy and information from place to place within the organism. Energy and information can be delocalized or spread over all systems, and energy and information from all systems can be focused or concentrated into any single system.

Ch'i as an example. Acupuncture is one aspect of a larger system that includes the martial arts. The phenomena involved in the martial arts reveal aspects of biological communication and energy flow that are of great interest and of great medical importance.

It can be recalled from Chapter 5 that Albert Szent-Györgyi believed that life is too rapid and subtle to be explained by slow molecular and neural processes. He was looking for a faster means of communication and realized that electrons and protons would be likely carriers of energy and information. Exactly what was he talking about, and how can it possibly relate to clinical medicine or human performance?

The research mentioned earlier dramatically shows how the *impasses* that have slowed scientific progress in the past have been resolved, paving the way for the emergence of a new medicine. The implications are staggering. We are witnessing a major step in the evolution of our species as we become cognizant of the organizing and energizing principles of life and of our cognition of our aliveness.

PROSPECTS FOR THE FUTURE

Based on the information gathered here, it now would be timely to replicate the electromyographic studies done by Valerie Hunt some decades ago, focusing on the neurophysiological correlates of extraordinary states of awareness and movement.

It also would be extremely interesting to have an update on the 1993 symposium on the mechanism of energy transduction in biological systems. This symposium was held at the New York Academy of Sciences and was published in 1974 (see Chapter 13). To the extent possible, it would be desirable to assemble the participants from that earlier symposium and/or their students, as well as others who have engaged in bioenergetic research in the past quarter century. The conference would revolve around updating the same three basic questions that were discussed at the earlier symposium: (1) Does a crisis in bioenergetics persist? (2) If so, what is the nature of the proposed crisis? (3) How can the crisis be resolved? As before, it would be desirable to have a blend of experimentalists and theorists. The goal would be a statement of the current status of bioenergetics, a delineation of the current unresolved issues, and suggestions on the means to expand our knowledge on this topic. It would be desirable to have experts in the field evaluate the soliton and quantum coherence models of sensation and action.

The acupuncture model of regulation involves a set of channels or meridians that function in communication. The meridians comprise "an invisible network that links together all the fundamental substances and organs" of the body (Kaptchuk 1983). The channels are unseen, but they convey nourishment, strength, and communications that unify all the parts of the organism. Acupuncture theory aims to discern the interconnections among substances, organs, and meridians.

AN INTERVIEW WITH MAE-WAN HO

We conclude with a question-and-answer interview with one of the individuals who has thought deeply about the material presented in this book (Box 25-1). This is not an interview that actually took place; instead it is based on Mae-Wan Ho's writings, the questions they raise, and the ways she begins to answer them. There emerges from this "interview" an appreciation of the deeper implications of the material assembled in this book. The "interview" is presented here with her permission and agreement. It is based on several of her classic publications: "Quantum coherence and conscious experience" (Ho 1997), "The biology of free will" (Ho 1996b), "Bioenergetics and biocommunication" (Ho 1996a), and "Coherent energy, liquid crystallinity and acupuncture" (Ho 1999). The interested reader will find these articles invaluable and inspiring.

Box 25-1 "Interview" with Mae-Wan Ho

Question: Which part of the body is in control?

Mae-Wan: Nothing is in control, yet everything is in control. Each part is as much in control as it is sensitive and responsive. Choreographer and dancer are one and the same. Global and local, whole and part, are indistinguishable. The living matrix network is a *molecular democracy* of distributed control. More important than control is the source of the integration that gives rise to large-scale actions that are coordinated in a continuum from the macroscopic to the molecular.

Question: Does consciousness control matter or vice versa?

Mae-Wan: An organic sentient whole is an entangled whole. In the ideal, the self and the environment are domains of coherent activities. This is a pure state that permeates the whole of the being with no definite localizations or boundaries. The coherent self couples coherently to the environment so that one becomes as much in control of the environment as one is responsive.

Question: Where in the brain are consciousness and memory located?

Mae-Wan: Consciousness and memory will not be found at some definite location in the brain. Instead, both are distributed and delocalized throughout the system.

Question: Is there a regular pattern of neural activity related to consciousness?

Mae-Wan: In spite of decades of searching for repeatable patterns in brain activity, none have been found. Every perception is influenced by all that have gone before.

Question: Where do we find the evidence for this organicist perception of a unified theory of consciousness and systemic regulation?

Mae-Wan: The new *organicism* recognizes no boundaries between disciplines. It arises in the space between all disciplines. It is an unfragmented knowledge system by which one lives. It is a nondualist and holistic participatory knowledge system resembling those of traditional indigenous cultures all over the world.

Question: What about the laws of physics and genetics?

Mae-Wan: The laws of physics are not laid down once and for all. They especially do not dictate what we can or cannot think. The sentient coherent being is free of the laws of physics and is free to explore and create its possible futures. For example, living systems are not subject to the laws of thermodynamics. The genetic landscape has been found to be far more flexible and fluid than we thought 10 years ago. The epigenetic landscape has a fluid topography that includes multiple developmental pathways. The genetic paradigm was fatally undermined at least 10 years ago, when a plethora of "fluid genome" processes were first discovered, and many more have come to light since.

Box 25-1 "Interview" with Mae-Wan Ho—cont'd

Question: How does quantum coherence relate to bioenergetics?

Mae-Wan: Quantum coherence is far more dynamic than the biochemical models of energy flow. Quantum coherence explains how we can have energy at will, whenever and wherever it is needed. Quantum coherence recognizes a wide variety of possible methods or modes of energy flow and energy storage. Energy can be delocalized over all modes (lifting an object is a mode of energy utilization, as are thinking, running, and breathing) or concentrated into a single mode (such as a defensive, offensive, or protective movement). Energy is always available within the system and can be mobilized with maximum efficiency over all modes. Energy can be transferred directly and immediately to a place where it is needed, without heat loss or dissipation.

Question: Just exactly what is quantum coherence?

Mae-Wan: Organisms are made up of strongly dipolar molecules packed tightly together in regular, almost crystalline arrays. Large voltages are present. Electric and elastic forces cause the molecules in these arrays to vibrate. Because the structures are geometrically coherent, large collective modes or coherent excitations will develop. These are described as phonons or photons. When the coherence builds to a certain level, a large-scale Fröhlich wave is produced. In essence, the organism behaves as a single crystal. The anteroposterior axis is the optical axis for the whole organism. There is something very special about organic wholeness that is only describable in terms of quantum coherence.

Question: Is there a good metaphor for quantum coherence?

Mae-Wan: The laser is a good metaphor. Energy is pumped into a cavity containing atoms capable of emitting light. At low levels of pumping, the atoms emit randomly, as in an ordinary lamp. As the pumping energy is increased, a threshold is reached where all the atoms oscillate together in phase and send out a giant light track that is a million times stronger than that emitted by individual atoms. Energy pumping and dynamic order are intimately linked.

CONCLUSION

We conclude with a comparison of the conventional and the newly emerging schemes for the origin of living structure and consciousness (Figure 25-3). In the past it had been thought that the genes give rise to proteins that then spontaneously assemble into the living structures that carry out living processes, including consciousness. In the emerging quantum model, it is the action of quantum coherence that organizes the parts into living structures, and it is the action of quantum coherence that gives rise to consciousness as a distributed and emergent property of the assembled parts.

This would be yet another highly speculative theory to add to an already highly speculative field were it not for some remarkable studies documenting the presence of quantum coherence in living systems. In Chapter 9, brief mention was made of the work of Ross et al. (1997). These researchers developed an imaging technique:

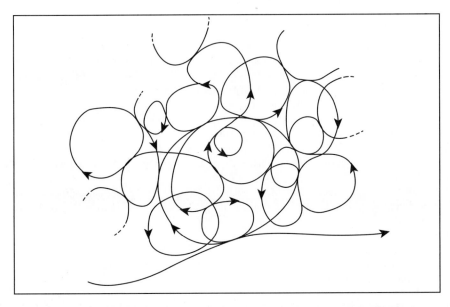

Figure 25-3 The many-fold cycles of life coupled to energy and information flows. Organisms can take advantage of a complete spectrum of coupled cycles, storing and mobilizing energy and information using many different kinds of efficient transfers. Energy and information input into any of the body's systems can be readily delocalized over all systems; conversely, energy and information from all systems can become concentrated into any single system. Energy coupling in living systems is symmetrical, which is why we can have energy at will, whenever and wherever required. (From Ho, M.-W. 1996, 'The biology of free will', *Journal of Consciousness Studies,* vol. 3, pp. 231-244. Used with permission of Mae-Wan Ho.)

This imaging technique enables us to literally see the whole organism at once, from its macroscopic activities down to the long-range order of the molecules that make up its tissues. The colors generated depend on the structure of the particular molecules—which differ for each tissue—and their degree of coherent order. For weakly birefringent material, the color intensity is approximately linearly related to both intrinsic birefringence and the order parameter. The principle is exactly the same as that used in detecting mineral crystals in geology; but with the important difference that the living liquid crystals are dynamic through and through. The molecules are all moving about busily transforming energy and material in the meantime, and yet they still appear crystalline. (Ross et al. 1997)

REFERENCES

del Guidice, E.S., Doglia, S. & Milani, M. 1982, 'Self-focusing of Fröhlich waves and cytoskeleton dynamics', *Physics Letters,* vol. 90A, pp. 104-259.

del Guidice, E.S., Doglia, S. & Milani, M. 1983, 'Order and structures in living systems', in *Nonlinear Electrodynamics in Biological Systems,* eds W.R. Adey & A.F. Lawrence, Plenum Press, New York, pp. 477-488.

del Guidice, E.S., Doglia, S. & Milani, M. 1988, 'Spontaneous symmetry breaking and electromagnetic interactions in biological systems', *Physica Scripta,* vol. 38, pp. 505-507.

del Guidice, E.S., Doglia, S., Milani, M. & Vitiello, G. 1985, 'A quantum field theoretical approach to the collective behavior of biological systems', *Nuclear Physics*, vol. B251, pp. 375-400.

del Guidice, E.S., Doglia, S., Milani, M. & Vitiello, G. 1986a, 'Electromagnetic field and spontaneous symmetry breaking in biological matter', *Nuclear Physics*, vol. B275, pp. 185-199.

del Guidice, E.S., Doglia, S., Milani, M. & Vitiello, G. 1986b, 'Solitons and Coherent Electric Waves in a Quantum Field Theoretical Approach', in *Modern Bioelectrochemistry*, eds F. Gutman & H. Keyzer, Plenum, New York.

del Guidice, E.S., Doglia, S., Milani, M. & Vitiello, G. 1988, 'Coherence of electromagnetic radiation in biological systems', *Cell Biophysics*, vol. 13, pp. 221-224.

del Guidice, E.S., Doglia, S., Milani, M., Vitiello, G., Smith, J.M. & Vitiello, G. 1989, 'Magnetic flux quantization and Josephson behavior in living systems', *Physica Scripta*, vol. 40, pp. 786-791.

del Guidice, E.S., Doglia, S., Milani, M. & Vitiello, G. 1991, 'Dynamic mechanism for cytoskeleton structures', in *Interfacial Phenomena in Biological Systems*, ed M. Bender, Marcel Dekker, New York.

Eiseley, L. 1957, *The Immense Journey*, Random House, New York.

Eremko, A.A. & Brizhik, L.S. 2000, 'Regulation of metabolic charge transport via self-induced and external microwave radiation', *Proceedings of the Fifth International Conference on Quantum Medicine*, eds I.P. Chervonyi & V.V. Kireev, Donetsk, Ukraine, 26-30 September, pp. 55-58.

Fröhlich, H. 1978, 'Coherent electric vibrations in biological systems and the cancer problem', *IEEE Transactions on Microwave Theory and Techniques*, vol. MTT-26, pp. 613-617.

Ho, M.-W. 1993, *The Rainbow and the Worm: The Physics of Organisms*, World Scientific, Singapore.

Ho, M.-W. 1996a, 'Bioenergetics and biocommunication', in *Computation in Cellular and Molecular Biological Systems*, eds R. Cuthbertson, M. Holcombe & R. Paton, World Scientific, Singapore, pp. 251-264.

Ho, M.-W. 1996b, 'The biology of free will', *Journal of Consciousness Studies*, vol. 3, pp. 231-244.

Ho, M.-W.1997, 'Quantum coherence and conscious experience', *Kybernetes*, vol. 26, pp. 265-276.

Ho, M.-W. 1999, Coherent energy, liquid crystallinity and acupuncture, presented to the British Acupuncture Society, 2 October 1999, Available: http://www.i-sis.org/acupunc.shtml.

Ho, M.-W. & Knight, D. 1998, 'The acupuncture system and the liquid crystalline collagen fibers of the connective tissues', *American Journal of Chinese Medicine*, vol. 26, pp. 251-263.

Ho, M.-W., Popp, F.A. & Warnke, U. 1994, *Bioelectrodynamics and Biocommunication*, World Scientific, Singapore.

Kaptchuk, T.J. 1983, *The Web That Has No Weaver*, Congdon and Weed, New York.

Oschman, J.L. 2000, *Energy Medicine: The Scientific Basis*, Churchill Livingstone/Harcourt Brace, Edinburgh.

Preparata, G. 1995, *QED Coherence in Matter*, World Scientific Publishing Company, Singapore. For a review of this book, see Flower, R.G. 1996, *Frontier Perspectives*, vol. 6, pp. 33-34.

Ross, S., Newton, R.H., Zhou, Y.M., Hafegee, J., Ho, M.W., Bolton, J. & Knight, D. 1997, 'Quantitative image analysis of birefringent biological materials', *Journal of Microscopy*, vol. 187, pp. 62-67.

Smith, C.W. 1988, 'Electromagnetic effects in humans', in *Biological Coherence and Response to External Stimuli*, ed. H. Fröhlich, Springer Verlag, Berlin, pp. 205-232.

Winfree, A.T. 1984, 'Wavefront geometry in excitable media', *Physica*, vol. 12D, pp. 321-332.

Young, J.Z. 1975, *The Life of Mammals: Their Anatomy and Physiology*, 2nd ed, Clarendon Press, Oxford.

Afterword

"I have now in my hands all the threads which have formed such a tangle. There are, of course, details to be filled in, but I am certain of all the main facts."

Sherlock Holmes, in *A Study in Scarlet*

A DETECTIVE STORY

We began with a plot for a detective novel: a set of strange, bewildering, and disconnected clues. Always confident, Sherlock is convinced that these peculiar bits can be fitted together and the mystery solved. Now, after 25 chapters of detail, we begin to fit the pieces together.

A major thread came from one of the most insightful and creative scientists of the century, Albert Szent-Györgyi. He noticed something fundamental: living systems respond to their environments with such speed and subtlety that they must have a previously unsuspected high-speed communication system. Energy and information seem to be conducted and semiconducted extremely rapidly throughout the organism. The process is much faster than nerve conduction or diffusion of chemical signals. Small and highly mobile units, such as electrons and protons, must be carrying energy and information from place to place. Szent-Györgyi shifts the focus of his research program to the subatomic realm, where he expects to find an explanation for cancer and other degenerative diseases.

A new field of *electronic biology* begins to emerge. Materials scientists and electronic engineers readily grasp its significance. They begin a research effort that gradually builds into an international enterprise, with generous funding from the American public through the National Foundation for Cancer Research. For mainstream biologists and biomedical researchers, however, the whole topic is "premature." This is the term used in a much-discussed paper by molecular biologist Gunther Stent published in *Scientific American* in 1972. Stent defined the obstacle to scientific discovery that occurs when a claim or hypothesis is "premature" because its implications cannot be connected to canonical or mainstream knowledge by a simple series of logical steps. To this day, it is difficult for many biologists and biomedical researchers to connect the submolecular realm with generally accepted knowledge in their fields.

While focusing intently on the very small, Prof reminds us that *"all levels of organization are equally important and we have to know something about all of them if we want to approach life"* (Szent-Györgyi 1974). His studies show that water is intimately involved in the communication process.

The importance of water is confirmed (Corongiu & Clementi 1981). A web of water molecules, held together by hydrogen bonds, binds the components of deoxyribonucleic acid (DNA) and proteins together. Without water, the molecules of the living body literally would fly apart.

A NOBEL EMBARRASSMENT

The compartmentalization of science and other intellectual endeavors into separate disciplines has prevented those who study high-speed organic processes from having a conversation with each other, even though their researches are all pointing in the same direction. This was illustrated dramatically by the 2000 Nobel Prize award in Chemistry (Heeger, MacDiarmid & Shirakawa 2000). The Nobel Committee recognized three scientists *"for the discovery and development of conductive polymers,"* or "plastic that conducts electricity."

The embarrassing problem, apparently unknown to the scientists involved and to the Nobel Committee, was that the phenomenon of electronic conduction in polymer films had already been discovered many years earlier by several other groups (Eley et al. 1953; McGinness, Corry & Proctor 1974). In addition, credit for a much earlier "discovery" of the same phenomenon must be given to natural evolution, which incorporated conductive polymers in a network that extends throughout the living body.

The Nobel Committee's lack of acknowledgement of the earlier work gave rise to discussions of "the disregard syndrome" (Ginsburg 2001) and "bibliographic negligence" (Garfield 1991). However, this unfortunate mistake probably was not due to dishonesty or negligence but rather to the way we have separated and insulated academic disciplines from cross-fertilization. In any case, it helps explain why this extremely interesting and important discovery (organic semiconduction) has not become more widely known and discussed. It helps explain why the *electronic biology* developed by Albert Szent-Györgyi is seldom mentioned by biologists or biomedical researchers. Current research on biological semiconduction has huge therapeutic implications that are leading to dramatic advances in biomedicine.

A CONTINUUM

A British scientist notices that a major protein spans the membranes of the red blood cell, linking the cellular and extracellular environments (Bretscher 1971). The scientific community initially is skeptical (as is usually the case with great discoveries), but it is soon found that such links between cells and their surroundings are not only ubiquitous, but they are *vital* to life and health. The *integrins* and related molecules prove to be key links in a system-wide communication network. They serve as energetic and informational pathways that convey external stimuli into the cell and nucleus, and in the opposite direction. Szent-Györgyi's concept of a system-wide communication network has acquired a tangible form.

The connective tissue system in the body, which consists of fascia, ligaments, tendons, bones, and cartilage, has long been recognized as an all-pervasive connecting system that determines the overall form of the body, as well as the architecture of each system and organ. The microscopic design of individual cells and tissues, in turn, is largely determined by cytoskeletal architecture. The discovery that the connective tissue and cytoskeletal and nuclear matrices are interconnected has led to a *continuum* concept for the organization of living tissues. The entire molecular network, which extends into every cell and every nucleus everywhere in the body, is termed *a tissue-tensegrity matrix system* (Pienta & Coffey 1991) or simply *the living matrix* (Oschman & Oschman 1993). A noted acupuncturist has a poetic name, "the web that has no weaver" (Kaptchuk 1983).

Historic discussions between two Austrian PhD students, Ludwig von Bertalanffy and Paul A. Weiss, pave the way for a whole systems theory and its application to biology (Weiss 1973). The systemic interconnectedness of the living matrix provides a substantive basis for a wide range of therapeutic techniques that have been given a rather nebulous classification, *wholistic*. Complexity theory and nonlinear dynamics now have a tangible system-wide living substrate to explore.

We begin to see that the operation of the whole network, that life itself, depends on the integrated activities of all of the components. The living matrix is a cooperative or collective or synergistic system, with both linear and nonlinear aspects. Living structure and consciousness are seen as emergent properties of the whole matrix, properties that are not observable or understandable from studying the parts. Many of the seemingly miraculous observations in hands-on bodywork, movement therapies, and transcendent athletic and artistic performances arise because of the capacity of the entire system to undergo dramatic nonlinear shifts or "phase changes."

THE LIVING MATRIX VERSUS THE NERVOUS SYSTEM

Figure A–1 summarizes the situation. When we think of communications taking place throughout the living body, we first think of the nervous system. This is a serious bias that we need to overcome. The nervous system is but one part of a much more pervasive system, the living matrix continuum, that reaches all parts of the organism, without exception, including the innermost parts of every cell in the body and the cells making up the nervous system. The nervous system does not extend into every part of the living matrix, but the living matrix reaches into every part of the nervous system.

Most biologists know about the discovery of the integrins, but they are not nearly as excited as are complementary therapists or cutting-edge performers. Those who work daily with the human body as a dynamic and interconnected energetic system have long had an intuitive appreciation of the continuity and energetics of the living matrix system.

Our bias that the nervous system is the most important communication system in the body emerges simply because we use our nervous system when we think, or at least we think we do. We have this bias because it is the nervous system that gives rise to our immediate conscious experience. It is only in rare moments of crisis, transcendence,

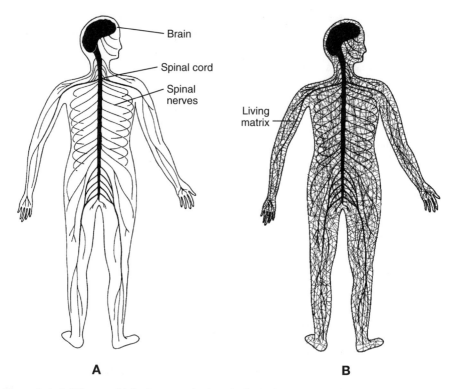

Figure A–1 **A,** When we think of communications in the body we think of the nervous system. **B,** There is an even more pervasive system, the living matrix, which extends into every part of the body, including the cells comprising the nervous system. Nerves do not reach into every part of the living matrix, but the living matrix extends into every part of the nervous system, as well as the other great systems of the body, such as the circulatory system, digestive tract, lymphatic system, and integument.(Drawing of the nervous system is Figure 27.1, page 137 from Mackean D.G. 1973 *Introduction to Biology*, John Murray Ltd, London, used with kind permission of DG Mackean.)

or exceptional performance that we get glimpses of other systems in operation. This book is an introduction to those other systems.

Neurological consciousness

The nervous system consists of a vast number of neurons and neuronal connections. For a long time it was accepted that conscious experience arises from decisions made within the brain by integrating sensory information with the memories stored within the hundreds of billions of nerve cells in the cortex and their myriad synaptic terminals. As we saw in Chapter 15, the whole system came to be portrayed as a computer-like switching circuit, leading to a view of brain = mind = computer. The neuron doctrine states that neuroscience alone ultimately will provide the entire framework for understanding the mind, both conscious and unconscious.

There is no question that the nervous system functions in sensation, perception, thought, and movement. However, the repair of injuries, either physical or emotional, probably has little to do with nerve functioning. Pain and altered gait (to immobilize the injured area during the repair process) are neurological phenomena that assist in the healing process, but the actual repair of tissues takes place in the domain of the connective tissues and the cells within it. There are many vital processes that seem to have little to do with nerves.

Generations of neurophysiologists have made impressive contributions to our understanding of the brain, but we still are at a loss with regard to some fundamental questions about the origins of what Sir John Eccles referred to as "non-physical and transcendent properties," such as feelings, thoughts, memories, intentions, and emotions. As he approached the end of his distinguished career, Eccles concluded, in 1993, that the methods and theories he had pioneered, and for which he had received the Nobel Prize, were inadequate for the task he had set for himself. It would be necessary to take this investigation to a smaller level of scale, to the ultramicroscopic aspects of the synapse, and its quantum properties.

More threads

Another distinguished scientist, Herbert Fröhlich, finds that the whole living network is continuously vibrating at light frequencies (Fröhlich 1988). This occurs because of the huge electrical potentials and the high degree of molecular order (crystallinity) within tissues. The living matrix is simultaneously a mechanical, vibrational, energetic, photonic, and informational network. The entire composite of physiological and regulatory processes we refer to as *the living state* takes place within the context of a continuously interconnected living matrix.

Perhaps the degenerative diseases do not really have "causes" in the usual sense; instead, they may be the consequences of a loss of "systemic cooperation." One of the accomplishments of complementary medicine is to slow or reverse the accumulation of subtle disorders or imbalances or communication breakdowns that compromise our immune defenses and repair systems. Structural and movement approaches stimulate the body's repair systems to repair themselves, maintaining or restoring "systemic cooperation" with its many beneficial consequences for health and performance.

Many hands-on, energetic, and movement therapists, including acupuncturists, work with a system-wide information/energy system corresponding to the acupuncture meridians and their fine branches. A group in Irvine, California, demonstrated that needling acupuncture points in the foot, used to treat vision problems, rapidly activate neurons in the visual cortex in the brain (Cho et al. 1998). Independently, M.E. Hyland at the University of Plymouth, United Kingdom, used extended network and macro-quantum entanglement theories to predict the existence of a hyperfast communication system within the body that accounts for some of the effects of various complementary and alternative therapies (Hyland 2003).

Clues that living matter possesses a primordial high-speed network capable of sensing and moving come from a number of areas in athletics, performing and martial arts, and physiology (Hunt, Conrad, Murphy; see Chapters 6 and 7). Several scientists independently suggest that there is a reversible linkage between the sensation and movement systems in cells and tissues. In other words, there is a basis for a direct, non-neural pathway from sensors to muscles (Atema, Lowenstein and colleagues, Gray; see Chapter 16).

A consequence is that sensation may not be what it seems to be, based on neuro-physiological measurements or conscious experience. Survival in predator-prey situations has forced living systems to evolve sensors with extraordinary sensitivity. Typically one records the action potentials established when a sensory cell responds to a stimulus. By reducing the strength of the stimulus until there is no response, one can obtain a "threshold" value for the sensor. Stimuli weaker than the threshold do not result in a nerve impulse. It is possible, however, that a "subthreshold" stimulus could nonetheless trigger a conformational wave that propagates throughout the entire living matrix. In this way, the entire body is "alerted" to subtle aspects of the sensory environment, even though the nervous system is not.

Synchrony in conversation

In Boston, W.S. Condon describes the simultaneous "dance" of conversation in which speakers and listeners share synchronous micromovements. This phenomenon, now called *synchrony,* begins within 20 minutes of birth. The neonate's movements become entrained by and synchronized with the speech of nearby adults. By the time the child begins to talk, he or she has already internalized the form and structure of the language and body language system of the culture (Condon & Sander 1974). Sensation and movement in the living matrix are an intimate part of the development of verbal and emotional behavior.

During the same period (early 1970s), while Condon is studying millions of feet of film of people in conversation, biochemists around the world are confronted with a mystery: how energy gets from place to place in living systems (see Chapter 13 and Appendix C). The Ukrainian biophysicist A.S. Davydov develops an answer that bears his name, the *Davydov soliton,* a coherent matter wave that is capable of transferring large amounts of energy and information from place to place without dissipation or loss (see Chapter 19). His work is applied in the development of fiberoptic cables carrying the huge volume of Internet traffic between large cities.

Davydov also suggests that solitons entering the protein lattice in a muscle can directly initiate contraction (see Chapter 20). His non-neural model of muscle activation draws little attention.

From time to time, just about everyone notices special moments when "time seems to slow down" or when extraordinary healings or memorable performances take place. Jazz musicians, martial arts masters, and performers share an interest in a state of consciousness characterized by transcendent clarity and coordination. Therapists look for a "zero point" in which extraordinary healing takes place. Certain spiritual and contemplative practices reveal states of "pristine" or "atemporal" awareness.

TRAUMA

For those who have been injured, either physically or emotionally, the resolution of trauma has deep significance. This is as true for individuals as it is for communities and nations. For as travel and communication bring us closer together on our little globe, the resolution of historic ethnic and national animosities becomes increasingly more vital to our collective future.

The insights of therapists working at the cutting edge of trauma therapy have been invaluable in shaping the content of this book. The living matrix concept sheds light on methods such as those being developed by Levine (1997), Porges (1995), Scott (1997), Rosenberg (2003), Redpath (1995), Callahan and Callahan (1996), Fleming (1996), Chitty (2001), Gallo (1999), and many others.

The living matrix is obviously the "terrain" in which injuries and traumas take place. Not only is this system a high-speed communication network, it also has enormous capacity to store information. This "information" takes the form of memories and the literal energy blockages resulting from tissue damage. The ways stored information and energy blockages affect our moment-to-moment experience is coming into focus and has practical significance for the therapist or performer who must deal with the effects of physical or emotional trauma and injury on a daily basis.

The biologist views the living body as a remarkably designed system of systems, the culmination of millions of years of evolutionary experimentation with the laws of nature. Given the multitude of possible situations in which the organism may find itself, what is the ideal design and strategy for survival? How is this strategy reflected in the structure and rhythmic functions of every organ and tissue? When survival comes down to making a life-or-death choice in a fraction of a second, what are the most likely and appropriate decisions and compromises? Precisely how do traumatic experiences affect the future potential of the organism? Can the internalized memories of these experiences be resolved in a way that leaves the organism free to continue on its path toward its own perfection and authenticity? This is one of the most profound goals of all therapies.

JARGON CAN OBSCURE

Scientific jargon can obscure important phenomena. Event-related desynchronization (ERD) is an example. This is the technical name for one of the most fascinating discoveries in neurophysiology. Few people, including scientists, even know about it. The subject has been discussed and debated by neuroscientists since 1949, and it has great interest for the theme of this book.

Desynchronization

In 1949 Canadian neurophysiologists reported that the normal sensorimotor brain rhythms (such as the well-known alpha rhythm) recorded with corticograms (from electrodes touching the brain surface) become desynchronized or blocked during and even *before* movements (Jasper & Penfield 1949).

These studies were performed over a period of 10 years. Electrodes were placed on the brain surface during operations on the cerebral cortex. The subjects were consenting conscious patients with focal epilepsy. When a patient was asked to clench his or her fist, there was a blocking of some of the brain wave rhythms in the cortical region controlling hand movements. In some cases the blocking was recorded *before* the initiation of the movement in response to the command, "get ready to move your fingers" (Jasper & Penfield 1949). Figure A–2 shows the original recordings and a more recent demonstration of the phenomenon from Durka et al. (2000).

Readiness potentials and fields

Years after the study by Jasper and Penfield, German scientists made a remarkable discovery. Electrical and magnetic activity begins as much as 3 to 4 seconds prior to a movement. Figure A–3 shows an electrical recording from electrodes on the scalp (Kornhuber and Deecke 1964) along with the corresponding magnetic field recorded with a magnetometer in the space around the head (Deecke, Weinberg & Brickett 1982; Hari et al. 1983).

The electrical field presaging a movement originally was called the *bereitschaftspotential,* in English the *readiness potential.* In essence, it is a shift in the brain's electrical activity associated with preparing for a movement. These electrical events take place in a part of the brain known as the *secondary motor area of the cerebral cortex* (Deecke, Grözinger & Kornhuber 1976).

What does this mean? Stated simply, we "think" that when we "think" of moving our body, it moves. I ask my finger to move, and it moves. There is no perceptible delay between the will to move and the experience of the movement.

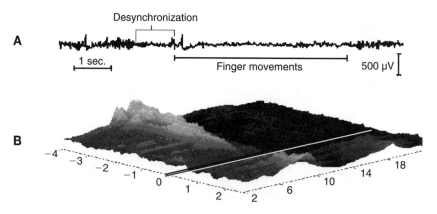

Figure A-2 A, First recording showing event-related desynchronization in brain waves. The regular electrical brain rhythms such as alpha, are reduced in intensity prior to a movement. (Based on Jasper H.H. & Penfield W. 1949 'Electrocortiograms in man: effect of the voluntary movement upon the electrical activity of the precentral gyrus', Archiv fur Psychiatrie und Nervenkrankheiten, vereinigt mit Zeitschrift für de gesampte Neurologie und Psychiatrie, vol. 83, pp. 163-174. Used with permission from Springer-Verlag GmbH & Co. KG, Heidelberg.) **B,** A modern recording of event-related desynchronization, showing how two of the primary rhythms are depressed prior to a movement. A white line has been drawn across the image to show the point at which the movement takes place. (From Durka PJ, Ircyha D, Neuper C, & Pfurtscheller, G 2000, 'Time-frequency microstructure of event-related desynchronization and synchronization', *Medical and Biological Engineering and Computing,* vol. 39, pp. 315-321. Copyright IEE.)

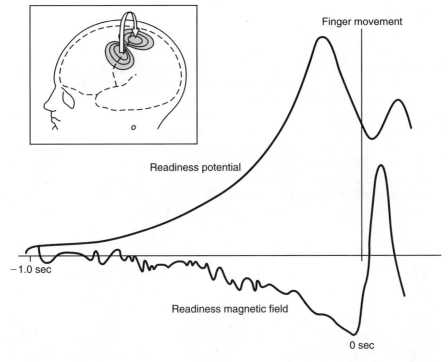

Figure A–3 Electrical and magnetic activity beginning prior to a movement. The upper trace is a recording from electrodes on the scalp showing the readiness potential (Bereitschaftspotential) prior to a finger movement. (From Kornhuber, H.H. & Deecke, L. 1964, 'Hirnpotentialänderungen beim Menschen vor und nach Willkürbewegungen, dargestellt mit Magnetbandspeicherung und Rückwärtsanalyse', *Pflügers Archiv fur Gesamte Physiologie,* vol. 261, p. 52. Used with permission from Springer-Verlag GmbH & Co.KG, Heidelberg.) The lower trace is the corresponding readiness magnetic field (Bereitschaftsmagnetfeld) recorded with a magnetometer in the space around the head. (From Deecke, L., Weinberg, H. & Brickett, P. 1982, 'Magnetic fields of the human brain accompanying voluntary movement: Bereitschaftsmagnetfeld', *Experimental Brain Research,* vol. 48, pp. 144–148. Used with permission from Springer-Verlag GmbH & Co.KG, Heidelberg.) The inset shows a neuromagnetic pulse in the space near the head. (From Okada, Y. 1983, 'Neurogenesis of evoked magnetic fields', in *Biomagnetism: An Interdisciplinary Approach,* eds S.J. Williamson, G.L. Romani, L. Kaufman & I. Modena, NATO Advanced Study Institute Series, Plenum Press, New York, 1983, p. 409, Figure 12.6.1.)

We know that the brain has to send a neural impulse to the muscle before it con-tracts. It obviously takes a finite length of time for the nerve impulse to travel along the motor nerve to the muscle; however, this is such a fast process that there seems to be essentially no delay between thought and action, or so we think. The research shows that a variety of neural events take place a significant time before a movement.

It was not too surprising to find neural activity preceding an act, as the brain must obviously prepare itself for any movement. It was, however, quite surprising that this activity begins a second or even more before the action. This gives rise to the fascinat-ing question, "When do we consciously decide to initiate an act such as a movement?"

A perceptual delay

The half-second delay applies to sensations and to the initiation of movements. Indications are that it may take as much as half a second to become aware of a sensation, but the brain plays a trick. The experience is assigned to an earlier point in time, that is, to the moment at which the stimulation actually occurred in the "real world."

Hameroff (2002) describes the situation as follows. Consider watching your feet while you are walking. You visually see your feet impacting the ground, and you sense the pressures on the bottoms of your feet. The two sensations seem simultaneous. However, the signals from the bottoms of your feet must travel through the long nerves and spinal cord to reach the brain. Because of conduction times and synaptic delays, these sensations should reach the brain significantly after the visual image is formed, although we do not experience this delay. We see and feel our feet touching the ground simultaneously, even though it may take a half-second or more for the sensory inputs to be processed. The subjective sensations are referred back in time to the moment of the actual physical impact. In this way the brain cleverly conceals from us the elaborate tactics it is using to create a coherent experience of the present moment. In other words, there appears to be a substantial delay between "reality" and our perception of reality. Stated differently:

> *Consciousness lags behind what we call reality. It takes half a second to become conscious of something, though this is not how we perceive it. Outside of our conscious awareness, an advanced illusion rearranges events in time.*   NØRRETRANDERS (1998)

Extensive research on this subject has been carried out by Benjamin Libet and his colleagues (see Libet 1985 and the critical discussion that follows). The results and conclusions are as fascinating as they are controversial. The advanced or sophisticated illusion Nørretranders refers to rearranges the timing of the flows of sensory information so that one has a smoothly synchronized experience of the present moment of consciousness, even though the various sensory systems are operating at different speeds.

Popper and Eccles (1985) stated, "This antedating procedure does not seem to be explicable by any neurophysiological process." Roger Penrose (1990) presented the logical view that consciousness cannot be pinpointed in time at all. He wrote, "I suggest that we may actually be going badly wrong when we apply the usual physical rules for time when we consider consciousness!"

Another view is the "multiple drafts model" of Dennett (1992). The idea is that there is no unequivocal flow of time in consciousness, but there are many different "drafts" present concurrently, and the draft of consciousness experienced is selected from them, while the others are discarded.

Microgenesis

Dennett's scheme has some similarity to a previous model called *microgenesis* developed by Jason Brown (1988) and further elaborated in a subsequent book (Brown 2000) (Figure A–4). In Brown's model, each conscious moment arises from a "bottom-up" unfoldment, a series of steps that retrace the evolution of the brain, the personality structure, and sensory inputs. Normally each pulse that leads to a moment of consciousness lasts about one tenth of a second, or the duration of a single brain wave. Each wave begins deep in the brain, in the thalamus, which serves as the clock or "pacemaker." From the thalamus the wave of electrical activity spreads upward, into the evolutionarily newer brain structures, ultimately reaching the surface of the cortex, where perceptions and actions come together to form the conscious moment.

The conscious moment is written on a "magic writing pad" on the surface of the cortex (Figure A–4). This image gradually fades, from the top down, as a new wave arrives to take its place. The continual replacement of the image occurs so smoothly that consciousness appears to flow seamlessly from moment to moment. At the same time as the conscious moment is perceived, the nearby cortical "motor keyboard" initiates actions through the musculoskeletal system. Perceptions of the moment and decisions to act are smoothly integrated at the top of the microgeny.

In Brown's model, sensations entering the microgeny at successive points do not provide the building blocks for construction of the image of the world. Instead, sensation constrains, sculpts, or selects the developing world from deep "pre-objects" formed from the personality structure. This includes the traumatic memories and attitudes developed from past experiences. Hence, *meaning* is actually determined *prior* to awareness. The "world out there" is what survives a transit through the microgenetic sequence.

TRAUMA TIME

These considerations have obvious and profound implications for the experience of trauma and for trauma therapy. In *Trauma Energetics,* Redpath (1995) noted that the trauma of an event is set in place in the fraction of a second *before* our self-awareness can notice it. Years later, energetic regulatory systems continue to scan this section of "held-energy" or traumatic memory roughly 10 times per second, with each brain wave. What Redpath calls *serious action* is impossible. Serious action is defined as movement and experience that are not referenced to, or motivated by, traumatic patterns, either within ourselves or within the culture around us.

Now we can build a hypothesis around these concepts. Because the energetic "signature" of a trauma is recorded prior to conscious awareness of the event, the signature resides outside of the thought and speech centers of the brain and, possibly, entirely outside of the nervous system (Figure A–1). Successful trauma resolution depends less on recalling the neurological "records" of an event than it does on finding the changes incorporated in the tissues the instant before the event was consciously experienced. In terms of bodywork, energetic, and movement therapies, we seek methods that can directly access these patterns.

Practical ways of interacting with the preconscious aspects of a trauma involve appropriate tapping into the structure/energy/movement system of the organism. Again, Nørretranders (1998) provides some excellent clues about how to do this. Given the nature of the situation, the following would seem to be good advice:

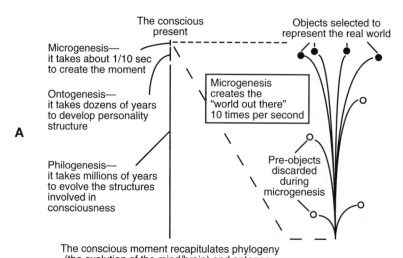

A

The conscious present

Microgenesis—
it takes about 1/10 sec
to create the moment

Ontogenesis—
it takes dozens of years
to develop personality
structure

Philogenesis—
it takes millions of years
to evolve the structures
involved in
consciousness

Objects selected to
represent the real world

Microgenesis
creates the
"world out there"
10 times per second

Pre-objects
discarded
during
microgenesis

The conscious moment recapitulates phylogeny
(the evolution of the mind/brain) and ontogeny
(personality development)

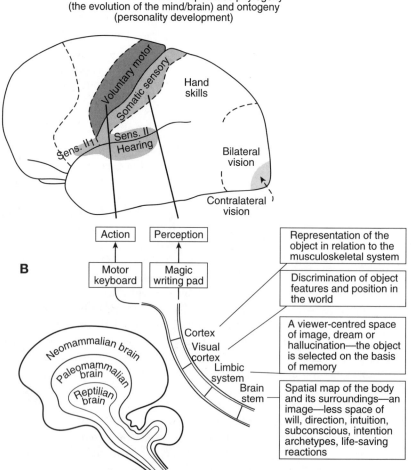

B

Voluntary motor

Somatic sensory

Hand
skills

Sens. III

Sens. II
Hearing

Bilateral
vision

Contralateral
vision

Action

Perception

Motor
keyboard

Magic
writing pad

Neomammalian brain

Paleomammalian
brain

Reptilian
brain

Cortex

Visual
cortex

Limbic
system

Brain
stem

Representation of the
object in relation to the
musculoskeletal system

Discrimination of object
features and position in
the world

A viewer-centred space
of image, dream or
hallucination—the object
is selected on the basis
of memory

Spatial map of the body
and its surroundings—an
image—less space of
will, direction, intuition,
subconscious, intention
archetypes, life-saving
reactions

Figure A–4 A, Formation of the conscious moment according to the microgenetic concept developed by Jason Brown (1988) and further elaborated in a subsequent book (Brown 2000). Each conscious moment arises from a "bottom-up" unfoldment, a series of steps that retrace the evolution of the brain, the development of the personality structure, and sensory inputs. Normally each pulse leading to a moment of consciousness lasts about one tenth of a second, or the duration of a single brain wave. In Brown's model, sensations do not provide building blocks for construction of the image of the world. Instead, sensation constrains, sculpts, or selects the developing world from deep "pre-objects" formed from the personality structure. Hence, meaning is actually determined prior to awareness. The "world out there" is what survives a transit through the microgenetic sequence. B, Each conscious moment begins with the formation of a wave of electrical activity deep within the brain, probably in the thalamus, which serves as the clock or "pacemaker." From the thalamus the wave of electrical activity spreads upward, into the evolutionarily newer brain structures, ultimately reaching the surface of the cortex, where perceptions and actions come together to form the conscious moment. The conscious moment is written on a "magic writing pad" on the surface of the somatic sensory cortex. This image gradually fades, from the top down, as a new wave arrives to take its place. As the conscious moment is perceived, the nearby cortical "motor keyboard" or somatic motor area initiates actions through the musculoskeletal system. Perceptions of the moment and decisions to act are smoothly integrated at the top of the microgeny. (Drawing of the brain modified from Guyton, A.C. 1971, *Textbook of Medical Physiology,* 4th edn, WB Saunders Company, Philadelphia, p. 717, Figure 61–4; scheme of microgenesis modified from Oschman, J.L. 2000, *Energy Medicine: The Scientific Basis,* Churchill Livingstone, Edinburgh, p. 115, Figure 8.2.)

Trust your hunches and intuitions—they are closer to reality than your perceived reality, as they are based on far more information.

Consciousness gives us a picture of the world and of ourselves acting in that world; however, both pictures are heavily edited. Nørretranders explains that the sensory picture is edited in a way that prevents consciousness from knowing that the entire organism has already been affected by the sensation. Similarly, the picture of one's actions is also distorted to create an illusion of a moment of conscious decision. Enormous amounts of information from the environment and the preparations for movement are edited out or removed from both pictures, and it takes time to do this.

Nørretranders noted that developments in a variety of fields construct a picture of a world far more complicated than consciousness ordinarily is able to deal with. There is far more information coming in from "out there" than we can possibly become aware of. What we are actually conscious of is winnowed down extraordinarily before it is registered as sensation.

Each second, our consciousness reveals to us a tiny fraction of the 11 million bits of information our senses pass on to our brains. Most of the information from our senses goes to our unconscious. Trust your hunches and intuitions—they are closer to reality than your perceived reality, as they are based on far more information.

NØRRETRANDERS (1991)

The unconscious

Precisely where in the organism will we find the "unconscious" to which Nørretranders refers? Where is the repository for the vast majority of our sensory input, amounting to some millions of bits per second? How is sensory information stored in this repository, and how do these "records" affect our moment-to-moment experiences and behavior? Is this "unconscious" neurological, or is there some other system in the body that "stores" traumatic memories? Answers to these questions are emerging.

> *The unconscious is the matrix out of which consciousness grows; for consciousness does not enter the world as a finished product, but is the end-result of small beginnings. . . . The conscious rises out of the unconscious like an island newly risen from the sea.*
>
> JUNG (1954)

The living matrix is a major part of the "sea" from which our conscious experience emerges. The "unconscious" has long been a nebulous term from the perspective of basic science because, like learning and memory, the heroes of the scientific era, the neuroscientists, have not been able to find them. With the living matrix we have an enormous and elaborate microcircuitry that has virtually unlimited capacity to store and process information while the nervous system is plodding along creating the narrow window on the world we call *conscious experience.*

I am suggesting that the "unconscious" probably has less to do with the nervous system than we have thought in the past. Neurological concepts of the unconscious, almost by definition, must be incomplete. A different hypothesis, worthy of exploration and testing, is that all of the tissues and cells in the body, including but not limited to neurons, are capable of "remembering." If this is correct, no amount of therapeutic inquiry or discussion at the level of neurological consciousness will access the bulk of the traumatic residue. Other approaches are needed, and the new "energy psychology" techniques listed earlier may actually operate by accessing or affecting this other system.

A vast information storage system

Reasons for looking for an information storage system in the living matrix were summarized in a 1961 paper written by the distinguished molecular biologist and neuroscientist Francis O. Schmitt (1903–1995) from the Massachusetts Institute of Technology. Schmitt appreciated that the problems of memory and learning are beyond electrophysiological measurement, however detailed. He understood "that much of the higher activity of the brain eludes detection by conventional electrophysiological methods . . . remembering, learning, and thinking may be sub-served by phenomena that are electrically silent to those instruments." Memory, he suggested, will be found in giant macromolecular polymers such as proteins, ribonucleic acid (RNA), and DNA. Importantly, Schmitt pointed out that there are other informational storage systems in the body, including the chromosomes (DNA) and the antigenic codes in the immune system. These represent vast reservoirs of memory of which we are never consciously aware.

Schmitt also pointed out that information coded in the sequence of components in a molecule may function at spectacularly low levels of energy. As an example, he mentioned the homing of fish to their streams where they were spawned due to the recognition of extremely low concentrations of molecules *remembered* as taste or smell. "Some kind of *amplifier action* other than that manifested whenever a stimulus elicits a bioelectric response in a network of neurons would seem to be required to explain the data." This was more than 30 years before Gilman (1997) received the Nobel Prize for his work on cellular regulations, including amplification.

Prophetically, Schmitt went on in his 1961 article to consider the transfer of electrons, protons, and light in the retrieval of long-term memory traces. These aspects of Schmitt's 1961 report have become part of the current discussions of the energetic mechanisms involved in memory and consciousness.

What Schmitt left out of his thoughtful paper was a far more pervasive informational system represented by the collagen of connective tissue. Collagen is the most abundant protein in nature and forms the bulk of the animal body. The number of collagen molecules in the body (counted in hundreds of trillions) far exceeds the number of neurons (hundreds of billions) and DNA molecules (also hundreds of billions). Moreover, each collagen molecule is a very long and thin polymer (about 215 times longer than it is thick) composed of about 3000 amino acid subunits (monomers). Information is not encoded in the amino acid sequence of collagen because the sequence is relatively invariant from molecule to molecule. Instead, the real potential for information storage is in the arrangement of the surrounding water molecules, in the myriad interactions with the ground substance matrix, and in the huge number of ionic and hydrogen bonds joining each collagen molecule with its neighbors within the collagen array. Taken together, the pervasive system of collagen molecules, ground substance, and water has the potential to form an incomprehensibly large repository of information about body experience.

Holography is a very sophisticated way to store information, and Karl Pribram saw immediately that the brain could exploit holographic principles (see Chapters 22 and 23). Perhaps physical structures responsible for memory are elusive for the same reason that the patterns on a holographic plate are unintelligible and bear no relation to the images they encode. Perhaps brain structures and patterns of nerve impulses and vibrations in the living matrix contain no first-order information (like a regular photograph) about memory and learning. Perhaps memory is to be found not in the patterns of neural activity or vibrations within the living matrix but in their Fourier transforms (Pribram 1977).

By focusing on the role of the nervous system in memory, neuroscientists have neglected a vast information storage system that forms virtually the bulk of the body. This is the connective tissue system, which is the very system in the body, the terrain, which is injured or traumatized.

So vast is the storage capacity of this system that it is my belief that it is capable of "remembering" every movement the body makes and every position held, whether in normal activities or during trauma, during the entire lifetime of the organism.

Cellular memory

In the meantime, others recognized that the elusive property we call *memory* is not confined to the nervous system, but that virtually all cells in the body can store information. For example, individuals with transplanted organs often acquire "memories" from donors of transplanted tissues. This phenomenon can best be explained by "cellular memory" (Pearsall 1998; Sylvia & Novak 1997). The mechanisms involved have been researched by a number of scientists. One cellular component that has been implicated in a variety of studies is the microtubule (Hameroff 1993), although there probably are others.

Excitable media

Neurons are capable of transmitting information because the neuronal membrane is excitable. The entire living matrix is also excitable. Waves of energy travel at high velocity from place to place through the entire matrix.

This idea has been expressed a number of times by researchers working in very different fields of study. The most dramatic examples are in the martial arts. An individual with a well-organized and "educated" connective tissue system is capable of sensing the environment, making choices, and moving faster than, and ahead of, neurological consciousness. When the martial arts master is attacked, he or she often becomes consciously aware of the attacker *after* the attacker has been thrown. The entire process of sensing and responding to the attack takes place *before* the event has reached the master's neurological awareness.

Multiple forms of consciousness

These findings give rise to the possibility that living organisms have two or more sorts of "consciousness." The oldest form, in terms of evolution, is the consciousness of living matter itself, as exemplified by the "lower" organisms from which we are descended: bacteria, protozoa, and sponges (see Chapters 15 and 16). These organisms lack anything resembling a nervous system, yet they are able to quickly sense and respond to their environments. Nerves are a more recent evolutionary development, but it is certain that the older system persists in all living matter and still is present in all animals with a nervous system.

A classic example is the white blood cell, which resembles an ameba in its form and function. Both the ameba and the white blood cell are sophisticated scavengers, moving about and looking for something to eat. White cells actually "patrol" particular areas in the body, referred to as *area codes*. The white cell is ready at a moment's notice to attack and destroy bacteria (Springer 1994). The built-in intelligence and decision making of these cells has nothing to do with neurology, because they lack anything resembling a nervous system.

The older form of consciousness, which we could term *connective tissue consciousness* or *matrix consciousness* (in single-celled organisms lacking a true connective tissue), continues to have vital functions in the higher organisms, but we usually are not aware of this system because our primary consciousness is neurological. We usually are not aware of our white blood cells fighting an infection or of skin cells migrating toward a cut that has to be repaired.

Neurological consciousness enables us to contemplate and philosophize about nature, the meaning of our place within it, and the meaning of our personal history. Matrix consciousness has different functions. It enables us to store and process the massive amounts of information our senses take in each second, only a small portion of which reaches consciousness. It is the "operating system" of the body, silently working in the background, controlling the myriad of physiological events taking place each instant of our lives. It is capable of performing life-saving reactions faster than the reactions mediated by the nervous system. It is in this matrix that our traumatic history shapes our experience of every conscious moment.

Time, trauma time, therapeutic time, and performance time

We have mentioned the possibility of two or more kinds of "consciousness." One is ancient in terms of evolution, residing in the living matrix; the other is the more modern neurological consciousness.

There are others. For example, if you sit on a tack, you get up very quickly before you have a chance to contemplate or philosophize about the nature of the stimulus or who is to blame. Another fast reaction occurs when you pull your hand away from a hot surface you have accidentally touched. These are examples of reflexes: responses that take place before you are consciously aware of the stimulus. In a reflex the sensory signal goes to the spinal cord, where it directly triggers a motor nerve that stimulates an automatic reaction. Only later does a signal go up the spinal cord to the brain so that you can become conscious of the tack or the hot surface.

Notice how sitting on the tack leads to a signal that propagates along a sensory nerve network that branches into two separate pathways. One branch stimulates muscle contraction, and the other branch takes the information upward to the brain. Another pathway is even faster and more direct. It is through the living matrix. The sensory information arriving at the surface of the sensory receptor travels through the cell matrix to the connective tissue and thence throughout the organism. This proposed branching of sensory flows was shown in Figure 7–2, B.

Because these information-processing schemes operate at different transmission velocities, there arises the possibility of separate "consciousnesses" taking place within the body. I place the word "consciousness" in quotation marks because we usually refer to neurological consciousness as the one and only kind of consciousness we experience. We are generally not aware of the faster schemes, but we get glimpses of them from time to time, as when we sit on a tack, we have a traumatic experience, or when "time seems to slow down," as when we watch the martial arts master responding to an attacker or watch a gold medal performance in the Olympics. Figure A–5 shows these two pathways and a speculative scheme for the ways they may interact during a traumatic event.

SOME CONCLUSIONS

Sensation, perception, and movement are obviously fundamental components of our conscious life. For the therapist, athlete, or performer of any kind, and for those who

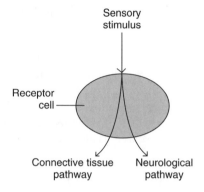

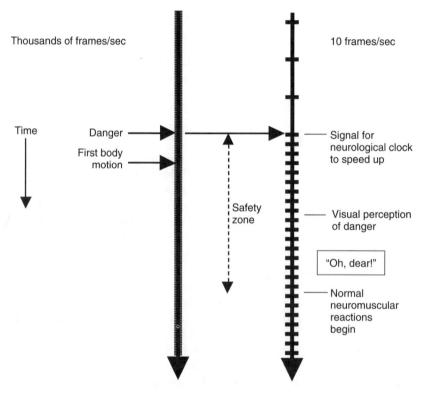

Figure A–5 Two systems responding to a potentially traumatic situation. It is proposed that there is a split in the pathway followed by a signal generated within sensory receptor cells. A precise location for this split in the visual system was proposed in Figure 16–4. The neurological pathway is shown to the right. This system can lead to a conscious perception of danger, an analysis of the significance of the situation, and neuromuscular avoidance or life-saving reactions. The connective tissue pathway, shown to the left, rapidly conducts the energy and information from the stimulus throughout the body. Both the propagation velocity and the "clock" or pacemaker for this system are far faster than the nervous system. Although we do not know the velocity of transmission through the connective tissue system, it could be extremely fast, on the order of the quantum

Legend continued

Figure A–5 Continued
coherence oscillations shown in Figure 25–2C. In times of danger, extreme trauma, or emergency, the "awareness" of the situation and body reactions (movements) arise first in the connective tissue system. Signals are propagated from the receptor cell directly to the contractile lattice within the muscle cells, as shown in Figures 20–1 and 24–2. The connective tissue system signals the clock or pacemaker in the nervous system (possibly in the thalamus) to speed up, increasing the number of conscious frames per second. Because more frames are occurring per second, one has the experience of "time slowing down." It takes fraction of a second for the visual system to form an image of the impending danger and a bit longer for the seriousness of the situation to reach conscious awareness, shown as "Oh, dear!" There is a further delay before the onset of normal neuromuscular reactions to the situation. In *Trauma Energetics,* Redpath (1995) noted that the trauma of an event is set in place in the fraction of a second before we are consciously aware of it. Years later, this energetic "signature" of the event continues to be referenced in the formation of every conscious moment. What Redpath calls *serious action* is impossible. Serious action is defined as movement and experience that are not referenced to, or motivated by, one's traumatic history. It is suggested that the energetic "signature" of a trauma is recorded prior to conscious awareness of the event, in the connective tissue, outside of the thought and speech centers of the brain. Successful trauma resolution depends less on recalling and verbalizing the neurological "records" of an event than it does on finding the record laid down in the connective tissue the instant before the event was consciously experienced. In terms of bodywork, energetic, and movement therapies, we seek methods that can directly access these patterns.

coach them or treat their injuries, the ways we sense and respond to the world around us, and within us, are of vital interest.

The clues gathered together here provide examples of exceptional sensation and action and their possible scientific basis. The book was written to stimulate thinking about technique and accomplishment in both therapy and performance in general.

You will not find here specific suggestions about therapeutic technique or how to train for performances. I have not suggested that you "put your right hand here and your left hand there" or "exercise until your heart rate reaches this level." Instead, we have developed a context or foundation for future explorations, experimentation, research, and experience. We have explored a common substrate involved in sensation of the environment, processing of the resulting information, and action through movement.

Several themes stand out. One is that living systems are far, far more sensitive to the energetic environment than we realized. Physicists repeatedly used their powerful theories to calculate the limits of biological sensitivity, but the behavior of living systems consistently defied those calculations. One reason is that cells contain molecular amplifiers and signal processors that are far more sophisticated than anything physicists or electronic engineers have dared dream of. An individual cell is the most powerful supercomputer in the world. The assembly of cells that constitutes the human organism represents a technology of unimaginable sophistication and capabilities.

It is tremendously exciting and rewarding to be at the interface between various fields of inquiry related to energy medicine. Many kinds of sophisticated electronic healing devices are being developed. These devices work because they interact with the organism as an electronic/protonic/electromagnetic/photonic/acoustic/thermal/vibrational system. These devices are educating us about how the organism operates as an energetic network. Moreover, the devices are teaching therapists about what can be accomplished with the appropriate input of energy. It is my belief, and I state this as a hypothesis for testing, that any healing that can be accomplished using an electronic device also can be accomplished by a human being. In workshops in Denmark (at the Stanley Rosenberg Institute) and Switzerland (at the Colorado Cranial Institute and the Polarity Therapie Zentrum Schweiz), I have witnessed a number of remarkable examples of this.

Resonance is an important mechanism involved in energy healing, whether it involves electronic devices or energy therapists. It is not widely appreciated that molecules are resonant antennas, emitting characteristic signals and responding to tiny signals of the appropriate frequency, much like a radio transmitter emits frequencies that entrain rhythmic oscillations in a distant receiver (see Figure 16–5).

A dramatic example of the remarkable capacity of resonance to transfer information over incredible distances took place in March 2002. To celebrate the thirtieth anniversary of its launch, the National Aeronautics and Space Administration (NASA) beamed a message to Pioneer 10 across 7.4 billion miles of space. The signal was sent from a radio telescope in the desert east of Los Angeles, California, and a radio telescope in Spain received a response 22 hours 6 minutes later.

In 1983, Pioneer 10 became the first manmade object to leave the solar system. Its transmitter has a power of a few watts, comparable to a small flashlight (NASA 2002) or to the output of the human heart (see Foreword). A tiny amount of power sent a return signal some 3.7 billion miles back to the earth. The signal traveled at the speed of light for some 11 hours 3 minutes. This phenomenon hints at the effectiveness of resonant electromagnetic interactions taking place between living organisms over the much shorter distances involved in hands-on or energy healing or even distant healing (prayer).

Because of amplification and resonance, the general trend in energy medicine is toward devices or techniques using increasingly less energy and effort on the part of therapists. Once the appropriate frequency and amplitude are discovered, virtually any physiological process in the body can be influenced with a tiny signal.

Given the extraordinary "electronic instrument" we all have within us, what are the limits to human achievement? A valuable perspective came from a remarkable book by John Lilly (1985), *The Center of the Cyclone*. Lilly's dolphin research and his explorations of consciousness revealed to him that "What we believe to be true is true, within certain limits, which are, themselves, beliefs. In the province of the mind, there are no real limits." He introduced the concept of meta-beliefs, which are beliefs about belief systems. The statement just quoted, that there are no real limits, is a meta-belief, that is, one can transcend what one thinks is possible by examining and transcending one's beliefs.

This concept ties in with what many philosophers and explorers have concluded about the connection between mind and reality, and there is a well-documented physiological connection. Performers, therapists, and patients alike can benefit from mental rehearsals or internal imaging of an activity, without physically doing anything. Mental imaging is being applied in training for athletic events, dance, theatre, music, combat, and healing work. Mental rehearsal sets up "anticipatory fields" in the motor cortex. These fields include the readiness potential and readiness magnetic field shown in Figure A–3. These fields spread through the tissues and into the space around the body. This can lead to a "preconditioning" of biochemical pathways, energy reserves, and patterns of information flow. Cells throughout the body are then poised to work together at the instant of demand (Oschman 2000, p. 227).

Given the great sensitivity of living systems to tiny energy fields, the resonance phenomenon described earlier, the vast memory storage and information processing power of the living matrix, and the fields produced as the brain prepares for a movement, it is not surprising that one can anticipate the movements another person is going to make even before that person is aware that he or she has decided how to move. It is not surprising that micromovements of a speaker entrain micromovements of a listener, as Condon so carefully demonstrated. It is no longer a mystery that a well-prepared athlete can anticipate the movements of an opponent, or that the performances of members of a team, dance troupe, or jazz band can become totally integrated and coordinated. There are some therapeutic techniques that can pinpoint and treat the real cause of an ailment. Knowing that there is an internal substrate involved in these phenomena can lead to advances in performance and therapeutic methodology.

Nørretranders (1991) recommended that hunches and intuitions be trusted because they are closer to reality than perceived reality, given that they are based on far more information. The process of accessing one's hunches and intuitions comes naturally to some individuals and must be learned by others. Lilly's advice can be incorporated into one's intentions with regard to accessing the vast wealth of internal wisdom we all possess. A meta-intention is an intention about intentions. By believing in and intending to access internal wisdom, what Redpath calls *serious action* is facilitated. Where this becomes extremely relevant is in the treatment of deep trauma. With the appropriate preparation, the therapist will "know" how to resolve the issue with minimal effort and without the need for discussion of the traumatic event.

The approach is well stated in what is known as the *Captain Kirk principle:* intellect is driven by intuition, intuition is directed by intellect (Star Trek 1966). The rationale for these concepts is finding support from an emerging field of scientific inquiry summarized by Myers (2002) in his book, *Intuition: Its Powers and Perils.* Myers demonstrates through numerous well-replicated experiments that intuition—"our capacity for direct knowledge, for immediate insight without observation or reason"—is as much a component of our thinking as is analytical logic.

Intuition is subtle perception and learning—knowing without knowing that you know. Although intuition has often been shunned by scientists, Myers points out that intellect and intuition are complementary, not competitive. Without intellect, intuition drives us into chaos; without intuition, we are unable to resolve issues that are too complex or that are happening too quickly for logical analysis (Shermer 2002).

I have seen a demonstration of the Trager method (by Deane Juhan) that began with a pause, during which the intention was to simply connect with the patient. In workshops in Denmark, Stanley Rosenberg demonstrated the contrast between putting hands on a patient and starting to massage him or her, and putting hands on and pausing for a few moments. It is during this interval that enormous amounts of information can flow between the two individuals. The rich source of wisdom and guidance that Nørretranders refers to as "hunches and intuition" can set in place, and the result may be surprisingly different from the well-thought-out technical approach. Training in technique is vital, but sooner or later every therapist learns that there are ways of transcending method, of allowing the information within the patient to govern what takes place in the therapy session.

Albert Szent-Györgyi spent the last decades of his scientific career studying electronic and protonic conduction in proteins. All of his research, plus that of Ho (1993, 1997, 1999), supports the idea that energy and information can move about within the body by a mechanism that differs from nerve transmission and hormonal regulations. These energy flows are maintained by quantum coherence, as described in Chapter 25.

The velocity of energy and information transfer through the living matrix is unknown, but there are indications that it is extremely rapid, certainly far faster than nerve conduction. The primary channels or *superhighways* of this information flow appear to be the acupuncture meridians.

These concepts begin to account for the psychotherapeutic techniques that involve tapping on particular acupuncture or acupressure points, for example. I suggest that tapping interacts directly with the electronic system within the body that stores traumatic and other memories and that references these memories during every conscious moment. By asking a patient to focus on his or her emotional discomfort while tapping on the points, the therapist is directly interacting with the flows of energy and information related to the original emotional state or phobia. To be more specific about this idea, tapping or applying pressure to any tissue creates electrical fields because of the piezoelectric (pressure electricity) effect.

Information stored in the matrix shows up in subtle yet profound phenomena we refer to as *intuition, intention,* and *the unconscious.* Dreams, hypnosis, the placebo effect, subliminal perception, and spontaneous healing (Weil 1995) may be better understood through study of the living matrix system.

Ideally this book has opened up some possibilities for exploration. I have aimed to respect and recall the nearly forgotten works of pioneering scientists whose ideas can shape the future of medicine and human performance and the realization of the vast potentials of the human spirit.

Our time longs for a new synthesis—it waits for science to satisfy our higher needs for a view of the world that shall give unity to our scattered experience.

HUGO MUNSTERBERG (1899)

REFERENCES

Bretscher, M.S. 1971, 'A major protein which spans the human erythrocyte membrane', *Journal of Molecular Biology,* vol. 59, pp. 351–357.

Brown, J.W. 1988, *The Life of the Mind. Selected Papers,* Lawrence Erlbaum Associates, Hillsdale, NJ.

Brown, J.W. 2000, *The Self-embodying Mind,* Barytown Ltd., Barytown, NY.

Callahan, R.J. & Callahan, J. 1996, *Thought Field Therapy and Trauma: Treatment and Theory,* Callahan, Indian Wells, CA.

Chitty, J. 2001, *Polyvagal Theory, the Triune Autonomic Nervous System, and Therapeutic Applications,* [Online], Available: http://www.polaritycolorado.com/writings/triune_autonomic_article.PDF.

Cho, Z.H., Chung, S.C., Jones, J.P., Park, J.B., Park, H.J., Lee, H.J., Wong, E.K. & Min, B.I. 1998, 'New findings of the correlation between acupoints and corresponding brain cortices using functional MR', *Proceedings of the National Academy of Sciences of the USA,* vol. 95, pp. 2670–2673. An update on this research was presented by J.P. Jones at a conference, Bridging Worlds and Filling Gaps in the Science of Spiritual Healing, held in Kona, Hawaii, November 29-December 3, 2001.

Condon, W.S. & Sander, L.W.W. 1974, 'Neonate movement is synchronized with adult speech. Integrate participation and language acquisition', *Science,* vol. 183, pp. 99–101.

Corongiu, G. & Clementi, E. 1981, 'Simulations of the solvent structure for macromolecules. I. Solvation of B-DNA double helix at T=300 K', *Biopolymers,* vol. 20, pp. 551–557.

Deecke, L., Grozinger, B., & Kornhuber, H.H. 1976, 'Voluntary finger movement in man: cerebral potentials and theory', *Biological Cybernnetics,* vol. 23, pp. 99–119.

Deecke, L., Weinberg, H. & Brickett, P. 1982, 'Magnetic fields of the human brain accompanying voluntary movement: Bereitschaftsmagnetfeld', *Experimental Brain Research,* vol. 48, pp. 144–148.

Dennett, D.C. 1992, 'Multiple drafts versus the Cartesian theatre', in *Consciousness explained,* Little Brown, Boston, pp. 101–138

Doyle, A.C. 1887, 'A study in scarlet', *The Original Illustrated "Strand" Sherlock Holmes, The Complete Facsimile Edition,* Mallard Press, New York, 1990, pp. 37–38.

Durka, P.J., Ircyha, D., Neuper, C. & Pfurtscheller, G. 2000, 'Time-frequency microstructure of event-related desynchronization and synchronization', *Medical and Biological Engineering and Computing,* vol. 39, pp. 315–321.

Eccles, J. 1993, 'Keynote', in *Rethinking neural networks: Quantum fields and biological data,* ed. K.H. Pribram, Lawrence Erlbaum Associates, Hillsdale, NJ, pp. 1–28.

Eley, D.D., Parfitt, G.D., Perry, M.J. & Taysum, D.H. 1953, 'Semiconductivity of organic substances I', *Transactions of the Faraday Society,* vol. 49, pp. 79–86.

Fleming, T. 1996, *Reduce Traumatic Stress in Minutes: The Tapas Acupressure Technique (TAT),* Workbook, Torrance, CA.

Fröhlich, H. 1988, *Biological Coherence and Response to External Stimuli,* Springer-Verlag, Berlin.

Gallo, F.P. 1999, *Energy Psychology,* CRC Press, Boca Raton, FL.

Garfield, E. 1991, 'Bibliographic negligence: A serious transgression', *The Scientist,* vol. 5, p. 14.

Gilman, A.G. 1997, 'G proteins and regulation of adenylyl cyclase', Nobel Lecture presented December 8, 1994, in *Nobel Lectures Physiology or Medicine 1991–1995,* ed. N. Ringertz, World Scientific, Singapore, pp. 182–212.

Ginsburg, I. 2001, 'The Disregard Syndrome: A menace to honest science?', *The Scientist,* vol. 15, p. 51.

Hameroff, S. 1993, 'Nanoneurology and the cytoskeleton: Quantum signaling and protein conformational dynamics as cognitive substrate', in *Rethinking neural networks: Quantum fields and biological data,* ed. K.H. Pribram, Lawrence Erlbaum Associates, Hillsdale, NJ, pp. 317–376.

Hameroff, S.R., Kaszniak, A.W., & Chalmers, D.J. (editors). 1999, 'The timing of conscious experience', in *Toward a science of consciousness III. The third Tucson discussions and debates.* MIT Press/Bradford Books, Cambridge, MA, p. 341.

Hari, R., Hamalainen, M., Kaukoranta, E., Reinkainen, K., & Teszner, D. 1983, 'Neuromagnetic responses from the second somatosensory cortex in man', *Acta Neurologica Scandanavica,* vol. 68, pp. 207–212.

Heeger, A.J., MacDiarmid, A.G. & Shirakawa, H. 2000, Nobel Prize in Chemistry.

Ho, M.W. 1993, *The Rainbow and the Worm. The Physics of Organisms,* World Scientific, Singapore.

Ho, M.W. 1997, 'Quantum coherence and conscious experience', *Kybernetes,* vol. 26, pp. 265–276.

Ho, M.W. 1999, 'Coherent energy, liquid crystallinity and acupuncture', Talk presented to the British Acupuncture Society, October 2, 1999, [Online], Available: http://www.i-sis.org/acupunc.shtml.

Hyland, M.E. 2003, 'Extended network entanglement theory: Mechanisms for subtle therapy and empirical prediction. Submitted to *Journal of Alternative and Complementary Medicine.*

Jasper, H.H. & Penfield, W. 1949, 'Electrocortiograms in man: Effect of the voluntary movement upon the electrical activity of the precentral gyrus', *Archiv für Psychiatrie und Nervenkrankheiten, vereinigt mit Zeitschrifte für die gesamte Neurologie und Psychiatrie,* vol. 183, pp. 163–174.

Jung, C.G. 1954, in *Collected Works of CG Jung: Volume 17. Development of Personality,* Princeton University Press, Princeton, NJ, p. 52.

Kaptchuk, T.H. 1983, *The Web That Has No Weaver,* Congdon and Weed, New York.

Kornhuber, H.H. & Deecke, L. 1964, 'Hirnpotentialänderungen beim Menschen vor und nach Willkürbewegungen, dargestellt mit Magnetbandspeicherung und Rückwärtsanalyse', *Pflügers Archiv fur Gesamte Physiologie,* vol. 261, p. 52.

Levine, P.A. 1997, *Waking the Tiger: Healing Trauma: The Innate Capacity to Transform Overwhelming Experiences,* North Atlantic Books, Berkeley, CA.

Libet, B. 1985, 'Unconscious cerebral initiative and the role of conscious will in voluntary action', *The Behavioral and Brain Sciences,* vol. 8, pp. 529–566.

Lilly, J.C. 1985, *The Center of the Cyclone. An Autobiography of Inner Space,* The Julian Press, New York.

McGinness, J.E., Corry, P., & Proctor, P.H. 1974, 'Amorphous semiconductor switching in melanins', *Science,* vol. 183, p. 853.

Munsterberg, H. 1899, *Psychology and Life,* Boston, Houghton Mifflin.

Myers, D.G. 2002, *Intuition: Its Powers and Perils,* Yale University Press, New Haven, CT.

NASA. 2002, *USA Today,* March 4, 2002, p. 3A.

Nørretranders, T. 1998, *The User Illusion,* Viking/Penguin, New York.

Oschman, J.L. 2000, *Energy Medicine: The Scientific Basis,* Churchill Livingstone, Edinburgh.

Oschman, J.L. & Oschman, N.H. 1993, 'Matter, energy and the living matrix, *Rolf Lines,* vol. 21, pp. 55–64.

Pearsall, P. 1998. *The Heart's Code,* Broadway Books, New York.

Penrose, R. 1990, *The Emperor's New Mind,* Oxford University Press, New York.

Pienta, K.J. & Coffey, D.S. 1991, 'Cellular harmonic information transfer through a tissue tensegrity-matrix system', *Medical Hypotheses,* vol. 34, pp. 88–95.

Popper, K. & Eccles, J. 1985, *The Self and Its Brain,* Springer International, Berlin, p. 364.

Porges, S.W. 1995, 'Orienting in a defensive world: Mammalian modifications of our evolutionary heritage. A polyvagal theory', *Psychophysiology,* vol. 32, pp. 301–318.

Redpath, W.M. 1995, *Trauma Energetics,* Barberry Press, Lexington, MA.

Rosenberg, S. 2003, '*Can You Test for Physiological Causes for Chronic Stress?*' [Online], Available: http://www.stanleyrosenberg.com.

Schmitt, F.O. 1961, 'Molecule-cell, component-system reciprocal control as exemplified in psychophysical research', The Robert A. Welch Foundation Conferences on Chemical Research. V. Molecular Structure and Biochemical Reactions, Houston, Texas, December 4–6, pp. 33–37.

Scott, M. 1997, 'Deep story: An investigation of the use of the embodied voice in the treatment of traumatic memory', Ph.D. Thesis, University of Massachusetts, Amherst, MA.

Shermer, M. 2002, 'The Captain Kirk principle', *Scientific American,* vol. 287, no. 6, p. 39.

Springer, T.A. 1994, 'Traffic signals for lymphocyte recirculation and leukocyte emigration: The multistep paradigm', *Cell,* vol. 76, pp. 301–314.

Star Trek, 1966, Episode 5, The Enemy Within.

Stent, G.S. 1972, 'Prematurity and uniqueness in scientific discovery', *Scientific American,* vol. 227, pp. 84–93.

Sylvia, C. & Novak, B.B. 1997, *A Change of Heart,* Little Brown, Boston.

Szent-Györgyi, A. 1974, 'Drive in living matter to perfect itself,' *Synthesis,* vol. 1, pp. 14–26.

Weil, A. 1995, *Spontaneous Healing: How to Discover and Enhance Your Body's Natural Ability to Maintain and Heal Itself,* Alfred A Knopf, New York.

Weiss, P.A. 1973, *The Science of Life: The Living System—A System for Living,* Futura Publishing Co., Mount Kisco, NY.

Appendix A: John McDearmon Moore, Jr., D.O., F.A.C.G.P. (1916–2000)

A COUNTRY DOCTOR

A substantial part of this book is about the work of the dedicated scientists who research the foundations of life, in laboratory and theoretical settings. Equally important in the search for wellness are the practices and findings of the dedicated doctors, nurses, and therapists whose healing touch eases the journey for those who are sick or injured. The best of these practitioners combine the discoveries of modern science with the intuitive wisdom of the heart. As a tribute to these deeply committed individuals, we look at the life of one exemplary figure, a country doctor from Tennessee.

When "Mack" Moore retired from 48 years of medical practice, homage from a cross-section of politicians, dignitaries, and ordinary folk came pouring in from all over the country to his hometown, Trenton, Tennessee. They were paying tribute to a man who had become almost a member of the family to all these grateful people. A personal letter from President George H. Bush, Sr., spoke of Mack's many contributions to the national community. And there were tears of farewell from patients for a man they had always been glad to see coming down the road, on a horse or wagon, or on foot when the path was washed out. Yet little did his admiring patients know of the struggles he endured for many of those 48 years—struggles against the ridicule and innuendos of many in the then-accepted medical profession who felt threatened by the subtlety and seeming complexity of the notion of wholistic health—a model now on its way to becoming the new worldwide paradigm for wellness.

As often happens, Mack's choice of profession came from a moving personal experience. In the fall of 1935, he entered the United States Naval Academy at Annapolis, Maryland. Soon thereafter he began to suffer from a debilitating kidney illness for which he was hospitalized repeatedly during the next three years. Conventional "treat the symptoms" allopathic care at the Academy failed to cure the painful condition. He was advised that he would not be able to withstand the rigors of duty at sea. Noticing Mack's interest in medicine, his attending physician encouraged him to leave the Academy and begin a course of study to become a physician himself.

Mack's family doctor had begun his career as an osteopathic physician, so when Mack's kidney flared up again, he sought an osteopath. The idea of helping the body restore itself to wholeness by ministering to the entire body was out of synch with the times for many, if not most, of the physicians practicing in 1939. The treatment Mack's osteopath recommended seemed too simple and not interventional enough to many practitioners of the healing arts at the time; yet the recommended treatment, an alternation of acid-ash diet, over a short period of time, coupled with musculoskeletal manipulation, permanently cured his "chronic" condition. Mack was now beginning to deeply embrace the idea of a physician as a mediator, rather than as a "God-like" interventionist.

Mack realized early on that engaging the mind in the purposeful envisioning of its own perfect health was a crucial element of healing. Because of this approach, which seemed to prevent illness before it started, Mack often said that psychology was the most important part of his training. The precepts of osteopathy—the integrity of the whole body and every element related thereto—from structural, to electrochemical, and, yes, even to psychological, found expression in the healing touch of this humble man.

Graduating from the Kirksville College of Osteopathic Medicine and Surgery in 1943, Mack returned to his hometown and took over the practice of the same osteopath who had delivered him into the world in 1916. Within 7 years, he and his wife, Agnes, opened the Trenton Osteopathic Hospital, the first hospital in the county. Mack recruited an osteopathic surgeon for surgical cases.

The little eight-bed hospital was almost always full. A dedicated team of nurses and staff held regular study sessions to stay abreast of new treatments and innovations in the health world. Way ahead of his time was Mack's theory that nutrition had a very important relationship to health and that obesity contributed directly to diabetes and heart-related issues.

The family quarters were located above the hospital and clinic, so the doctor was available for emergencies at any hour. It was commonplace for a call to come from the emergency room at 2 or 3 in the morning. "Doc" would awaken from a deep sleep and rush down to try to save someone's life. A few hours later he would be making rounds in the hospital and preparing for a grueling day of osteopathic manipulative treatments, sewing lacerations, operating on someone with acute appendicitis, delivering an overdue baby, and so on. He had to intensely love his work, otherwise he could not have kept up the pace.

One of the hallmarks of a successful life is humility. Mack always knew that God was doing the healing—something that practitioners sometimes forgot and that patients sometimes misunderstood. From the days of the witch-doctors all the way up to the present era, one of the occupational hazards of "doctoring" is the tendency for "who is doing what" to get lost in the shuffle. Patients want their physicians to be infallible, of course, and that is an impossible demand for fallible human beings, regardless of their training. Mack had a humbleness about him that moved people to admire and respect him even when he didn't have the pat answers they were hoping for. His faith got him through the difficult moments that every physician faces, and it endeared him to others, even to those who hoped against hope that he would perform a miracle that was not within his power to bring about.

Mack's generosity and compassion were legendary. Thousands of times he simply wrote off the fees he would normally charge for this services and supplies for patients who could not pay for the treatments they desperately needed. Others would bring cakes or wine or chili sauce to pay for treatments.

Ever true to the spirit of the General Practitioner, Mack was never one-dimensional or specialized. He loved the great out of doors and enjoyed doing things with his hands besides administering healing touch to his patients. He was an able carpenter and electrician, and an avid gardener and arborist. He also owned a dairy farm, which he took a great interest in and, when he wasn't quail hunting on a rare day off, he was busy plowing or putting up fences or terracing his pastures.

Over the years, Mack and Agnes served in many demanding capacities in state and national offices of professional associations and their auxiliaries. Mack was named a Fellow of the American College of General Practitioners in 1969. He and Agnes were involved in many civic and religious activities within their community. In the late 1960s and early 1970s national attention was being focused on the need for health care in rural areas throughout the country. Being a member of the search team of the Brookings Institute and the Memphis Regional Medical Program—a cooperative effort of the time—Mack worked with officials of local, state, and federal governments to fund a rural Health Access Station, located in a fairly remote area of West Tennessee, for diagnosis, treatment or referral of patients living in that area. Mack sponsored an R. N. from the area for what was then a newly conceived nursing profession—the Nurse Practitioner. This young lady—now bearing License#1—became the point person of the clinic working under a protocol of a panel of doctors with Mack's monitoring. Due in part to the lower health care costs for those served by the station, the area gradually became more prosperous until the clinic was replaced by more traditional facilities eighteen years later. The Nurse Practitioner, whose role at first was scorned by the medical profession, now was much sought after.

When the threat of nuclear attack seemed a possibility during the Cold War, Mack, as Civil Defense Director for the county, was alert to the need for state and county government officials to have a safe place in which to work together in the case of such an emergency. Working with government officials, Mack inspected similar centers in other states and planned one to be built in his county. After much leg work and many hours with county board members the facility became a reality with a near 100% protection factor against nuclear radiation. In addition to the shelter and a protocol for those who might use it, there was a decontamination center, equipped with supplies, along with operational equipment and facilities for a prolonged stay, as well as a morgue. Fortunately the Cold War did not materialize, but the Center today is one of busiest places in the county, helping handle emergence situations all over the country.

By 1978 it was no longer economically feasible to maintain a small hospital. Regulatory requirements were increasing at a staggering rate. Economy of scale dictated a much larger facility just to break even. The little hospital that had inspired the whole community needed to be replaced with something larger. Mack was asked to join with several allopathic physicians in a 50-bed community hospital. After much soul searching, Mack closed "the little hospital that could" and agreed to practice at the larger hospital, which he did until 1985. It was another milestone for an osteopathic physician to practice in a conventional hospital with MDs.

By 1985, Mack had been practicing for 42 years. Realizing that obesity was becoming a major health problem in American, he left the community hospital and transformed his practice into a bariatric center. Once again he was well ahead of his time.

Obesity and attendant illness are reaching epidemic proportions. Until his retirement in 1991, Mack brought hope of living a normal healthy life to thousands of individuals.

Never one to stop learning, Mack continued to study new ideas in wholistic health, such as cranial-sacral therapies, even in retirement. In 1999, one year before his passing, after having his first cranial adjustment, he remarked that the cranial bones *can* move despite what his professors had told him almost 60 years before. He understood well the old adage: "It's not what we don't know that hurts us, it's what we know that just ain't so." Mack's whole life was a celebration of that truth. He was never a "yes man," despite a healthy respect for institutions and traditions. He knew his *own* truth, that he was able to take action and do meaningful things without undue hesitation and fear of contradiction from others. A lifetime of deeds like Mack Moore's is the greatest legacy and monument a person can leave behind. This book is dedicated to this rare and pioneering spirit—and to all those who bravely follow such a path.

Appendix B: Magnetobiology and electro-dynamic fields in therapeutics and human performance

More than 50 years ago, the distinguished Yale Professor Harold Saxton Burr recognized the direct link between all body functions and the energy fields produced within and around the body. He referred to the phenomenon as the *electro-dynamic field*. Based on a series of studies conducted between 1932 and 1956, Burr asserted that all disturbances, physical or emotional, show up in the field long before any symptoms or pathological structure can be detected by ordinary diagnostic methods. Moreover, correcting or normalizing the energy field reverses the degenerative process. This research has not been given the attention it deserves. (For a list of Burr's publications, see Burr 1957.)

Burr's discoveries are of profound significance in complementary medicine and human performance. From the biological perspective, magnetic and electromagnetic senses are a built-in feature of organisms ranging from bacteria to mammals (Loeb 1918; Fraenkel & Gunn 1961; Barnothy 1964, 1969; Palmer 1967; Presman 1970; Dubrov 1978; Kholodov & Lebedeva 1992). Evolution has provided animals with this sensitivity for navigating; finding food and identifying predators; and predicting the weather, earthquakes, and other earth changes.

Biological applications of electro-dynamic fields include the dramatic synchronized movements of schools of fish and flocks of birds. These behaviors are thought to be anti-predator adaptations. Predators are confused by multiple synchronously moving targets that are tightly packed together. Maintaining synchrony and spatial position within groups requires the sophisticated functioning of complex sensory and motor systems. The phenomenon is also dramatically demonstrated in some martial arts schools in which there is training in sensing and responding to an attacker without using the usual sensory information.

The great sensitivity of organisms to environmental rhythms has been debated for a long time because the extreme sensitivities involved have seemed beyond belief. It is only recently that the extensive behavioral evidence has been supported by convincing research on the biophysical and molecular mechanisms involved (reviewed by Binhi 1999, 2002).

Burr's findings on electro-dynamic fields were passed over by medical science in spite of knowledge of the electrocardiogram and electroencephalogram. These well-known diagnostic tools demonstrate that the body produces measurable electrical fields. Ampère's Law (see Figure 1-2) requires that these electrical fields must be accompanied by magnetic fields in the space around the body. These biomagnetic fields have been documented (see Figures 1-4 and 2-4, *A*). The mechanisms by which magnetic fields affect cell functions have been worked out in detail (see Figure 1-6 and Binhi 2002). All elements of Burr's thesis have been confirmed.

Electro-dynamic fields provide the basis for a variety of therapies in which information is obtained from the energy field of a patient, and the field of the therapist is used to connect with the patient to enhance physical and emotional health.

In terms of athletic and artistic performances, the electro-dynamic fields Burr discussed provide information about bodily movements *and about the planning of movements before they occur* (readiness magnetic fields; see Figure A-3 [Afterword]). This information is projected *instantaneously* into the space around the body.

Whereas disease and disorder "advertise" themselves to the sensitive observer through the energy field, the signals developed by the neuromuscular system are much larger. They also should be easy to "read" once one knows they are there and that there are ways of detecting and responding to them.

I suggest that these electro-dynamic fields play a key role in peak athletic and artistic performances in which participants must function together in a highly coordinated or synchronized manner. *Anticipatory awareness* enables the individual or a whole team to prepare for an opponent's intended movements.

For the athlete or performer, energy fields can be used in several ways both to prevent injuries and to accelerate the healing process when injuries occur. Sports injuries, for example, usually are addressed by treating the injured part of the body. Complementary medicine teaches us that an injury to a particular part of the body may have occurred because of weakness or imbalance in another area. For example, knee or ankle problems often arise because the muscles around the pelvis or lower back are hypertoned, hypotoned, or out of balance. The sophisticated athletic trainer knows that both prevention and repair involve whole-body flexibility and balance.

Energy medicine has dramatic applications in situations where the quick restoration of function to an individual player or performer is vital to the team effort. When a player is injured, as often happens in football games for example, the other team members stand around waiting for the medical team to assess the injury and determine what should be done about it (Figure B-1, *A*). This is an opportunity for application of the group's energy to assist the medical staff. A therapist familiar with energy methods will be able to use his or her own energy and the energy of the group to "jump-start" the healing process in several ways.

The medical team will obviously ask the injured player where the pain is located. However, the therapist who understands the whole-system aspects of therapeutics knows that the location of a pain is only part of the story. There may be other areas that have been strained but that will not become painful until sometime after the injury. Tensions in these areas may further exacerbate the primary injury if the player returns to the game. The skilled energy therapist can determine the system-wide aspects of the injury, including the emotional impacts, and bring in all of the body's assets to jump-start healing. This can be done without interfering with the medical

Figure B-1 Energy medicine has dramatic applications when quick restoration of function to an injured player or performer is vital to the group effort. **A,** an example from football, in which an injured player is down on the field. Teammates stand around waiting for the medical staff to assess the injury and determine a treatment protocol. **B,** A therapist who appreciates energy medicine kneels opposite the medical staff, focusing on the systemwide aspects of the injury, including the emotional impact, with an intention of jump-starting the healing process. Other team members contribute by focusing their attention on the injured performer, forming an "energy circuit" that not only assists the healing process, but that also enhances the "systemic cooperation" within the whole group, an effect that will enhance performance when play resumes.

team's efforts in bandaging, splinting, or applying cold to an injured joint, for example. Moreover, the other members of the team can contribute to this process by forming an energy "circuit" that includes the therapist, the medical staff, and the injured performer. A suggested arrangement is shown in Figure B-1, *B*. From the material discussed in this book, we predict that this energetic process will both jump-start the healing in the injured player and enhance the systemic cooperation within the entire team, an effect that will show itself when play resumes.

Caring for an injured performer is greatly facilitated if the therapist saw the injury occur so that he or she is precisely aware of what happened with the performer's body mechanics during the incident.

Finally, we consider the magnetic sense that enables the therapist to detect the locus of physical or emotional issues within a patient's body and that enables well-integrated performers to sense and respond to their environment, including the fields of other participants in an event. After considering what is known about the properties of water and the quantum coherence phenomenon described in Figure 25-2, I propose that magnetic sensitivity involves the influence of the external magnetic field on the spin axes of the water molecules associated with the living matrix. If this hypothesis is correct, the magnetic sense experienced by many energy therapists goes to the core of the living processes involved in creating structure and consciousness, as discussed in Chapter 25.

Access to the wealth of information taken in through the energy field is not accomplished consciously because the information is extraneural. It is here that insight and intuition must be allowed to prevail. This requires pausing, quieting the mental analysis of the situation, and having an intention to connect. This was discussed in more detail in the Afterword.

These concepts go beyond helping performers achieve their personal best, world records, and gold medals. Transcendent performance in any area lifts the spirits of all involved, whether performers or observers. It is the basic stuff of human evolutionary progress.

REFERENCES

Barnothy, M.F. 1964, 1969. *Biological Effects of Magnetic Fields, Volumes 1 and 2*, Plenum Press, New York.

Binhi, V.N. 1999, An analytical survey of theoretical studies in the area of magnetoreception, in *Electromagnetic Fields: Biological Effects and Hygienic Standardization*, eds M.H. Repacholi, N.B. Rubtsova & A.M. Muc, World Health Organization, Geneva, Switzerland, pp. 155–170.

Binhi, V.N. 2002, *Magnetobiology: Underlying Physical Problems*, Academic Press, New York.

Burr, H.S. 1957, 'Harold Saxton Burr', *Yale Journal of Biology and Medicine*, vol. 30, pp. 161–167.

Dubrov, A.P. 1978, *The Geomagnetic Field and Life: Geomagnetobiology*, Plenum Press, New York.

Fraenkel, G.S. & Gunn, D.L. 1961, *The Orientation of Animals. Kinesis, Taxes and Compass Reactions*, Dover Publications, Inc., New York.

Kholodov, Y.A. & Lebedeva, N.N. 1992, *Reactions of the Human Nerve System on the Electromagnetic Fields*, Nauka, Moscow [in Russian].

Loeb, J. 1918, *Forced Movements, Tropisms, & Animal Conduct*, J.B. Lippincott Company, Philadelphia.

Palmer, J.D. 1967, 'Geomagnetism and animal orientation. Do the guidance systems of animals respond to the earth's magnetic forces?', *Natural History*, vol. 76, pp. 54–57.

Presman, A.S. 1970, *Electromagnetic Fields and Life*, Plenum Press, New York.

Appendix C: A crisis in bioenergetics

MECHANISM OF ENERGY TRANSDUCTION IN BIOLOGICAL SYSTEMS: NEW YORK ACADEMY OF SCIENCES CONFERENCE

Some 60 participants drawn from all parts of the world and representing all facets of bioenergetics gathered in New York during 7 to 9 February 1973 to engage in a dialogue directed to the theme of the principles that can unify bioenergetics. The organizer, D. Green (University of Wisconsin), attempted to achieve the right blend of experimentalists and theorists and of the different disciplines (biology, chemistry, and physics) so that no one point of view would dominate the proceedings.

The conference revolved around three basic questions: Is there a crisis in bioenergetics?; What is the nature of the postulated crisis?; and How can the crisis be resolved? The traditionalists, drawn largely from the ranks of the experimentalists, took the position that within the present conceptual framework of bioenergetics, the solution of the outstanding problems was inevitable given the necessary time and intensity of effort. H. Huxley and B. Hartley (both of the Medical Research Council Laboratory of Molecular Biology, Cambridge University) and F. Harold (National Jewish Hospital, Denver), in their outstanding presentations, implicitly supported this thesis of the inevitability of progress along classical lines. Perhaps it would be more precise to say that the traditionalists, while open to persuasion, were unaware of compelling reasons for any drastic change in the conceptual framework. C. McClare (Kings College, London) was the spearhead of the group of both experimentalists and theorists who challenged the adequacy of the present approach to bioenergetics. This challenge was directed to the mechanism of muscular contraction [McClare and S. Ji (University of Wisconsin)], to the mechanism of enzymic catalysis (R. Lumry, University of Minnesota; Green; and Ji), to the mechanism of electron transfer (M.E. Winfield, Commonwealth Scientific and Industrial Research Organisation, Melbourne), to the mechanism of nerve transmission (L. Wei, University of Waterloo, Ontario; and D. Nachmansohn, Columbia University), and finally to the mechanism of energy transduction (McClare; Green; Ji; G. Weber, University of Illinois; R. Williams, Oxford University; F. Cope, Naval Air Development Center, Warminster, Pennsylvania; and

A. Bennun, Rutgers University). While there was by no means agreement within this group as to the nature of the crisis and the corrective measures required to resolve the crisis, the group were convinced that unless new dimensions were added to the conceptual framework, progress would grind to a halt.

McClare developed the thesis that there was a fundamental misconception about energy transduction in biological systems. In all of the classical mechanisms, the notion of macroscopic constrained equilibrium machines has been invoked, but this notion is in-applicable to biological transducing machines that operate at the molecular level. According to McClare, biological transducing machines are molecular machines, and only molecular machines can achieve the observed high efficiency of biological energy transductions. Intrinsic to molecular machines are the concepts of the generation of a vibrationally excited state by one of the reactants in the exergonic reaction, the transfer of energy by resonance, and the relaxation of the energized state by a work performance. Since energization of molecular machines depends upon resonance phenomena, the efficiency of energization can be theoretical.

Green and Ji have systematically developed the concept of molecular machines in their electromechanochemical model of energy transduction and have extended this model to enzymic catalysis, oxidative phosphorylation, active transport, and muscular contraction. In their thesis there is a set of fundamental laws that underlies the performance of all biological energy-transducing systems, and from these laws the mechanistic principles of the transducing systems can be deduced. The laws of bioenergetics, like the laws of thermodynamics, can only be deduced by the a priori method although their validity has to be established by experiment.

The conference was marked by debates that swirled around each of several central issues. Weber challenged the viability of the notion of vibrationally excited states crucial to the molecular machine concept, contending that the lifetime of such states (about 10^{-12} second) would be too short to be useful in biological energy-transducing systems. McClare emphasized that under appropriate conditions, the lifetime of vibrationally excited states, such as that of carbon monoxide, can extend into the second range. L. Shohet (University of Wisconsin), from the analysis of computer models of the alpha helix, concluded that by the appropriate selection of resonance conditions, transfer of vibrational energy through the helix can take place with minimal energy loss at the speed of sound. Ji developed the notion that protein structure could play a key role in stabilizing and extending the lifetime of vibrationally excited states generated in supermolecules. K.D. Straub (University of Arkansas Medical School), the first to propose energy transfer in biological systems by means of phonons, made some incisive comments about the stability question.

The chemiosmotic model of P. Mitchell (Glynn Research Laboratories, Bodmin, England) was the focus of greatest attention and most extended debate in the conference. The model was defended in masterly presentations by Harold and by V.P. Skulachev (State University, Moscow). Williams was the principal protagonist of the view that a transmembrane proton gradient could not be a driving force in energy coupling; only an intramembrane charge separation or potential would meet the energetic requirements. Two dramatic experimental developments which supported Williams' position were revealed at the conference. First, H.T. Witt (Technische

Universiteţ, Berlin) showed that the intrinsic membrane potential in chloroplasts, but not the pH gradient, could be correlated with the capacity for photosynthetic phosphorylation. Second, mention was made of evidence obtained in several different laboratories that coupling could be achieved in nonmembranous suspensions of macromolecules (reported by M.I.H. Aleem, University of Kentucky; and T. Ozawa, Nagoya University; and published by D.R. Sanadi, Retina Foundation, Boston). If there was one issue on which a consensus was reached at the conference, it was that the Mitchell model could only be viable with an intrinsic membrane potential as the driving force. In that revised form, the Mitchell model shares common ground with the electromechanochemical model.

E.N. Moudrianakis (Johns Hopkins University) has opened wide the door to a reevaluation of the first acceptor for activated phosphate in photosynthetic phosphorylation. He presented unequivocal evidence that the first acceptor was adenosine monophosphate and that formation of adenosine triphosphate (ATP) depended on a myokinase-like transfer of a phosphoryl group between two molecules of bound adenosine diphosphate (ADP). A similar interpretation, based on results obtained with submitochondrial particles, was first proposed by Ozawa several years ago, but eventually came in for severe criticism published by P. Boyer (University of California, Los Angeles) and M.E. Pullman (Public Health Research Institute, New York). What this demonstration by Moudrianakis probably means is that the hydrolysis of ATP to ADP and inorganic phosphate is not the microscopic reversal of oxidative or photosynthetic phosphorylation.

The sticky problem of how electrons from reduced complex III can find their way into the heme group in a crevice in the interior of cytochrome c was considered by R. Dickerson (California Institute of Technology), B. Chance (University of Pennsylvania), and Winfield. Despite the elegant x-ray studies of Dickerson and the exhaustive spectroscopic studies of Chance, there was no consensus about the mechanism of electron transfer, although the cloud of conjecture was thick. Winfield, in one of the most penetrating presentations of the conference, laid bare the sorry state of our present ignorance of this crucial problem. If evidence were needed for the crisis in bioenergetics, the problem posed by the mechanism or electron transfer involving the components of the mitochondrial electron transfer chain can provide food for thought.

The generation of accurate structural information about energy-transducing systems is undoubtedly the crowning achievement of contemporary bioenergetics. The surveys of progress in muscle (Huxley; M. Morales, University of California School of Medicine, San Francisco; and J. Gergely, Retina Foundation), the mitochondrion (Y. Hatefi, Scripps Clinic and Research Foundation, La Jolla, California; L. Packer, University of California, Berkeley; and A. Tzagoloff, Public Health Research Institute), the sarcoplasmic reticulum (D. MacLennan, University of Toronto; and A. Martonosi, St. Louis University), and the chloroplast (Moudrianakis; L. Vernon, Brigham Young University; and R. Park, University of California, Berkeley) were among the high points of the conference. The x-ray crystallographic studies of chymotrypsin (Hartley) and of cytochrome c (Dickerson) pointed up the stark contrast between the precision and beauty of the accumulated structural information about these two macromolecules and the pitifully limited understanding of the mechanisms of both catalysis and electron transfer. Clearly, knowledge of the structural information is a necessary but not sufficient condition for adumbrating the functional principles.

The number of new concepts presented at the conference was high. These included enzymes as transducers of thermal to electromechanochemical potential energy (Green and Ji), force-generating mechanisms for muscular contraction (R. Dowben, Southwestern University Medical School, Dallas; McClare; and Ji), thermodynamic models for oxidative phosphorylation (Bennun and Weber), and molecular models for nerve transmission (R. Keynes, Agricultural Research Council Institute of Animal Physiology, Babrahan, Cambridge, England; Wei; I. Tasaki, National Institute of Mental Health, Bethesda, Maryland; and Nachmansohn). Cope developed a solidstate model for mitochondrial function.

A gap of at least 100 years separates parallel developments in physics and biology. The cooperative and symbiotic relation between theory and experiment has by now a long tradition in physics, but this relation is only in its formative stage in biology. If this conference has helped to convey the message to workers in bioenergetics that experiment without theory is as sterile as theory without experiment, it will have served a useful and historic purpose.

H. Baum (Chelsea College, London), played the role of Solomon in a brilliant summary of the conference. He pointed up the close analogies between the prolonged debate over atomism that raged in the late 19th century and the present dilemma in bioenergetics.

Mitchell, the cosponsor of the conference, was unfortunately unable to be present, but, nonetheless, the power of his approach and insights dominated the proceedings. The notion of vectorial electron flow and vectorial coupling, which he has championed, would now appear to be one of the greatest achievements in bioenergetics.

DAVID E. GREEN
Institute for Enzyme Research,
University of Wisconsin,
Madison 53706

Index

Page numbers followed by f indicate figures; t, tables; b, boxes.